FADS & FALLACIES

In The Name of Science

Martin Gardner

FADS & FALLACIES

In The Name of Science

THE MASTERPIECES
OF SCIENCE

NEW AMERICAN LIBRARY

The Masterpieces of Science

Published by New American Library,
1633 Broadway,
New York, N.Y. 10019

First Printing 1986

PRINTED IN THE UNITED STATES OF AMERICA

To my Mother and Father

CONTENTS

Preface   ix
Preface to the Second Edition   xi
  1. *In the Name of Science*   1
  2. *Flat and Hollow*   13
  3. *Monsters of Doom*   23
  4. *The Forteans*   35
  5. *Flying Saucers*   46
  6. *Zig-Zag-and-Swirl*   58
  7. *Down with Einstein!*   68
  8. *Sir Isaac Babson*   78
  9. *Dowsing Rods and Doodlebugs*   86
 10. *Under the Microscope*   99
 11. *Geology versus Genesis*   106
 12. *Lysenkoism*   121
 13. *Apologists for Hate*   132
 14. *Atlantis and Lemuria*   143

15.	The Great Pyramid	151
16.	Medical Cults	162
17.	Medical Quacks	178
18.	Food Faddists	192
19.	Throw Away Your Glasses!	201
20.	Eccentric Sexual Theories	212
21.	Orgonomy	219
22.	Dianetics	230
23.	General Semantics, Etc.	246
24.	From Bumps to Handwriting	255
25.	ESP and PK	261
26.	Bridey Murphy and Other Matters	275
	Appendix and Notes	284
	Index of Names	313

Preface

NOT MANY BOOKS have been written about modern pseudo-scientists and their views. I found only two general surveys that provided leads or useful material—Daniel W. Hering's *Foibles and Fallacies of Science,* 1924, and *The Story of Human Error,* 1936, edited by Joseph Jastrow.

David Starr Jordan, the first president of Stanford and a renowned authority on fish, wrote in 1927 a book called *The Higher Foolishness.* In it, he coined the word "sciosophy" (meaning "shadow wisdom") to stand for what he termed the "systematized ignorance" of the pseudo-scientist. The book is infuriating because although Jordan mentions the titles of dozens of crank works, from which he quotes extensively, he seldom tells you the names of the authors.

Most of my research was done in the New York Public Library, which has a magnificent collection of crank literature. Unfortunately, only a tiny portion of it is identified as such (under headings like "Science—curiosa," "Impostors," "Quacks," "Eccentric Persons," etc.). Consequently it had to be exhumed by devious, obscure, and often intuitive methods.

Friends too numerous to mention have in various ways aided this research, but I wish specifically to thank Everett Bleiler, Prof. Edwin G. Boring, and Mr. and Mrs. David B. Eisendrath, Jr. for suggestions and favors relating to the book as a whole; and the following individuals for assistance on certain chapters: Dr. Alan Barnert, John Boyko, Arthur Cox, Charles Dye, Bruce Elliott, James H. Gardner, Thomas Gilmartin, Zalmon Goldsmith, Gershon Legman, Dr. L. Vosburgh Lyons, Robert Marks, Prof. H. J. Muller, and Allen W. Read.

I also wish to express thanks to Paul Bixler, editor of the *Antioch Review,* for permission to use in the first chapter portions of my article "The Hermit Scientist" (*Antioch Review,* Winter, 1950–51), and to my literary agent, John T. Elliott, for his insistence that this article could be expanded into a book.

And special thanks to Charlotte Greenwald for help in proofing and revising.

THE AUTHOR

Preface to Second Edition

THE FIRST EDITION of this book prompted many curious letters from irate readers. The most violent letters came from Reichians, furious because the book considered orgonomy alongside such (to them) outlandish cults as dianetics. Dianeticians, of course, felt the same about orgonomy. I heard from homeopaths who were insulted to find themselves in company with such frauds as osteopathy and chiropractic, and one chiropractor in Kentucky "pitied" me because I had turned my spine on God's greatest gift to suffering humanity. Several admirers of Dr. Bates favored me with letters so badly typed that I suspect the writers were in urgent need of strong spectacles. Oddly enough, most of these correspondents objected to one chapter only, thinking all the others excellent.

Some readers, however, liked the entire book and were kind enough to call my attention to occasional errors and to suggest new material that might be worth mentioning if the book were ever revised. Thanks to Hayward Cirker, president of Dover Publications, such a revision has now become possible. I have left the text unaltered save for corrections of unimportant errors, and added as much as I could in the way of documentation and fresh material in a lengthy appendix. One new chapter has been written to cover the recent Bridey Murphy mania and to discuss further the difficult question of when a reputable publisher is justified in bringing out a work of "unorthodox" science.

THE AUTHOR
1956

CHAPTER 1

IN THE NAME OF SCIENCE

SINCE THE BOMB exploded over Hiroshima, the prestige of science in the United States has mushroomed like an atomic cloud. In schools and colleges, more students than ever before are choosing some branch of science for their careers. Military budgets earmarked for scientific research have never been so fantastically huge. Books and magazines devoted to science are coming off the presses in greater numbers than at any previous time in history. Even in the realm of escape literature, science fiction threatens seriously to replace the detective story.

One curious consequence of the current boom in science is the rise of the promoter of new and strange "scientific" theories. He is riding into prominence, so to speak, on the coat-tails of reputable investigators. The scientists themselves, of course, pay very little attention to him. They are too busy with more important matters. But the less informed general public, hungry for sensational discoveries and quick panaceas, often provides him with a noisy and enthusiastic following.

In 1951, tens of thousands of mentally ill people throughout the country entered "dianetic reveries" in which they moved back along their "time track" and tried to recall unpleasant experiences they had when they were embryos. Thousands of more sophisticated neurotics, who regard dianetics as the invention of a mountebank, are now sitting in "orgone boxes" to raise their body's charge of "orgone energy." Untold numbers of middle-aged housewives are preparing to live to the age of 100 by a diet rich in yoghurt, wheat-germ, and blackstrap-molasses.

Not only in the fields of mental and physical health is the spurious scientist flourishing. A primitive interpretation of Old Testament miracle

1

tales, which one thought went out of fashion with the passing of William Jennings Bryan, has just received a powerful shot in the arm. Has not the eminent "astrophysicist," Dr. Immanuel Velikovsky, established the fact that the earth stopped rotating precisely at the moment Joshua commanded the sun and moon to stand still? For fifty years, geologists and physicists have been combining forces to perfect complex, delicate instruments for exploring underground structures. They've been wasting their time according to Kenneth Roberts, the well-known novelist. All you need is a forked twig, and he has written a persuasive and belligerent book to prove it.

Since flying saucers were first reported in 1947, countless individuals have been convinced that the earth is under observation by visitors from another planet. Admirers of Frank Scully's *Behind the Flying Saucers* suspect that the mysterious disks are piloted by inhabitants of Venus who are exact duplicates of earthlings except they are three feet tall. A more recent study by Gerald Heard makes out an even stronger case for believing the saucers are controlled by intelligent bees from Mars.

In the twenties, newspapers provided a major publicity outlet for the speculations of eccentric scholars. Every Sunday, Hearst's *American Weekly* disclosed with lurid pictures some outlandish piece of scientific moonshine. The pages of the daily press were spotted with such stories as unconfirmed reports of enormous sea serpents, frogs found alive in the cornerstones of ancient buildings, or men who could hear radio broadcasts through gold inlays in their teeth. But gradually, over the next two decades, an unwritten code of science ethics developed in the profession of news journalism. Wire services hired competent science writers. Leading metropolitan dailies acquired trained science editors. The American Medical Association stepped up its campaign against press publicity for medical quackery, and disciplined members who released accounts of research that had not been adequately checked by colleagues. Today, science reporting in the American press is freer of humbug and misinformation than ever before in history.

To a large extent, the magazine and book publishing firms shared in the forging of this voluntary code. Unfortunately, at the turn of the half-century they began to backslide. *Astounding Science Fiction,* until recently the best of the science fantasy magazines, was the first to inform the public of the great "Dianetic Revolution" in psychiatry. *True* boosted its circulation by breaking the news that flying saucers came from another planet. *Harper's* published the first article in praise of Velikovsky's remarkable discoveries, and similar pieces quickly followed in *Collier's* and *Reader's Digest. The Saturday Evening Post* and *Look* gave widespread publicity to Gayelord Hauser's blackstrap-molasses cult during the same month that the Pure Food and Drug Administration seized copies of his best-seller, *Look Younger, Live Longer*. The government charged that displays of the book next to jars of blackstrap constituted, because of the book's sensational claims, a "mislabeling" of the product.

Many leading book publishers have had no better record. It is true that L. Ron Hubbard's *Dianetics* was too weird a manuscript to interest the larger houses, but Velikovsky's equally preposterous work found two highly reputable publishers. Kenneth Roberts' book on the art of finding water with a dowsing rod, Scully's saucer book, and Heard's even more fantastic study also appeared under the imprints of major houses.

When book editors and publishers are questioned about all this, they have a ready answer. It is a free country, they point out. If the public is willing to buy a certain type of book in great quantities, do they not, as public servants, have every right—perhaps even the obligation—to satisfy such a demand?

No one with any respect for independent thinking would propose that a publishing house or magazine be *compelled,* by any type of government action, to publish only material sanctioned by a board of competent scientists. That, however, is not the issue. The question is whether the voluntary code of ethics, so painstakingly built up during the past two decades, is worth preserving. Velikovsky's book, for example, was widely advertised as a revolutionary astronomical discovery. The publisher, of course, had every legal right to publish such a book. Likewise, the scientists who threatened to boycott the firm's textbooks unless it dropped Velikovsky from its list, were exercising their democratic privilege of organized protest. The issue is not a legal one, or even a political one. It is a question of individual responsibility.

Perhaps we are making a mountain out of a molehill. It is all very amusing, one might say, to titillate public fancy with books about bee people from Mars. The scientists are not fooled, nor are readers who are scientifically informed. If the public wants to shell out cash for such flummery, what difference does it make? The answer is that it is not at all amusing when people are misled by scientific claptrap. Thousands of neurotics desperately in need of trained psychiatric care are seriously retarding their therapy by dalliance with crank cults. Already a frightening number of cases have come to light of suicides and mental crack-ups among patients undergoing these dubious cures. No reputable publisher would think of releasing a book describing a treatment for cancer if it were written by a doctor universally considered a quack by his peers. Yet the difference between such a book and *Dianetics* is not very great.

What about the long-run effects of non-medical books like Velikovsky's, and the treatises on flying saucers? It is hard to see how the effects can be anything but harmful. Who can say how many orthodox Christians and Jews read *Worlds in Collision* and drifted back into a cruder Biblicism because they were told that science had reaffirmed the Old Testament miracles? Mencken once wrote that if you heave an egg out of a Pullman car window anywhere in the United States you are likely to hit a fundamentalist. That was twenty-five years ago, and times have changed, but it is easy to forget how far from won is the battle against religious superstition. It is easy to forget that thousands of high school teachers of biology, in many of our southern states, are still afraid to teach the theory of evolution for fear of losing their jobs. There is no question but that informed and enlightened Christianity, both Catholic and Protestant, suffered a severe body blow when Velikovsky's book was enthusiastically hailed by the late Fulton Oursler (in *Reader's Digest*) as scientific confirmation of the most deplorable type of Bible interpretation.

Flying saucers? I have heard many readers of the saucer books upbraid the government in no uncertain terms for its stubborn refusal to release the "truth" about the elusive platters. The administration's "hush hush policy" is angrily cited as proof that our military and political leaders have lost all faith in the wisdom of the American people.

An even more regrettable effect produced by the publication of scientific rubbish is the confusion they sow in the minds of gullible readers about what is and what isn't scientific knowledge. And the more the public is confused, the easier it falls prey to doctrines of pseudo-science which may at some future date receive the backing of politically powerful groups. As we shall see in later chapters, a renaissance of German quasi-science paralleled the rise of Hitler. If the German people had been better trained to distinguish good from bad science, would they have swallowed so easily the insane racial theories of the Nazi anthropologists?

In the last analysis, the best means of combating the spread of pseudo-science is an enlightened public, able to distinguish the work of a reputable investigator from the work of the incompetent and self-deluded. This is not as hard to do as one might think. Of course, there always will be borderline cases hard to classify, but the fact that black shades into white through many shades of gray does not mean that the distinction between black and white is difficult.

Actually, two different "continuums" are involved. One is a scale of the degree to which a scientific theory is confirmed by evidence. At one end of this scale are theories almost certainly false, such as the dianetic view that a one-day-old embryo can make sound recordings of its mother's conversation. Toward the middle of the scale are theories advanced as working hypotheses, but highly debatable because of the lack of sufficient data—for example, the theory that the universe is expanding. Finally, at the other extreme of the scale, are theories almost certainly true, such as the belief that the earth is round or that men and beasts are distant cousins. The problem of determining the degree to which a theory is confirmed is extremely difficult and technical, and, as a matter of fact, there are no known methods for giving precise "probability values" to hypotheses. This problem, however, need not trouble us. We shall be concerned, except for a few cases, only with theories so close to "almost certainly false" that there is no reasonable doubt about their worthlessness.

The second continuum is the scale of scientific competence. It also has its extremes—ranging from obviously admirable scientists, to men of equally obvious incompetence. That there are individuals of debatable status—men whose theories are on the borderline of sanity, men competent in one field and not in others, men competent at one period of life and not at others, and so on—all this ought not to blind us to the obvious fact that there is a type of self-styled scientist who can legitimately be called a crank. It is not the novelty of his views or the neurotic motivations behind his work that provide the grounds for calling him this. The grounds are the technical criteria by which theories are evaluated. If a man persists in advancing views that are contradicted by all available evidence, and which offer no reasonable grounds for serious consideration, he will rightfully be dubbed a crank by his colleagues.

Cranks vary widely in both knowledge and intelligence. Some are stupid, ignorant, almost illiterate men who confine their activities to sending "crank letters" to prominent scientists. Some produce crudely written pamphlets, usually published by the author himself, with long titles, and pictures of the author on the cover. Still others are brilliant and well-educated, often with an excellent understanding of the branch of science in which they are speculating. Their books can be highly deceptive imitations of the genuine article—well-written and impressively learned. In spite of these wide variations, however, most pseudo-scientists have a number of characteristics in common.

First and most important of these traits is that cranks work in almost total isolation from their colleagues. Not isolation in the geographical sense, but in the sense of having no fruitful contacts with fellow researchers. In the Renaissance, this isolation was not necessarily a sign of the crank. Science was poorly organized. There were no journals or societies. Communication among workers in a field was often very difficult. Moreover, there frequently were enormous social pressures operating against such communication. In the classic case of Galileo, the Inquisition forced him into isolation because the Church felt his views were undermining religious faith. Even as late as Darwin's time, the pressure of religious conservatism was so great that Darwin and a handful of admirers stood almost alone against the opinions of more respectable biologists.

Today, these social conditions no longer obtain. The battle of science to free itself from religious control has been almost completely won. Church groups still oppose certain doctrines in biology and psychology, but even this opposition no longer dominates scientific bodies or journals. Efficient networks of communication within each science have been established. A vast cooperative process of testing new theories is constantly going on—a process amazingly free (except, of course, in totalitarian nations) from control by a higher "orthodoxy." In this modern framework, in which scientific progress has become dependent on the constant give and take of data, it is impossible for a working scientist to be isolated.

The modern crank insists that his isolation is not desired on his part. It is due, he claims, to the prejudice of established scientific groups against new ideas. Nothing could be further from the truth. Scientific journals today are filled with bizarre theories. Often the quickest road to fame is to overturn a firmly-held belief. Einstein's work on relativity is the outstanding example. Although it met with considerable opposition at first, it was on the whole an intelligent opposition. With few exceptions, none of Einstein's reputable opponents dismissed him as a crackpot. They could not so dismiss him because for years he contributed brilliant articles to the journals and had won wide recognition as a theoretical physicist. In a surprisingly short time, his relativity theories won almost universal acceptance, and one of the greatest revolutions in the history of science quietly took place.

It would be foolish, of course, to deny that history contains many sad examples of novel scientific views which did not receive an unbiased hearing, and which later proved to be true. The pseudo-scientist never tires reminding his readers of these cases. The opposition of traditional psychology to the study of hypnotic phenomena (accentuated by the fact that Mesmer was both a crank and a charlatan) is an outstanding instance. In the field of medicine, the germ theory of Pasteur, the use of anesthetics, and Dr. Semmelweiss' insistence that doctors sterilize their hands before attending childbirth are other well known examples of theories which met with strong professional prejudice.

Probably the most notorious instance of scientific stubbornness was the refusal of eighteenth century astronomers to believe that stones actually fell from the sky. Reaction against medieval superstitions and old wives' tales was still so strong that whenever a meteor fell, astronomers insisted it had either been picked up somewhere and carried by the wind, or that the persons who claimed to see it fall were lying. Even the great French *Académie des Sciences* ridiculed this folk belief, in spite of a number of early studies of meteoric phenomena. Not until April 26, 1803, when several thousand small meteors fell on the town of L'Aigle, France, did the astronomers decide to take falling rocks seriously.

Many other examples of scientific traditionalism might be cited, as well as cases of important contributions made by persons of a crank variety. The discovery of the law of conservation of energy by Robert Mayer, a psychotic German physician, is a classic instance. Occasionally a layman, completely outside of science, will make an astonishingly prophetic guess—like Swift's prediction about the moons of Mars (to be discussed later), or Samuel Johnson's belief (expressed in a letter, in 1781, more than eighty years before the discovery of germs) that microbes were the cause of dysentery.

One must be extremely cautious, however, before comparing the work of some contemporary eccentric with any of these earlier examples, so frequently cited in crank writings. In medicine, we must remember, it is only in the last fifty years or so that the art of healing has become anything resembling a rigorous scientific discipline. One can go back to periods in which medicine was in its infancy, hopelessly mixed with superstition, and find endless cases of scientists with unpopular views that later proved correct. The same holds true of other sciences. But the picture today is vastly different. The prevailing spirit among scientists, outside of totalitarian countries, is one of eagerness for fresh ideas. In the great search for a cancer cure now going on, not the slightest stone, however curious its shape, is being left unturned. If anything, scientific journals err on the side of permitting *questionable* theses to be published, so they may be discussed and checked in the hope of finding something of value. A few years ago a student at the Institute for Advanced Studies in Princeton was asked how his seminar had been that day. He was quoted in a news magazine as exclaiming, "Wonderful! Everything we knew about physics last week isn't true!"

Here and there, of course—especially among older scientists who, like everyone else, have a natural tendency to become set in their opinions— one may occasionally meet with irrational prejudice against a new point of view. You cannot blame a scientist for unconsciously resisting a theory which may, in some cases, render his entire life's work obsolete. Even the great Galileo refused to accept Kepler's theory, long after the evidence was quite strong, that planets move in ellipses. Fortunately there are always, in the words of Alfred Noyes, "The young, swift-footed, waiting for the fire," who can form the vanguard of scientific revolutions.

It must also be admitted that in certain areas of science, where empirical data are still hazy, a point of view may acquire a kind of cult following and harden into rigid dogma. Modifications of Einstein's theory, for example, sometimes meet a resistance similar to that which met the original theory. And no doubt the reader will have at least one acquaintance for whom a particular brand of psychoanalysis has become virtually a religion, and who waxes highly indignant if its postulates are questioned by adherents of a rival brand.

Actually, a certain degree of dogma—of pig-headed orthodoxy—is both necessary and desirable for the health of science.[1] It forces the scientist with a novel view to mass considerable evidence before his theory can be seriously entertained. If this situation did not exist, science would be reduced to shambles by having to examine every new-fangled notion that came along. Clearly, working scientists have more important tasks. If someone announces that the moon is made of green cheese, the professional astronomer cannot be expected to climb down from his telescope and write a detailed refutation. "A fairly complete textbook of physics would be only part of the answer to Velikovsky," writes Prof. Laurence J. Lafleur, in his ecellent article on "Cranks and Scientists" (*Scientific Monthly*, Nov., 1951), "and it is therefore not surprising that the scientist does not find the undertaking worth while."

The modern pseudo-scientist—to return to the point from which we have digressed—stands entirely outside the closely integrated channels through which new ideas are introduced and evaluated. He works in isolation. He does not send his findings to the recognized journals, or if he does, they are rejected for reasons which in the vast majority of cases are excellent. In most cases the crank is not well enough informed to write a paper with even a surface resemblance to a significant study. As a consequence, he finds himself excluded from the journals and societies, and almost universally ignored by the competent workers in his field. In fact, the reputable scientist does not even know of the crank's existence unless his work is given widespread publicity through non-academic channels, or unless the scientist makes a hobby of collecting crank literature. The eccentric is forced, therefore, to tread a lonely way. He speaks before organizations he himself has founded, contributes to journals he himself may edit, and—until recently—publishes books only when he or his followers can raise sufficient funds to have them printed privately.

A second characteristic of the pseudo-scientist, which greatly strengthens his isolation, is a tendency toward paranoia.[2] This is a mental condition (to quote a recent textbook) "marked by chronic, systematized, gradually developing delusions, without hallucinations, and with little tendency toward deterioration, remission, or recovery." There is wide disagreement among psychiatrists about the causes of paranoia. Even if this were not so, it obviously is not within the scope of this book to discuss the possible origins of paranoid traits in individual cases. It is easy to understand, however, that a strong sense of personal greatness must be involved whenever a crank stands in solitary, bitter opposition to every recognized authority in his field.

If the self-styled scientist is rationalizing strong religious convictions, as often is the case, his paranoid drives may be reduced to a minimum. The desire to bolster religious beliefs with science can be a powerful motive. For example, in our examination of George McCready Price, the greatest of modern opponents of evolution, we shall see that his devout faith in Seventh Day Adventism is a sufficient explanation for his curious geological views. But even in such cases, an element of paranoia is nearly always present. Otherwise the pseudo-scientist would lack the stamina to fight a vigorous, single-handed battle against such overwhelming odds. If the crank is insincere—interested only in making money, playing a hoax, or both—then obviously paranoia need not enter his make-up. However, very few cases of this sort will be considered.

There are five ways in which the sincere pseudo-scientist's paranoid tendencies are likely to be exhibited.

(1) He considers himself a genius.

(2) He regards his colleagues, without exception, as ignorant blockheads. Everyone is out of step except himself. Frequently he insults his opponents by accusing them of stupidity, dishonesty, or other base motives. If they ignore him, he takes this to mean his arguments are unanswerable. If they retaliate in kind, this strengthens his delusion that he is battling scoundrels.

Consider the following quotation: "To me truth is precious. . . . I should rather be right and stand alone than to run with the multitude and be wrong. . . . The holding of the views herein set forth has already won for me the scorn and contempt and ridicule of some of my fellowmen. I am looked upon as being odd, strange, peculiar. . . . But truth is truth and though all the world reject it and turn against me, I will cling to truth still."

These sentences are from the preface of a booklet, published in 1931, by Charles Silvester de Ford, of Fairfield, Washington, in which he proves the earth is flat. Sooner or later, almost every pseudo-scientist expresses similar sentiments.

(3) He believes himself unjustly persecuted and discriminated against. The recognized societies refuse to let him lecture. The journals reject his papers and either ignore his books or assign them to "enemies" for review. It is all part of a dastardly plot. It never occurs to the crank that this opposition may be due to error in his work. It springs solely, he is convinced, from blind prejudice on the part of the established hierarchy—the high priests of science who fear to have their orthodoxy overthrown.

Vicious slanders and unprovoked attacks, he usually insists, are constantly being made against him. He likens himself to Bruno, Galileo, Copernicus, Pasteur, and other great men who were unjustly persecuted for their heresies. If he has had no formal training in the field in which he works, he will attribute this persecution to a scientific masonry, unwilling to admit into its inner sanctums anyone who has not gone through the proper initiation rituals. He repeatedly calls your attention to important scientific discoveries made by laymen.

(4) He has strong compulsions to focus his attacks on the greatest scientists and the best-established theories. When Newton was the outstanding name in physics, eccentric works in that science were violently anti-Newton. Today, with Einstein the father-symbol of authority, a crank theory of physics is likely to attack Einstein in the name of Newton. This same defiance can be seen in a tendency to assert the diametrical opposite of well-established beliefs.[3] Mathematicians prove the angle cannot be trisected. So the crank trisects it. A perpetual motion machine cannot be built. He builds one. There are many eccentric theories in which the "pull" of gravity is replaced by a "push." Germs do not cause disease, some modern cranks insist. Disease produces the germs. Glasses do not help the eyes, said Dr. Bates. They make them worse. In our next chapter we shall learn how Cyrus Teed literally turned the entire cosmos inside-out, compressing it within the confines of a hollow earth, inhabited only on the inside.

(5) He often has a tendency to write in a complex jargon, in many cases making use of terms and phrases he himself has coined. Schizophrenics sometimes talk in what psychiatrists call "neologisms"—words which have meaning to the patient, but sound like Jabberwocky to everyone else. Many of the classics of crackpot science exhibit a neologistic tendency.

When the crank's I.Q. is low, as in the case of the late Wilbur Glenn Voliva who thought the earth shaped like a pancake, he rarely achieves much of a following. But if he is a brilliant thinker, he is capable of developing incredibly complex theories. He will be able to defend them in books of vast erudition, with profound observations, and often liberal portions of sound science. His rhetoric may be enormously persuasive. All the parts of his world usually fit together beautifully, like a jig-saw puzzle. It is impossible to get the best of him in any type of argument.[4] He has anticipated all your objections. He counters them with unexpected answers of great ingenuity. Even on the subject of the shape of the earth, a layman may find himself powerless in a debate with a flat-earther. George Bernard Shaw, in *Everybody's Political What's What?*, gives an hilarious description of a meeting at which a flat-earth speaker completely silenced all opponents who raised objections from the floor. "Opposition such as no atheist could have provoked assailed him"; writes Shaw, "and he, having heard their arguments hundreds of times, played skittles with them, lashing the meeting into a spluttering fury as he answered easily what it considered unanswerable."

In the chapters to follow, we shall take a close look at the leading pseudo-scientists of recent years, with special attention to native specimens. Some British books will be discussed, and a few Continental eccentric theories, but the bulk of crank literature in foreign tongues will not be touched upon. Very little of it has been translated into English, and it is extremely difficult to get access to the original works. In addition, it is usually so unrelated to the American scene that it loses interest in comparison with the work of cranks closer home.

With few exceptions, little time will be spent on theories which come under the broad heading of "occult." Astrology, for example, still has millions of contemporary followers, but is so far removed from anything resembling science that it does not seem worth while to discuss it. The theory that sunspots cause depressions (popular among conservative businessmen who like to think of booms and busts as natural phenomena to be blamed on something remote) is the last respectable survival of the ancient view that human affairs are linked with astronomical phenomena. This literature, however, belongs more properly to economics than to astronomy. The social sciences have, of course, their share of eccentric works, but for many reasons they form a separate subject for study.

Our survey will begin with curious theories of astronomy, the science most removed from the human landscape. It will proceed through physics and geology to the biological sciences, then into human affairs by way of anthropology and archeology. Four chapters will be devoted to medical quasi-science, followed by discussions of sexual theories, psychiatric cults, and methods of reading character. Finally, we shall make a serious appraisal of the reputable work of Dr. Rhine, with quick and not so serious glances at a few other venturers into the psychic fields.

The amount of intellectual energy that has been wasted on these lost causes is almost unbelievable. It will be amusing—at times frightening—to witness the grotesque extremes to which deluded scientists can be misled, and the extremes to which they in turn can mislead others. As we shall see, their disciples are often intelligent and sometimes eminent men— men not well enough informed on the subject in question to penetrate the Master's counterfeit trappings, and who frequently find in their devotion an outlet for their own neurotic rebellions. More important, we shall have impressed upon us the traits which these "scientists" hold in common. The atmosphere in which they move will become familiar to us as we begin to breathe the air of their fantastic worlds.

Just as an experienced doctor is able to diagnose certain ailments the instant a new patient walks into his office, or a police officer learns to recognize criminal types from subtle behavior clues which escape the untrained eye, so we, perhaps, may learn to recognize the future scientific crank when we first encounter him.

And encounter him we shall. If the present trend continues, we can expect a wide variety of these men, with theories yet unimaginable, to put in their appearance in the years immediately ahead. They will write impressive books, give inspiring lectures, organize exciting cults. They may achieve a following of one—or one million. In any case, it will be well for ourselves and for society If we are on our guard against them.

CHAPTER 2

FLAT AND HOLLOW

EVERY SCHOOLBOY knows that the earth is a solid ball, slightly flattened at the poles, and surrounded by a cosmos of inconceivable immensity. Since Magellan sailed around the globe in 1519, few have doubted that the earth is round. Yet it is precisely because these views are universally accepted that the shape of the earth is such a happy field of speculation for the pseudo-scientist.

Three eccentric theories of the earth have each won a surprising number of adherents in the present century: Voliva's flat earth, the view that the world is hollow and open at the poles, and—most incredible of all—that we are living on the *inside* of a hollow sphere.

It is hard to believe that any literate American, living in the first decade of the Atomic Age, would doubt that the earth is round; yet there are several thousand such persons. Most of them live in a drab little town called Zion, Illinois, on the shore of Lake Michigan about forty miles north of Chicago. They are the remnant of what at one time was a flourishing religious sect called The Christian Apostolic Church in Zion, founded in 1895 by a Scottish faith-healer named John Alexander Dowie.

Rev. Dowie was forcibly expelled from office as "General Overseer" of Zion in 1905. For the next thirty years, the community of 6,000 was ruled by the iron hand of Wilbur Glenn Voliva. Most of its citizens worked for Zion Industries, a million-dollar corporation which turned out an amazing variety of goods from fine lace to fig bars. No town in America had stricter Blue Laws. Motorists along the Lake front soon learned to avoid stopping in the village; they were likely to be arrested and fined for smoking cigarettes, or whistling on Sunday.

Voliva was a paunchy, baldish, grim-faced fellow who wore a rumpled frock coat and enormous white cuffs. Throughout his life he was profoundly convinced that the earth is shaped like a flapjack, with the North Pole in the center and the South Pole distributed around the circumference. For many years, he offered $5,000 to anyone who could prove to him the earth is spherical, and in fact made several trips around the world lecturing on the subject. In his mind, of course, he had not circumnavigated a globe; he had merely traced a circle on a flat surface.

According to Voliva, a huge wall of snow and ice prevents ships from sailing off the edge and falling into Hades. Below Hades is a sub-basement where live the spirits of a race that flourished on earth before the time of Adam. The stars are much smaller than the earth and rotate around it. The moon is self-luminous. The sun? Here is what Voliva has to say about the sun:

"The idea of a sun millions of miles in diameter and 91,000,000 miles away is silly. The sun is only 32 miles across and not more than 3,000 miles from the earth. It stands to reason it must be so. God made the sun to light the earth, and therefore must have placed it close to the task it was designed to do. What would you think of a man who built a house in Zion and put the lamp to light it in Kenosha, Wisconsin?"

A special May 10, 1930, issue of the sect's periodical, *Leaves of Healing,* is devoted entirely to astronomy. This 64-page number of the magazine is the most complete statement in print of Voliva's scriptural and scientific reasons for thinking the earth flat and motionless. "Can anyone who has considered this matter seriously," one article asks, "honestly say that he believes the earth is traveling at such an impossible speed? If the earth is going so fast, which way is it going? It should be easier to travel with it than against it. The wind always should blow in the opposite direction to the way the earth is traveling. But where is the man who believes that it does? Where is the man who believes that he can jump into the air, remaining off the earth one second, and come down to the earth 193.7 miles from where he jumped up?"

One of the best known proofs of the earth's rotation makes use of a device called the Foucault pendulum. This consists of a heavy weight suspended on a long wire. As the weight swings back and forth, inertia causes it to stay in the same plane of swing while the earth turns beneath it. The result is that the plane of swing seems to rotate slowly. The article quoted above disposes neatly of this proof. "If the earth's motion has anything to do with the movement of the pendulum," the author asks, "why must you start it going? The real fact is, and everybody who gives it a serious thought must see, that if the earth were whirling around with the speed astronomers say it is, the pendulum would fly straight out in space and stay there."

But the magazine's *pièce de résistance* is a double-spread photograph showing twelve miles of the shoreline of Lake Winnebago, Wisconsin. *The camera used was an 8 by 10 Eastman view camera . . .* reads the explanation. *The lens was exactly three feet above the water . . .* ANYONE CAN GO TO OSHKOSH AND SEE THIS SIGHT FOR THEMSELVES ANY CLEAR DAY. *With a good pair of binoculars one can see small objects on the opposite shore, proving beyond any doubt that the surface of the lake is a plane, or a horizontal line. . . . The scientific value of this picture is enormous.*

Because of Voliva's incredible ignorance, it is easier to see the psychological drives behind his outlandish views than in the case of cleverer cranks who conceal their motives under erudition and shrewd polemics. Voliva's drives were two in number—a desire to defend a religious dogma, and a paranoid belief in his own greatness so far removed from reality as to border on the psychotic. The first hardly needs elaboration. "We are fundamentalists," Voliva once declared. "We are the only *true* fundamentalists." And of course he was right. There are many passages in the Bible which if taken with extreme literalness suggest a flat rather than round earth, and one of the cardinal doctrines of the Dowie cult—one might say its essence—was to regard every word of the Bible as literally true.

However, to explain Voliva's astronomy as no more than a rationalization of a way of interpreting Scripture is to tell only part of the story. In past ages, perhaps, it was the full story. During the early centuries of faith, before the evidence for a round earth became overwhelming, it is easy to see how intelligent and well-informed theologians would prefer a literal interpretation of Old Testament passages. We can understand, for example, Saint Augustine's or Martin Luther's arguing that no human beings could live on the underside of the earth because they would be unable to see Christ descend from heaven at his Second Coming. But what are we to think of a man of the twentieth century who refused to admit the earth was round?

The answer lies in Voliva's delusions of grandeur. He regarded all astronomers as "poor, ignorant, conceited fools," and once boasted, "I can whip to smithereens any man in the world in a mental battle. I have never met any professor or student who knew a millionth as much on any subject as I do." Once during a courtroom legal wrangle he shouted, "Every man who fights me goes under. Mark what I say! The graveyard is full of fellows who tried to down Voliva. This other bunch will go to the graveyard too. God almighty will smite them." Although his sect never numbered more than ten thousand members he was able to assert, "I am just starting my real work. I shall evangelize the rest of the United States and Europe next."

Voliva often predicted the end of the world. As the years 1923, 1927, 1930, and 1935 rolled by—each of which he had set as the year of doom—it never occurred to him that the repeated failure of the skies to roll up like a scroll indicated any fallibility on his part. His death in 1942 must have been another surprise. He expected to live to 120 on a special diet of Brazil nuts and buttermilk.

Today, things have changed in Zion. Other churches have moved in. The Blue Laws have been repealed. Girls are wearing lipstick and nail polish, and in the summer, even shorts on Enoch Avenue will not cause their arrest. New York University, curiously, now owns controlling interest in Zion Industries. But there are several thousand elderly followers of Voliva, living quietly in the community, who still feel, in the words of their departed leader, that "the so-called fundamentalists . . . strain out the gnat of evolution and swallow the camel of modern astronomy."

Although it is difficult to find a flat-earth believer, in or out of Zion, who is not a fundamentalist, it would be a mistake to suppose that all eccentric views about the earth's shape have their origin in religious superstition. The best example in recent centuries of a non-religious theory is the hollow earth doctrine of Captain John Cleves Symmes, of the U.S. Infantry. After distinguishing himself for bravery in the War of 1812, Symmes retired from the Army and spent the rest of his life trying to convince the nation that the earth was made up of five concentric spheres, with openings several thousand miles in diameter at the poles.

He first announced his theory in 1818 by widely distributing a circular calling for one hundred "brave companions" to join him on a polar expedition to the northern opening—or "Symmes' hole" as it soon became known. It was the Captain's firm belief that the sea flowed through both polar openings, and that plant and animal life abounded on the concave interior as well as on the convex surface of the next sphere.

The more Symmes' theory was ridiculed, the angrier he became and the more energy he spent in finding "facts" to support his views. It became an obsession. For ten years he traveled about the country giving speeches in a stumbling, nasal voice, and trying to raise funds for his voyage. In 1822 and 1823, he petitioned Congress to finance the trip. The petitions were quietly tabled, although he was persuasive enough the second time to win 25 votes. In 1829, his health finally broke under the strain of lecturing. At Hamilton, Ohio, where he made his home at the time of his death, may be seen the weather-beaten monument raised to him by his son. A stone model of the hollow earth caps the memorial.

The most complete descriptions of Symmes' remarkable views are to be found in two books—*Symmes' Theory of Concentric Spheres*, written in 1826 by James McBride, the Captain's number one convert; and *The*

Symmes' Theory of Concentric Spheres, published in 1878 by his son, Americus Symmes. Hundreds of reasons are given for believing the earth hollow—drawn from physics, astronomy, climatology, the migration habits of animals, and the reports of travelers. Moreover, a hollow planet, like the hollow bones of the body, would be a sturdy and economical way for the Creator to arrange things. As one disciple put it, "A hollow earth, habitable within, would result in a great saving of stuff." "Reason, common sense, and all the analogies in the natural universe, conspire to support and establish the theory," the Captain's son concluded.

Symmes' beliefs made no dent whatever on the science of his day, but they did leave a strong impress on science fiction. In 1820, an anonymous writer using the name of Captain Seaborn published a fictional burlesque of the hollow earth under the title of *Symzonia.* It is a pleasantly told narrative about a steamship voyage to the southern polar opening where a strong current draws the ship over the "rim of the world." On the concave interior, Captain Seaborn finds a continent which he names Symzonia. There he meets a friendly race of people who wear snow-white clothes, speak in a musical tongue, and live in a socialist utopia. Edgar Allan Poe's unfinished *Narrative of Arthur Gordon Pym* was intended to describe a similar voyage. Jules Verne's *Journey to the Center of the Earth* probably was not influenced by Symmes' theory, but many later novels and dozens of short stories have been based directly on it.

Did Symmes derive his views from earlier speculations? There is no evidence that he did, although in 1721, Cotton Mather, in a book called *The Christian Philosopher,* defended a similar doctrine. Mather, in turn, had taken his theory from an essay published in 1692 by the famous English astronomer Edmund Halley (for whom the comet was named). Halley argued that the earth had a shell 500 miles thick, then two inner shells of diameters comparable to Mars and Venus, and finally a solid inner sphere about the size of Mercury. Each sphere, he believed, was capable of bearing life. Perpetual daylight could be provided by "peculiar luminaries," such as Virgil places above the Elysian fields, or perhaps the atmosphere between the shells was luminous. When a brilliant display of northern lights took place in 1716, Halley suggested it might be caused by an escape of this glowing gas. Since the earth was flattened at the poles, the outer shell would naturally be a trifle thinner at those points, he reasoned, and therefore likely to allow the gas to escape.

In 1913 an Aurora, Illinois, resident named Marshall B. Gardner—he was in charge of maintenance of machinery for a large corset company—published privately a small book titled *Journey to the Earth's Interior.* It described a hollow earth very much like Symmes', though the author became furious when anyone suggested he had based his thinking on the

earlier doctrine. In 1920, he enlarged the book to 456 pages. A frontispiece shows him to be a heavy man with a square face, pale eyes, and a
drooping black mustache.

Gardner rejected Symmes' "fantastic notion" of many concentric spheres.
Only the outer shell, he insisted, is known to exist. It is 800 miles thick.
Inside the hollow, a sun, 600 miles in diameter, gives perpetual daylight
to the interior. There are openings at both poles, each 1,400 miles across.
Other planets are constructed the same way. The so-called ice-caps on
Mars are really openings, and occasionally you can see gleams through
them from the inner sun. On the earth, light streaming out of the northern
opening produces the *aurora borealis*.

The frozen mammoths found in Siberia came from the interior of the
earth, and some may still be living there. The Eskimos also came from
inside, as indicated by their legends of a land of never-ending summer. A
chapter is devoted to an imaginary journey through the earth—into one
polar opening and out the other. A beautiful illustration in full color
shows the interior sun just above a watery horizon as the ship draws near
to the great rim. Seven chapters are devoted to various expeditions to the
North Pole. Gardner proves, of course, that no explorer ever really got
there.

The author admits he does not expect to get a "fair hearing" for his
views because of the "conservatism of scientists who do not care to
revise their theories—and especially when that revision is made necessary
by discoveries . . . made independently of the great universities." The
scientists, he writes bitterly, "have their professional free-masonry. If you
are not one of them, they do not want to listen to you." Ultimately,
however, he believes the public will accept his views and force the
scientists to do likewise.

He makes it quite clear that he does not wish to be confused with
scientific pretenders like Symmes who do not base their thinking on solid
facts. "Of course it is very easy for anyone to deny all the facts of
science and get up some purely private explanation of the formation of
the earth. The man who does that is a crank." Like all paranoid scientists, Gardner was incapable of seeing himself in any other light than as
an unappreciated genius, ridiculed at the moment, but destined to even-
tual honor. He makes the inevitable comparison of himself with Galileo.
It was the First World War, he feels, which diverted the world's attention
from his earlier book.

Ironically, it was less than six years after Gardner published the costly,
revised edition of his opus, that Admiral Richard Byrd flew over the pole
in a plane. There was, of course, no hole. Gardner ceased lecturing and
writing, although when he died in 1937, he was still convinced that his
theory had some merit.

Fantastic though Symmes' theory was, or Gardner's variation, an even more preposterous view was formulated in 1870 by another American— Cyrus Reed Teed. For 38 years, with unflagging energy, Teed lectured and wrote in defense of the theory that the earth was hollow and we were living on the inside!

Not much is known about Teed's early life. He was born in 1839 on a farm in Delaware County, New York, and in his youth was a devout Baptist. During the Civil War he served as a private with the Union Army, attached to the field hospital service. Later he graduated from the New York Eclectic Medical College, and established practice in Utica, N. Y. (Eclecticism was a popular medical cult of the last century. It relied mostly on worthless herb remedies.)

The Copernican theory, with its infinite spaces and gigantic suns, must have terrified young Teed. He longed to restore the cosmos to the small, tidy, womb-like character he found implied by Holy Scripture. That the earth was round he could not doubt, for mariners had sailed around it. But if this were so, where did space end? It seemed unthinkable that it would go on and on, without ever reaching a boundary.

One midnight in 1869, when Teed sat alone in a laboratory he had set up in Utica for the study of alchemy, he had a vision. His pamphlet, *The Illumination of Koresh: Marvelous Experience of the Great Alchemist at Utica, N. Y.*, describes this vision in detail. A beautiful woman spoke to him. She told him of his past incarnations and the role he was destined to play as a new messiah. And she revealed to him the key of the true cosmogony.

The key was a simple one. We are on the *inside* of the earth. The astronomers are right, in a way, only they have everything inside-out. Does not the Bible say God "hath measured the waters in the *hollow* of his hand?" (Isaiah 40:12) The more Teed meditated about this, the more he became convinced it was true. In 1870, under the pseudonym of Koresh (the Hebrew equivalent of Cyrus), he published *The Cellular Cosmogony* in which he outlined the new astronomical revelation.

The entire cosmos, Teed argued, is like an egg. We live on the inner surface of the shell, and inside the hollow are the sun, moon, stars, planets, and comets. What is outside? Absolutely nothing! The inside is all there is. You can't see across it because the atmosphere is too dense. The shell is 100 miles thick and made up of seventeen layers. The inner five are geologic strata, under which are five mineral layers, and beneath that, seven metallic ones. A sun at the center of the open space is invisible, but a reflection of it is seen as our sun. The central sun is half light and half dark. Its rotation causes our illusory sun to rise and set. The moon is a reflection of the earth, and the planets are reflections of "mercurial discs floating between the laminae of the metallic planes." The heavenly bodies we see, therefore, are not material, but merely focal points of light, the nature of which Teed worked out in great detail by means of optical laws.

The Foucault pendulum takes up an entire chapter. "The marvelous thing about this experiment," he writes, "is that any man possessing any claim whatever to the title of scientific should accept it." Teed's theory is that the turning of the pendulum is due to the influence of the sun. The whole thing, he concludes, "is the veriest nonsense, and later the 'scientists' will laugh at their own folly."

The earth, it is true, *seems* to be convex, but according to Teed, it is all an illusion of optics. If you take the trouble to extend a horizontal line far enough, you will always encounter the earth's *upward* curvature. Such an experiment was actually carried out in 1897 by the Koreshan Geodetic Staff, on the Gulf Coast of Florida. There are photographs in later editions of the book showing this distinguished group of bearded scientists at work. Using a set of three double T-squares—Teed calls the device a "rectilineator"—they extended a straight line for four miles along the coast only to have it plunge finally into the sea. Similar experiments had been conducted the previous year on the surface of the Old Illinois Drainage Canal.

Like most pseudo-scientists, who wish to impress the reader with their vast scientific knowledge, Teed has a tendency to let his words carry him into obscurities sometimes hard to follow. Planets, for example, are "spheres of substance aggregated through the impact of afferent and efferent fluxions of essence . . .." And comets are nothing less than "composed of cruosic 'force,' caused by condensation of substance through the dissipation of the coloric substance at the opening of the electro-magnetic circuits, which closes the conduits of solar and lunar 'energy.'"

Teed's paranoia emerges unmistakably in the bitterness with which he attacks "orthodox" scientists as "humbugs" and "quacks," who "palm off" their work as science on a "credulous public." An entire chapter is devoted to the "unreasonable opposition" and "stubborn resistance" of his enemies. He likens himself (as does almost every pseudo-scientist) to the great innovators of the past who found it difficult to get their views accepted. "The opposition to our work today is as unreasonable, absurd, and idiotic as that manifested against the work of Harvey and Galileo."

On another page: "We have devoted much energy and effort to bring the question of Koreshan Universology permanently before the people for public discussion. In this effort we have been held up to insolent ridicule and most bitter persecution, consonant with the invariable rule to which every innovation upon prevailing public sentiment is subject. . . . We have pushed our claims to a knowledge of cosmology until the advocates of the spurious 'sciences' begin to feel their insecurity. . . .''

And here is his most shocking and revealing declaration. ''. . . to know of the earth's concavity . . . is to know God, while to believe in the earth's convexity is to deny Him and all his works. All that is opposed to Koreshanity is antichrist.''

It was not surprising that Teed considered himself a messiah. Nor was it surprising that his medical practice in Utica, where he became known as the ''crazy doctor,'' declined rapidly. His wife, confused and ill, left him (his only child, Douglas Teed, later became a prominent southern artist and portrait painter). Finally ''Cyrus the Messenger,'' as he called himself in those days, abandoned medicine entirely and took to the road to preach his new revelation.

As an orator, he must have been magnificent. In Chicago, he attracted such a devoted following that he settled there in 1886, founding a ''College of Life'' and a periodical called *The Guiding Star* (soon succeeded by *The Flaming Sword*). Later he established Koreshan Unity, a small communal society housed in a building on Cottage Grove Avenue. *The Chicago Herald,* in 1894, credited him with 4,000 followers and said he had collected $60,000 from preaching engagements in California alone.

The paper described Koresh as ''an undersized, smooth-shaven man of 54 whose brown, restless eyes glow and burn like live coals. . . . He exerts a strange mesmerizing influence over his converts, particularly the other sex.'' In later years, he always wore a Prince Albert coat, black trousers, flowing white silk bow tie, and wide-brimmed black felt hat. Three out of four of his followers were women.

In Carl Carmer's book, *Dark Trees to the Wind,* 1949, there is an excellent chapter on Koreshanity which contains the testimony of one of its members about how he became converted to the movement. The testimony suggests that a desire to return to the womb may have played a considerable role in the cult's success. The man had been a barber in Chicago's Sherman Hotel. One day in 1900, when he was walking down State Street, his eye caught a huge sign which read, ''We live inside.'' A man was speaking to a small street crowd and selling copies of *The Flaming Sword*. The barber bought a copy. ''I read it in bed that night,'' he said. ''Before I went to sleep—I was inside.''

In the 1890's, Koresh obtained a tract of land about 16 miles south of Fort Meyers, Florida, where he founded the town of Estero. He called it ''The New Jerusalem'' and predicted it would some day be the capital of the world. Arrangements were made to accommodate eight million believers. Two hundred arrived. But they did manage to keep the colony going, in spite of ridicule from nearby newspapers.

Teed's death in 1908—resulting from a personal assault by the Marshal of Fort Meyers—was embarrassing. He had written a book called *The Immortal Manhood* in which he taught that after his "physical death," he would rise and take to heaven all the followers who had been faithful. When he died, on December 22, members of the colony stopped working and kept a constant prayer vigil over the body. After two days, Koresh began to show signs of corruption. Christmas came and went. The county health officer finally stepped in and ordered a burial.

Teed's followers buried their beloved leader in a concrete tomb on Estero Island, off the Gulf Coast. In 1921, a tropical hurricane pounded the island with giant waves that carried off the tomb. No trace of the body was ever found.

The cult's handsome little magazine, *The Flaming Sword,* made no mention of Teed's death in 1908. It continued to be published until as late as 1949 when the colony's printing plant was destroyed by fire. A 1946 issue pointed out that Teed's alchemical views had foreshadowed the atom bomb. In 1947, only a dozen members of the cult were left. And they were wrangling over property rights.

In Germany, Teed's writings provided the basis for a cult that flourished widely in the anti-cultural climate of the Nazi movement. It was known as the *Hohlweltlehre,* or Hollow Earth Doctrine, first proclaimed by Peter Bender, a German aviator badly wounded in the First World War. Bender corresponded at length with the Koreshans until his death in a Nazi prison camp. His work is still carried on in Germany, chiefly by Karl E. Neupert whose book *Geokosmos* is the most important textbook of the cult.[1]

In America, after 60 years of valiant battle, *The Flaming Sword* went down fighting. Its final issue contained an indignant article about a picture *Life* had published of some salt flats in Utah. The picture purported to show the earth's convexity. The editors had written *Life,* explaining the mistake. *Life* had answered, but *The Flaming Sword* editors felt that the reply was evasive.

CHAPTER 3

MONSTERS OF DOOM

EXACTLY HOW THE earth and its sister planets came into existence is a subject of much dispute among astronomers. A number of rival theories seem equally plausible. But all astronomers agree that whatever happened, it took place billions of years ago, and that within the tiny span of human history, the earth has been whirling on its axis and making its appointed journey around the sun with unperturbed regularity.

To the pseudo-scientist, any agreement among the "orthodox" is a welcome challenge. It is not surprising, therefore, to find as many eccentric theories about the earth's recent astronomical past as about its shape. In this chapter we shall glance at four such theories. Each of them maintains that the earth, since the time of man, has undergone one or more cosmic catastrophes. Each insists that the upheavals were caused by the earth's close contact with a heavenly body.

Worlds in Collision, by Dr. Immanuel Velikovsky, is the most recent of the four theories. When the book was published in 1950, after a shrewd publicity campaign, the first reaction of most professional astronomers was to regard it as a hoax. The second response—when it became clear that not only was Velikovsky in deadly earnest but publishers and editors seemed equally convinced he had something important to say— was one of rage. The flood of indignant letters to the publisher from scientists who threatened to boycott the firm's textbooks, led to the dismissal of the associate editor who brought the manuscript to the company's attention. Publication rights were turned over to Doubleday. By this time the book had jumped into the best-seller class, and Doubleday—which has no textbook department—was delighted to take it.

What are Velikovsky's theories? In brief, he believes that a giant comet once erupted from the planet Jupiter, passed close to the earth on two occasions, was tamed by Mars in a love affair that produced a brood of smaller comets, then finally settled down as the planet Venus. The comet's first encounter with the earth was during the Israelite Exodus of 1500 B.C. It caused the earth either to stop spinning entirely or to slow down suddenly—Velikovsky isn't sure which. In either case, the learned doctor correctly perceives that cataclysmic inertial results would have occurred. The surface of the earth literally burst—with mountains collapsing, others arising, floods, hurricanes, dust storms, fires, seas boiling, rivers the color of blood, a shower of meteoric rocks, the jumping of a huge spark from the earth to the comet's head, and a rain of petroleum. Modern cars and planes, he tells us, are propelled by fuel refined from "remnants of the intruding star that poured fire and sticky vapor. . . ."

One looks in vain for an explanation of the total absence of marked east-west orientations in geologic structures formed by these displacements, or for the fact that thousands of relatively delicate and ancient structures remained undisturbed by the great upheavals. The slightest shifting of the earth's crust, for example, would have altered the vertical lines of the long stalactites in Carlsbad, whose ages go far beyond the earliest of Velikovsky's catastrophes. The slow building of the Mississippi Delta, the imperceptible erosion that produced the Grand Canyon, the gradual recession of Niagara Falls, and countless other geologic processes have left formations that obviously have not been disturbed for tens of thousands of years.

According to Velikovsky, it was the stopping (or slowing) of the earth's spin which caused the Red Sea to divide precisely at the time Moses stretched out his hand, permitting the children of Israel to cross in safety before the waves engulfed the pursuing Egyptians. There was a momentary two-month retreat of the comet, then it returned just in time to provide the thunder, lightning, smoke, earthquakes, and trumpet blasts that accompanied the giving of the commandments to Moses on Mount Sinai. Several years later, a precipitate of carbohydrates that had been in the comet's tail fell in the form of Manna which kept the Israelites alive for forty years. Velikovsky does not explain why this Manna, if we are to believe the sixteenth chapter of *Exodus,* failed to precipitate on every seventh day.

After 52 years of quiet, the comet returned again. This visit coincided with Joshua's successful attempt to make the sun stand still upon Gibeon and the moon to rest in the Valley of Ajalon. Velikovsky is not certain whether this was due to a stopping of the earth, or an illusion created by a shifting of the earth's axis while it continued to turn. In any case, the geological upheavals occurred all over again, only this time there was no giant spark or a reversal of the magnetic poles. Hailstones which had remained close to the earth since the first visit of the comet, fell on top of the Canaanites.

For several years the earth cooled and settled, producing many local earthquakes. One of these quakes conveniently caused the walls of Jericho to topple and destroy the wicked city just as the priests blew their trumpets for the seventh time. "The Israelite tribes, believing in magic, thought that the sound of the earth came in response to the blowing of the ram's horn. . . ." Velikovsky reveals. Apparently the doctor's opinion is that God performed all the Old Testament miracles by means so cleverly contrived that all of them could be explained by natural laws.

Seven centuries after the time of Joshua, or about the middle of the eighth century before Christ, a new series of catastrophes occurred. They were due to the approach near the earth of the planet Mars, and account for numerous Old Testament events such as the sudden destruction of Sennacherib's army and sundry quakes prophecied by Amos, Isaiah, and others. The second half of the book is devoted to these Mars-produced minor upheavals. It lacks, however, the scope and flair of the first half, and its dreary details need not concern us. (Nor need we be concerned with the doctor's projected two-volume revision of the ancient history of the East. The first half of this fantasy, *Ages in Chaos,* was published in 1952.)

It may be of interest to note in passing that Velikovsky makes use of the close approach of Mars to explain what is perhaps the most astonishing scientific guess of all time. In *Gulliver's Travels,* Swift mentions casually that the astronomers of Laputa discovered that Mars had two satellites. Mars *does* have two moons, but they were not spotted until 156 years later. It was 100 years until a telescope was made large enough even to see the moons! Moreover, Swift's prediction about their periods of revolution corresponds closely to their actual periods. One of the moons, Phobos, goes around Mars in the same direction the planet rotates, but only in one-third the time so that it appears to rise in the west and set in the east. This is the only known body in the universe that revolves around a central body *faster* than the central body rotates, yet this fact also is included in Swift's brief description! In Velikovsky's opinion, Swift got his information from ancient manuscripts he had chanced upon, and which were based on actual observations of the moons at a time when Mars was close to the earth. The doctor thinks this was why Greek mythology had Mars' chariot drawn by two horses—the two horses, incidentally, for which the moons were later named.

Throughout *Worlds in Collision,* Velikovsky's evidence for his theories consists almost entirely of legends which he believes reflect memories of the ancient catastrophes. The task of assembling such a collection is much easier than one might suppose. Legends are difficult to date, and almost any imaginable type of nature miracle is likely to turn up somewhere in the varied folklore of a culture. All one has to do is comb the literature of mythology carefully, copy down the appropriate tales, and ignore the others. Even so, the ancient records often fail to come through to Velikovsky's satisfaction. When this is the case, he blames the silence on "collective amnesia."

Although *Worlds in Collision* is no longer taken seriously, it is astonishing how many who reviewed the book were caught off-guard. John J. O'Neill, science editor of the *New York Herald Tribune*, described the work as "a magnificent piece of scholarly research." Horace Kallen, co-founder of the New School for Social Research and a distinguished educator, wrote, "The vigor of the scientific imagination, the boldness of construction and the range of inquiry and information fill me with admiration." Gordon Atwater, at the time chairman and curator of New York City's Hayden Planetarium, declared that "the theories presented by Dr. Velikovsky are unique and should be presented to the world of science in order that the underpinning of modern science can be re-examined." Ted Thackrey, editor of the *New York Compass*, suggested that Velikovsky's discoveries "may well rank him in contemporary and future history with Galileo, Newton, Planck, Kepler, Darwin, Einstein. . . ." The book was also enthusiastically endorsed by Clifton Fadiman and Fulton Oursler.

Dr. Velikovsky is an almost perfect textbook example of the pseudo-scientist—self-taught in the subjects about which he does most of his speculation, working in total isolation from fellow scientists, motivated by a strong compulsion to defend dogmas held for other than scientific reasons, and with an unshakable conviction in the revolutionary value of his work and the blindness of his critics.

The doctor was born in Russia in 1895. After wandering about Europe and the Middle East, attending schools here and there, he obtained an M.D. from Moscow University and settled in Palestine as a general practitioner. Later he became a psychoanalyst, studying for a brief time under Wilhelm Stekel in Vienna. (For a sample of the doctor's writing as an analyst, see his article discussing Freud's dreams in the *Psychoanalytic Review*, October, 1941.) In 1939, he brought his wife and two daughters to New York. For the next nine years, to quote his own words, he daily "opened and closed the library at Columbia University." Presumably this was where he did his research on comets.

Velikovsky is a tall, slightly stooped, silver-haired man with sharp features, charming manners, and a persuasive way of speaking. His attitude toward those who disagree with him is one of disarming politeness. "If I had not been psychoanalytically trained," he told one reporter, "I would have had some harsh words to say to my critics." He is convinced, of course, that all the great scientists of the world are refusing to accept his work not because they see its mistakes, but because they are afraid to cut the umbilical cord which binds them to scientific orthodoxy.

That Velikovsky feels this strongly was brought out in his "Answer to My Critics," published in *Harper's,* June, 1951. He speaks of the "highly unscientific fury" with which scientists greeted his book, and coins the term "collective scotoma" (blind spot) for their inability to read it properly. He has little hope for the acceptance of his theories by the older generation of scholars who "have a vested interest in orthodox theories," and "are for the most part psychologically incapable of re-learning." He looks to the younger generation, whose minds are more flexible, for ultimate vindication.

Velikovsky's reply is followed by an article by Professor John Stewart, an astronomical physicist at Princeton. To Professor Stewart's devastating criticism, Velikovsky pens a brief rebuttal. Anyone familiar with modern science will find the doctor's two articles honeycombed with evasion. For example, Stewart points out that no known laws of gravity and motion can account for the ability of Velikovsky's comet to stop the earth from rotating, or to start it up again, or for Mars' ability to propel Venus into her present orbit. Velikovsky has a simple answer to all this. He invents electro-magnetic forces capable of doing precisely what he wants them to do. There is no scientific evidence whatever for the powers of these forces. They serve the same function for Velikovsky that curious optical laws served for Cyrus Teed. They explain the unexplainable. But so convinced is the hermit scientist that everyone is prejudiced except himself, that he can—with a straight face—belabor the "orthodox" for refusing to recognize these imaginary energies! "The reluctance to recognize the existence of electrical and magnetic forces in the celestial sphere," the doctor writes, ". . . is in danger of becoming a dogma, called upon to protect existing teachings in celestial mechanics."

Beneath Velikovsky's skillful writing and the torrent of alleged facts, references, footnotes, and quotations garnered from his nine-year hibernation in Columbia's library is one clear—though seldom-stated—emotional premise. "It was in the spring of 1940," he tells us, "that I came upon the idea that in the days of the Exodus, as evident from many passages of the Scriptures, there occurred a great physical catastrophe. . . ." The Old Testament is sacred scripture for both fundamentalism and traditional Judaism; and Velikovsky's theories, like those of Voliva, are no more than rationalizations of priorheld beliefs. Of course, the good doctor is much more learned and sophisticated than poor Voliva, so his fantasies are far less obvious.

If Velikovsky had been the first to develop a theory of comet-caused catastrophe, his book would have at least the merit of opening up a new field for pseudo-scientific exploration. Unfortunately, even this is denied him. He had two distinguished predecessors, with remarkably similar views. Each supported his case with the same myths and tales of folklore collected by the doctor, and each presented his theory in harmony with the same Old Testament texts.

William Whiston, a British clergyman and mathematician (he succeeded Newton as professor of mathematics at Cambridge) published his *New Theory of the Earth* in 1696. By that time it had become clear to everyone that the earth is round. Theologians had found an abundance of passages throughout the Bible which seemed to suggest a round earth. Even texts which had previously been used to prove flatness—such as "It is he that sitteth upon the circle of the earth" (Isaiah 40:22)—were now perceived to be descriptions of a globe. The Newtonian world-view was in the air, and there was an understandable desire to fit it as neatly as possible into the Biblical verses.

According to Whiston, the original "chaos" was the tail of a giant comet. "And the earth was without form, and void; and darkness was upon the face of the deep." Out of this chaos, the earth, planets, and their moons took shape. The orbits were perfect circles. The earth made its journey around the sun in exactly 360 days and the moon circled the earth in thirty. The earth, however, did not at first spin on its axis, so the "days" of creation were actually a year in length. It was not until Adam and Eve ate the forbidden fruit that the comet's force started the earth rotating. The earth's atmosphere in these early times was warm and clear. Moisture was so finely distributed that no rainbow could form.

On Friday, November 28, 2349 B.C., another comet "under the conduct of Divine Providence" visited the earth as God's instrument for punishing a wicked world. Water vapor from the comet's tail condensed and fell upon the land for forty days and nights. Fortunately, Noah was able to save his family and a shipload of animals from destruction. The earth's orbit was thrown into an elliptical shape, increasing the year to its present length. There was a slowing of the earth's rotation. At first, Whiston thought this was due entirely to the inertia of the large amount of water acquired by the earth, but in later editions of the book, he decided that magnetic forces also played a role. Eventually the skies cleared, a rainbow appeared for the first time, and the water slowly drained into the "bowels" of the earth. All this is worked out in great mathematical detail, with an abundance of diagrams and learned footnotes in Greek. Scores of legends from different cultures are cited to substantiate the Biblical account.

The climate of scientific opinion in seventeenth-century England was so favorable to Protestant orthodoxy, and knowledge of astronomy and geology in such rudimentary stages, that Whiston's work was well received by his colleagues. The great Newton and the English philosopher John Locke praised the book highly. One must not, therefore, think of Whiston as a pseudo-scientist to the same degree as later proponents of theories of comet-caused cataclysms.

After the time of Whiston, the notion that wandering comets were responsible for great earth catastrophes continued to fire the imagination of pseudo-scientists. It was not until 1882, however, that a really new twist was given to these speculations. This was the year a Minnesota Irishman named Ignatius Donnelly published a sensational work called *Ragnarok*.

Donnelly was one of America's most colorful political figures. Although he lacked college training, he won early admittance to the bar, and it was not long until his fiery oratory and agricultural reform views made him lieutenant-governor of Minnesota. He was then only twenty-eight. Four years later he was elected U.S. Congressman, and later state senator. A reformer with radical leanings, for years he edited a weekly paper called the *Anti-Monopolist*. His novel, *Doctor Huguet*, was a moving plea for racial tolerance, and a second novel, *Caesar's Column*, was the first American "It Can't Happen Here" fantasy. Shocked by the political corruption of his home state, he pictured in horrifying scenes the coming of a twentieth-century fascism. The novel was warmly praised by Hawthorne, and sold over a million copies. When Donnelly died in 1901 he was the Populist candidate for vice president.

Donnelly has rightly been dubbed the Prince of U.S. Cranks. Usually the pseudo-scientist is obsessed with but one central topic, but Donnelly was obsessed by three—the existence of Atlantis, a cipher message concealed in the plays of Shakespeare (proving they were written by Francis Bacon), and the catastrophic effects upon the earth of a visiting comet. In a later chapter we shall look into his *Atlantis. The Great Cryptogram* and *The Cipher in the Plays* fall outside our scope. Here we shall be concerned only with *Ragnarok*.

D. Appleton and Company brought out the book and it was an immediate success with the general public. Scientists ignored it, but the reviews in popular journals were favorable. Note how remarkably similar these excerpts are to the passages previously quoted from reviews of Velikovsky's work. "It is a bold enterprise," one critic declared, "and its very boldness gives it a peculiar fascination. The vast range of the survey and the multitude of witnesses, of every age and clime, which the author passes in review, yield the reader a decidedly new sensation. . . ." Another reviewer commented, "*Ragnarok* supplies a new theory . . . coherent in all its parts, plausible, not opposed to any of the teachings of modern science, and curiously supported by the traditions of mankind. If the theory is true, it will . . . revolutionize the present science of geology."

The word "Ragnarok" comes from an ancient Scandinavian legend and means a "rain of dust." Donnelly devotes two hundred pages to myths from all over the world telling of this early catastrophic event which he believes was caused by a giant comet. It swooped suddenly on the earth, with ". . . world-appalling noises, thunders beyond all earthly thunders, roarings, howlings, and hissings, that shook the globe . . . a storm of stones and gravel and clay dust . . . leveling valleys, tearing away and grinding down hills, changing the whole aspect of the habitable globe . . . through the drifts of debris glimpses are caught of the glaring and burning monster; while through all and over all is an unearthly heat. . . ."

After the comet had passed, and the great fires subsided, an Age of Darkness settled over the world and the Ice Age began. Donnelly rejected completely the theory, widely accepted by the geology of his time, that deposits of "till"—unstratified clay and gravel found here and there about the globe—were produced by moving glaciers. They came instead, he argued, from the congealed dust of the comet's tail. Flourishing cities were demolished by the earth's convulsion. The *Book of Job,* the story of Sodom and Gomorrah, and other Old Testament tales are memories of this event. Concerning Joshua's miracle he writes:

> And even that marvelous event, so much mocked at by modern thought, the standing-still of the sun at the command of Joshua, may be, after all, a reminiscence of the catastrophe. . . . In the American legends, we read that the sun stood still, and Ovid tells us that 'a day was lost.' Who shall say what circumstances accompanied an event great enough to crack the globe itself into immense fissures? It is at least, a curious fact, that in Joshua (Chapter X) the standing still of the sun was accompanied by a fall of stones from heaven by which multitudes were slain.

Donnelly, the reformer, could not resist the temptation to close his book with a blast at plutocracy. Addressing himself to "Dives"—the rich man—he pleaded: "Widen your heart. Put your intellect to work to so readjust the values of labor, and increase the productive capacity of Nature, that plenty and happiness, light and hope, may dwell in every heart. . . . And from such a world God will fend off the comets with his great right arm, and the angels will exult over it in heaven."

It is hard to say how much Donnelly actually believed the scientific claims he set forth in his book. There was a touch of the charlatan about him, and perhaps *Ragnarok* was written primarily to gain public attention and money. In any case, astronomical and geological opinion had advanced far beyond the days of Whiston, and so his comet whizzed by the scientists with scarcely an official notice.

Twelve years after Donnelly's death a Viennese mining engineer, Hans Hörbiger, in collaboration with an amateur astronomer, published his monumental, 790-page *Glazial-Kosmogonie*. It is one of the great classics in the history of crackpot science—filled with photographs and elaborate diagrams, heavy with the thoroughness of German scholarship, and from beginning to end, totally without value. It is almost as though the Germans, so superior in most fields of scientific learning, refused to be surpassed even in the field of pseudo-science.

When Hörbiger's book first appeared, it provoked bellows of rage from German astronomers, but in the mystical, anti-intellectual atmosphere of the rising Nazi movement, its fantastic theories soon acquired millions of fanatical followers. The cult became known by the abbreviation WEL, the initials of *Welt-Eis-Lehre* (Cosmic Ice Theory)—a term used by Hörbiger for describing his doctrine.

According to Willy Ley, the rocket authority who was living in Germany at the time, the WEL functioned almost like a political party. It issued leaflets, posters, and publicity handouts. Dozens of popular books were printed describing its views, and the cult maintained a monthly magazine called *The Key to World Events*. Disciples often attended scientific meetings, Ley reports (in an article, "Pseudo-Science in Naziland," *Astounding Science Fiction*, May, 1947), to interrupt the speaker with shouts of "Out with astronomical orthodoxy! Give us Hörbiger!"

The master's paranoia was not well concealed. In a letter to Ley he wrote, ". . . either you believe in me and learn, or you must be treated as an enemy." His "enemies" included, of course, all the leading astronomers of the world. The Vienna engineer was convinced that they rejected his views solely because he was not a recognized astrophysicist.

Moons, not comets, fascinated Hörbiger. Before the earth captured Luna, its present moon, it had at least six previous ones and maybe more. All the geologic upheavals that Donnelly blamed on comets, the Austrian pseudo-scientist blamed on these former satellites.

Chief among the scholarly satellites captured by Hörbiger was a British student of mythology, Hans Schindler Bellamy. After Hörbiger died in 1931, Bellamy wrote the definitive English account of WEL, backing it up with over 500 legends drawn from the folklore of the world. Titled *Moons, Myths, and Man*, the book was first published by the London firm of Faber and Faber. Harper brought out an American edition in 1936, and the following year Faber and Faber issued it again in revised form. It is chiefly from this work that the following brief account of WEL is drawn.

Space, according to Hörbiger, is filled with rarefied hydrogen. It offers enough resistance to planets and moons to cause them to move in slow spirals toward the central body. Eventually, all the planets will drop into the sun. Occasionally, small planets, on their spiral inward journey, get "captured" by larger planets and become moons. Bellamy is chiefly interested in the last moon before Luna because man existed during its cycle of birth and death, and so was able to hand down in the form of legends a highly accurate record of what took place. Myths are, he says, a kind of "fossil history." Their study is "a new science of pre-Lunar culture."

The tertiary moon, as Bellamy calls it, was slightly smaller than the present one. As it came closer to the earth, it moved faster and faster around it. The oceans were pulled into a "girdle tide"—a high, narrow belt near the equator. The rest of the earth entered a glacial age. To escape freezing, men migrated into "island refuges" or high mountain areas on either side of the girdle—Mexico, Tibet, Abyssinian highlands, the Bolivian Meseta. Huge in appearance, the moon was soon circling the earth six times a day, eclipsing the sun three times, and itself blotted out three times by the earth's shadow. Its pitted surface resembled scales and gave rise to legends of dragons and flying monsters. It was the "Devil" of the Hebrew-Christian folklore.

Eventually the pull of the earth became greater than the cohesion holding the moon together. It began to fall apart. A thick layer of ice on its surface cracked and melted, causing great rains and gigantic hailstones. Then came a torrent of rocks as the moon disintegrated completely. The earth had been twisted out of shape by the monster. Now it sprang back into spherical form with violent earthquakes and intense volcanic action. The "girdle tide" flowed back over the earth. This was the Deluge of Noah, and the basis of other myths of a great flood. Here and there, however, individuals managed to escape. Bellamy writes:

> Not without tenderness, we leave this chapter on Arks. The passage in Genesis has always appealed to our imagination; when we were young, at Sunday-school it was a favorite text; and when we grew up and doubted, for lack of a plain mythological explanation, many of the obscure statements of the Bible, it remained one of the chief passages which we felt might be based upon fact. And, indeed, at the close of the Tertiary Aeon there tossed on the rolling waves of the deluge many an ark, the cradle of a new race. . . . We ourselves may be descended from one of those deluge heroes— unless our forefathers found refuge on some mountain peak or escaped in some other way.

Legends of dragon slayers and battles of gods against monsters reflect the destruction of the satellite. A period of calm and mild climate descended on the moonless world. Myths of paradise recall this epoch. It came to an end when Luna, our present moon, was captured. The capture sent the earth into new paroxysms of quaking. The axis shlfted, the poles glaciated, the continent of Atlantis was submerged, and the Quaternary geologic period began.

Bellamy calculates that the capture of Luna took place about 13,500 years ago. Not only do we have myths describing it, but there are also racial memories buried in our subconscious. "When a boy," he writes, "I often dreamt vividly of a large moon . . . glaringly bright, and so near that I believed I could almost touch its surface. It moved quickly through the heavens. Suddenly it would change its aspect and—almost explosively— burst into fragments, which, however, did not fall down immediately. Then the ground beneath me would begin to roll and pitch, helpless terror would fall upon me—and I would awake with the sick feeling which one has after a terrible nightmare."

Later, when he had his first peep at the moon through a telescope, he found the pockmarked surface troublingly familiar. In 1921, he beame acquainted with Hörbiger's theory, and to his surprise, found the Austrian writer describing his dream. "Since that time I have often tried to coax my subconscious mind to give me another performance of my cosmological dream, but in vain. My mental efforts must have shocked that cell which had reproduced this memory of a dead age. Or, perhaps, the finding of a thoroughly satisfactory solution to that dream-picture had made its further repetition unnecessary."

Among Bellamy's many other books, two are quaint enough to deserve mention. *The Book of Revelation Is History* is a commentary on the Apocalypse to end all commentaries. He regards the vision of Saint John as a factual record of the close of the Tertiary Age! A commentary on *Genesis,* titled *In the Beginning God,* takes a similar approach. Genesis is the account, not of creation, but of recreation after the last catastrophe. He has a novel interpretation of the story about Eve's formation from Adam's rib. It reflects the memory of an early Caesarean birth involving a heroine of the deluge. Somehow the myth got the sex wrong!

Our present moon is, according to Hörbiger, now grimly spiraling its way toward us. A coating of ice 140 miles thick surrounds its surface. Similar coats of ice envelop Mercury, Venus, and Mars.

On Mars, the frozen sea extends 250 miles down. Cracks in this ice appear to us as the Martian "canals." The Milky Way is a ring of gigantic ice blocks. When someone reminded Hörbiger that photographs have proved the Milky Way consists of billions of stars, his answer was blunt and direct. The prints, he said, were faked.

The giant ice blocks also spiral toward the center of the solar system. When one falls into the sun, it causes a sunspot. The ice immediately vaporizes and the cloud is blown out into space where it freezes into a fine cosmic powder. It is this ice which coats all the planets except the earth. For reasons too complex to go into here, the cosmic ice does little more to the earth than produce high cirrus clouds and occasionally a violent hailstorm.

"To pick flaws in this theory," writes Willy Ley, "is about as easy—and as pleasant—as gathering Japanese beetles from an infested flowerbed." But the German astronomers stopped laughing when the cult began to grow. So successful were the members in combining WEL with Nazi political philosophy, that the Propaganda Ministry was obliged to state, "one can be a good National Socialist without believing in the WEL." From the cult's literature, Ley quotes the following statements:

"Our Nordic ancestors grew strong in ice and snow; belief in the World Ice is consequently the natural heritage of Nordic Man."

"Just as it needed a child of Austrian culture—Hitler!—to put the Jewish politicians in their place, so it needed an Austrian to cleanse the world of Jewish science."

"The Führer, by his very life, has proved how much a so-called 'amateur' can be superior to self-styled professionals; it needed another 'amateur' to give us complete understanding of the universe."

The above passages reveal with horrible clarity how easily the paranoia of the Master finds responsive echoes in the paranoid drives of fanatical disciples.

It is interesting to note that Velikovsky's book makes no mention of Hörbiger's rival theory or of Bellamy's ingenious folklore documentation, yet the doctor's followers are vastly outnumbered by believers in the WEL. In Germany and England today, it may still have more than a million adherents. The Hörbiger Institute continues to flourish in the Reich. A branch in England is busily turning out books, pamphlets, and periodicals. "Final proof of the theory," a recent London leaflet states, "awaits the conclusion of the first successful interplanetary flight, a matter in which the Institute is greatly interested."

Pseudo-scientific doctrines, like religious sects, are hard to kill. It took Byrd's flight over the North Pole to deal a death blow to "Symmes' hole." Perhaps the Cosmic Ice Theory will find disciples until the first spaceship lands on the cratered surface of an iceless moon.

CHAPTER 4

THE FORTEANS

SOONER OR LATER in this book, we shall have to come to terms with Charles Fort. Because Fort aimed his heaviest artillery at the astronomers and gave his imagination freest reign in proposing unorthodox views about the heavens, this seems the appropriate spot to introduce him.

Charles Hoy Fort was born in Albany, New York, in 1874. As a boy, his interest in science led him to collect minerals and insects, and occasionally to stuff a bird. He never went to college. After working for a time as a reporter, and trying his hand at a novel (*Outcast Manufacturers,* 1909), and a few short stories (which Theodore Dreiser published in his *Smith's Magazine*), Fort came into a small real estate inheritance. It was this income which freed him for an almost unbelievable stint of private research. For the remaining twenty-six years of his life, he pored over old magazines and newspapers, taking notes on every mysterious occurrence which did not jibe with established scientific notions. Most of this work was done in the British Museum. Later he returned to New York where he lived in the Bronx with his wife, Anna, and continued his studies at the New York Public Library.

Fort was a large, shy, bear-like man, with a brown walrus mustache, and thick glasses. His apartment was filled with shoe boxes crammed with notes and clippings. On the walls were framed specimens of spiders and butterflies, and under a glass, he kept a hunk of dirty, asbestos-like material which had dropped from the sky. For recreation he played a solitaire game he invented called "super checkers." It involved a thousand men on a huge board of several thousand squares. His wife, according to novelist Tiffany Thayer, never understood what went on in her husband's mind, and "never read his or any other books."

Fort had only two friends—Dreiser and Thayer. Convinced that Fort was a genius, it was Dreiser who persuaded his own publisher to bring out the first of Fort's four books, *The Book of the Damned*. By the "damned," Fort meant all those views which are excluded by dogmatic science—the "lost souls" of data. His self-appointed mission was the "undamning" of this data. The book was written in a curious, breathless style. At times, it broke into passages of profound wisdom, high humor, and beautiful phrasing.

Fort's second book, *New Lands*, was published in 1923 with an introduction by Booth Tarkington. By this time, a number of American writers had become fascinated by Fort's hilarious attack on what he called the scientific "priestcraft." In 1931, Thayer rounded up the writers for a historic banquet at the Savoy Plaza, on which occasion the Fortean Society was born. The founders included, besides Dreiser and Thayer, such literary lights as Alexander Woollcott, Tarkington, Ben Hecht, Burton Rascoe, and John Cowper Powys.

Fort's third book was titled *Lo!* "*Lo!* was my suggestion," writes Thayer, "bcause in the text the astronomers are forever calculating and then pointing to the sky where they figure a new star or something should be and saying 'Lo!'—and there's nothing whatever to be seen where they point. Fort agreed to *Lo!* at first hearing." *Wild Talents*, Fort's last book, was published a few weeks after his death in 1932.

In 1937, at his own expense, Tiffany Thayer began issuing the *Fortean Society Magazine*, now called *Doubt*. Fort had willed thirty-two boxes of unpublished notes to Thayer (a gesture which infuriated Dreiser) and one of the purposes of the magazine is to reprint these notes. Portions of them appear in each issue. The magazine's chief purpose, however, is to embarrass scientists as much as possible by printing news items they can't explain, or uncomplimentary stories about the scientists themselves. Thus when a British astronomer fell off his telescope on one occasion, *Doubt* ran gleeful accounts of the accident. These news stories are sent to Thayer by established Fortean "correspondents," such as George Christian Bump of Chicago, and by the magazine's readers.

A brochure issued by the Society defines its aims as follows:

> The Fortean Society is an international association of philosophers—that is, of men and women who would live no differently if there were no laws; of men and women whose behavior is not a sequence of reflex jerks caused by conditioning, but rather the result of some cerebration, or of some mystical whimsicality of their own. . . . Eminent scientists, physicists and medical doctors

are members—likewise chiropractors, spiritualists and Christians—even one Catholic priest. . . .

The Society provides haven for lost causes, most of which—but for our sympathy—might become quite extinct. . . . A good many adherents of a flat earth are members, anti-vivisectionists, anti-vaccinationists, anti-Wasserman-testers, and people who still believe disarmament of nations would be a good thing. . . .

These members embrace the only "doctrine" Forteanism has, that of suspended judgment, temporary acceptance and eternal questioning. . . .

In many ways, the Fortean Society resembles the Sherlock Holmes cult of Baker Street Irregulars. Just as the Irregulars keep up the elaborate pretense that Holmes was an actual person, so the Forteans keep up the elaborate pretense that Fort's wild speculations are as likely to be true as the "established preposterousness" (Fort's phrase) of accepted science. At bottom, the Society is a gigantic joke, but Thayer and most of the members appear to take it very seriously, and part of the joke consists of getting angry when anyone suggests that it is a joke. All Fortean correspondence, incidentally, is dated by a thirteen-month calendar, the year "one" being 1931—the date of the founding dinner. The thirteenth month has, naturally, been named "Fort."

Before examining Fort's attitude toward science and coming to some conclusion about it, perhaps it would be best first to take a look at his unique cosmology.

Fort had a passionate distrust of astronomers. The first half of *New Lands* is concerned almost entirely with proving that all astronomers are stumblebums, worse than astrologers in predicting events, making all their major discoveries by accident, and craftily concealing from the public the basic unreliability of their "medieval science."

"They computed the orbits of Uranus," Fort writes. "He went somewhere else. They explained. They computed some more. They went on explaining and computing, year in and year out, and the planet Uranus kept on going somewhere else." Finally, to save face, they decided another planet was "perturbing" Uranus. For the next fifty years they pointed their telescopes at different spots of the sky until by accident they found Neptune. Now Neptune has unpredictable movements. If the astronomers are as good as they think they are, Fort challenged, let them find another planet beyond Neptune. Unfortunately, this was written before Pluto was discovered in 1930, but Fort still had the last laugh. Pluto proved to be very much smaller than astronomers expected.

Fort did not work out a cosmology in detail. But he did offer a series of suggestions which he felt were no more absurd than the astronomers' solar system, ". . . a stricken thing that is mewling through space, shocking able-minded, healthy systems with the sores on its sun, its ghastly moons, its civilizations that are all broken out with sciences; a celestial leper, holding out doddering expanses into which charitable systems drop golden comets. . . ."

The earth, Fort suggested, is relatively stationary. "Perhaps it does rotate, but within a period of a year. Like everybody else, I have my own notions upon what constitutes reasonableness, and this is my idea of a compromise." He replies in some detail to traditional "proofs," like the Foucault pendulum, of the earth's daily rotation.

To explain the motion of the stars around the earth, Fort supposed the earth to be surrounded by an opaque shell, not very far away. The stars are holes in the shell through which light shines. Perhaps the twinkling is due to a "quivering" of this shell. The shell is not rigid. "There may be local vortices in the most rigid substance, and so stars, or pores, might revolve around one another. . . ." Now and then meteors spray through gelatinous portions of the shell, detaching lumps of the substance as they pass. Fort collected records of hundreds of occasions on which a jelly-like substance had fallen from the sky. He warned aviators they might someday find themselves "stuck like currants," but he admitted, "I think, myself, that it would be absurd to say that the whole sky is gelatinous: it seems more acceptable that only certain areas are."

Nebulae, Fort suggested, are glowing patches on the shell. Dark nebulae are opaque patches. Some may be projections that "hang like super-stalactites in a vast and globular cave." One of them, known as the Horse-head nebula, "stands out, as a vast, sullen refusal to mix into a frenzy of phosphorescent confetti. It is a solid-looking gloom, such as, some election night, the Woolworth Building would be, if Republican, and all the rest of Broadway hysterical with a Democratic celebration."

"There may be civilizations in the lands of the stars," Fort wrote in *Lo!*, "or it may be that, in the concavity of a starry shell, vast, habitable regions have been held in reserve for colonization from this earth." Here is Fort's poetic vision of coming space travel:

> The time has come
> The slogan comes—
> *Skyward ho!*

The treks to the stars. Flows of adventurers—and the movietone news—press agents and interviews—and somebody about to sail to Lyra reduces expenses by letting it be known what brand of cigarettes he'll take along—

Caravels with wings—and the covered planes of the sky—and writers of complaints to the newspapers: this dumping of milk bottles and worse from the expeditions is an outrage. New comets are watched from this earth—long trains of voyagers to the stars, when at night they turn on their lights. New constellations appear—the cities of the lands of the stars.

And then the commonplaceness of it all.

Personally conducted tours to Taurus and Orion. Summer vacations on the brink of Vega. ''My father tells me of times, when people before going to the moon, made their wills.'' ''Just the same there was something peaceful about those old skies. It's getting on my nerves, looking up at all those lipstick and soap and bathing suit signs.''

Somewhere floating above, Fort declared, is a Super-Sargasso Sea, with an island in it which he called Genesistrine. From these regions come the various objects and living things that often fall to the earth. Fort collected thousands of records about mysterious showers of worms, fish, dead birds, bricks, manufactured stone and iron objects, colored rain, little frogs (he was puzzled by the fact there has never been a record of tadpoles falling), and periwinkles. Most of these things are accumulations of rubbish blown into the Super-Sargasso Sea from the earth or other planets, recently or aeons ago.

There are well authenticated instances of red rains. The conventional explanation is that reddish-colored dust becomes mixed with the water. But Fort had better explanations:

Rivers of blood that vein albuminous seas, or an egg-like composition in the incubation of which the earth is a local center of development—that there are super-arteries of blood in Genesistrine: that sunsets are consciousness of them: that they flush the skies with northern lights sometimes. . . .

Or that our whole solar system is a living thing: that showers of blood upon this earth are its internal hemorrhages—

Or vast living things in the sky, as there are vast living things in the oceans—

Or some one especial thing: an especial time: an especial place. A thing the size of the Brooklyn Bridge. It's alive in outer space—something the size of Central Park kills it—

It drips.

Charles Fort had a thousand other equally colorful theories, and before the book is finished, we shall have occasion to refer to many of them. But now we must decide exactly what to make of his upsetting speculations. Was Fort a humorist, or was he a crackpot? Were his books, as

Hecht called them, a "Gargantuan jest," or did he really believe the theories he advanced?

Tiffany Thayer, who ought to know, has given a clear answer in his introduction to the 1941 one-volume edition of Fort's four books. "As an intimate of the man through a period of years, permit me to assure you that he believed nothing of the kind. . . . Charles Fort was in no sense a crank. He believed not one hair's breadth of any of his amazing 'hypotheses'—as any sensible adult must see from the text itself. He put his theses forward jocularly—as Jehovah must have made the platypus and, perhaps, man. . . ."

Earlier in the same essay Thayer writes that Fort ". . . packed a belly laugh in either typewriter hand . . . he roared at his subject, guffawed at the pretensions of its serious practitioners, chortled at their errors, howled at their inconsistencies, chuckled at his readers, snickered at his correspondents, smiled at his own folly for engaging in such a business, grinned at the reviews of his books and became hilarious at my expense when he saw that I was actually organizing the Fortean Society.

". . . Charles Fort had the most magnificent 'sense of humor' that ever made life bearable to a thoughtful man. Never forget that as you read him. If you do, he'll trick you. He'll make you hopping mad sometimes, but—as your choler rises—remember he's doing it purposely and that just when you're boiling he'll stick his head up and thumb his nose at you. . . ."

At this point, one may well ask why, if Fort didn't believe in his theories, did he spend twenty-six years on such "minor tasks"—as he once described it—of going through twenty-five years of the *London Daily Mail?* The answer is that more meaning than meets the eye lurks behind Fort's madness.

Fort was an Hegelian. In the last analysis, existence—not the universe we observe, but everything there is—is a unity. There is an "underlying oneness," an "inter-continuous nexus" which holds everything together. "I think we're all bugs and mice," he wrote, "and are only different expressions of an all-inclusive cheese." Fort was not a religious man, but he granted that the totality of things might be an organism with intelligence. There was no harm in calling it God. "Maybe he, or it, drools comets and gibbers earthquakes. . . ."

There is, then, a final reality and truth. But *for us,* the little bugs and mice, there are only the broken lights, the half-truths and the phantom realities. Everything is in a "hyphenated state of being." Fort never tired of such adjectives as "real-unreal," "likely-unlikely," "good-bad," "material-immaterial," "soluble-insoluble," and so on. Because everything is continuous with everything else, it is impossible to draw a line between truth and fiction. If science tries to accept red things and exclude yellow, then where will it put orange? Similarly, nothing is "included" by science which does not contain error, nor is there anything "damned" by science which does not contain some truth.

It was meditation on this continuity of all things that led Fort into skepticism with a vengeance. Like the ancient Greek skeptics who lived by the motto of "no more"—meaning one belief is "no more" true than another—Fort accepted nothing. "The Sickle of Leo," he wrote, ". . . gleams like a great question mark in the sky. . . . God knows what the answer to anything is." Again: "I believe nothing. I have shut myself away from the rocks and wisdoms of ages, and from the so-called great teachers of all time, and perhaps because of that isolation I am given to bizarre hospitalities. I shut the front door upon Christ and Einstein, and at the back door hold out a welcoming hand to little frogs and periwinkles."

Fort could write, "I believe nothing of my own that I have ever written," but the important thing to remember is that he did not believe anything he had ever read, either. "I am . . . obviously offering everything in this book as fiction," he said in *Wild Talents,* but he hastened to add that it was fiction in the same sense as Newton's *Principia,* Darwin's *Origin of Species,* mathematical theorems, and every printed history of the United States!

Fort doubted everything—including his own speculations. When his more astute admirers insist that he was not the arch-enemy of science he was reputed to be, but only the enemy of scientists who forget the ephemeral character of all knowledge, they are emphasizing the sound and healthy aspect of Forteanism. It is true that no scientific theory is above doubt. It is true that all scientific "facts" are subject to endless revision as new "data" are uncovered. No scientist worthy of the name thinks otherwise. But it is also true that scientific theories can be given high or low degrees of confirmation. Fort was blind to this elementary fact—or pretended to be blind to it—and it is this blindness which is the spurious and unhealthy side of Forteanism. If a Baker Street Irregular began to think Sherlock Holmes actually did exist, all the good clean fun would vanish. Similarly, when a Fortean seriously believes that all scientific theories are equally absurd, all the rich humor of the Society gives way to an ignorant sneer.

Fort himself admitted that although all things are continuous, there also is *discontinuity.* He expressed it in a characteristic way. It is impossible, he said, to tell whether certain microscopic forms of life are animal or plant, but this does not mean we cannot distinguish between such extremes as a hippopotamus and a violet. "No one . . . would send one a bunch of hippopotami as a token of regard." Apparently it never occurred to Fort that once this is granted, it becomes possible to draw a similar line between a theory which has a high degree of probable truth, and a theory of extremely low degree.

It is worth going into this matter in some detail because it has an important bearing on everything in this book. If we cannot distinguish between truth and error, between science and pseudo-science, then this work might just as well have been written about such men as Newton and Darwin. A good Fortean would have to say, "Of course!" But the obvious fact is that we *can* make the distinction. Naturally there are many kinds of border cases, like the orange between yellow and red—cases where we cannot say for sure whether a theory is respectable or shabby, sane or insane. But when we consider the extremes, like the hippo and the violet, we can distinguish between the scientific value of Einstein's work and the contributions of a Velikovsky. We can grant that Einstein may be wrong, and there is a faint (*very* faint) possibility that Velikovsky may be right, but the extremes of the continuum are so great that we are justified in labeling one a scientist and the other a pseudo-scientist.

Fort himself must have known that this line could be drawn. He explains carefully in one of his books why he has neglected Santa Claus. "I am particular in the matter of data, or alleged data," he writes. "And I have come upon no record, or alleged record, of mysterious footprints in snow, on roofs of houses, leading to chimneys. . . ." The lack of data, then, must have had a bearing on the probability that Santa Claus existed, and Fort was willing to "exclude" Santa Claus on that basis.

And there is the hilarious section of *Wild Talents* in which Fort rejects the newspaper account of a dog that said "Good morning!" then vanished in a thin, greenish vapor. It was not the talking that bothered Fort—he had many clippings about talking animals. It was the vanishing in the thin, greenish vapor. "You can't fool me with that dog story," he says. But he makes clear that he only draws the line because everybody has to draw lines somewhere. He was careful not to say the line was justified in terms of truth and error.

Perhaps we are taking Fort too seriously, and are only falling into another one of his traps. He was far from an ignorant man, and his discussions of such topics as the "principle of uncertainty" in modern quantum theory indicate a firm grasp of the subject matter. Opposition to the notion that electrons move at "random" is not in fashion at the moment. Nevertheless, Fort's jibes are in harmony with the more technical criticisms of Einstein and Bertrand Russell! Even when Fort made a scientific boner, as he did occasionally, it is hard to know whether he made it deliberately or without knowing better.

Curiously, Fort had little interest in science fiction. There is no indication he read a line of it. Perhaps this explains why his speculations, though amusing, are not particularly ingenious. His rotating shell of stars, for example, is an extremely dull construction and had, in fact, been earlier proposed by an Italian crank. It is often said that Fort had a strong influence on modern science fiction, but this seems an exaggeration. It is true that about a dozen novels and scores of short stories have been based on his ideas, but these works are more of the "weird tale" variety than science fiction. A few Fortean terms like "teleportation" have become staple science-fantasy property, but in general his ideas have proved too mundane to serve as useful story gimmicks. Dreiser once tried to convince H. G. Wells that there was science-fiction material in Fort's writings, but without success. Wells never regarded Fort's speculations as more than amusing scientific rubbish.

Why the Fortean Society continues to exist is hard to understand. If we lived in an age in which the majority of citizens had a clear comprehension of science, there might be some point in preserving an organization to remind scientists of their limitations. The astrology magazines on the stands and the sales of Velikovsky's books are sufficient reminders of how far we are from such an age.

It was all very amusing in 1931. Now, the Society's magazine, *Doubt,* has become a dreary prolongation of a joke that should have been buried with Fort. It does little more than chant the Fortean clichés, report unfunny news items, and print the utterly valueless notes which Fort left with Thayer. Particularly objectionable, and completely humorless, are its recent attacks on tonsillectomies (on the ground that they cause polio) and vivisection, not to mention the frequent un-Fortean injection into news reports of the editor's political prejudices.

Even the "scientists" promoted by the Society are, for the most part, dull and uninspired. Major General Alfred Wilkes Drayson, for example, has a rank of honor among the Forteans second only to Fort himself. Drayson was a professor at the Royal Military Academy, Woolwich, England, during the latter part of the last century, who explained the earth's ice age in terms of a tilting of the earth's axis. The "Drayson hypothesis" had a vogue in England, especially among the military. Drayson published a number of books and pamphlets, at considerable personal expense, and was extremely bitter about the opposition of orthodox astronomy to his views. A member of the Fortean Society, the late astrologer Alfred H. Bailey, published *The Drayson Problem* in 1922—a book currently sold by the Society, in the unlikely case you might wish to explore the Major General's theory. (Other valuable references on Drayson include *Draysonia,* 1911, by Admiral Sir Algernon F. R. de Horsey, and several booklets by Major R. A. Marriott.)

In recent years, on top educational levels, there has been a minor, but observable, Fortean trend. It is due, in part probably, to a revival of religious orthodoxy, and in part perhaps, to resentment against the atom bomb. Its subtlest manifestation is in certain sections of the Hutchins-Adler Great Books Movement. Nothing official, of course, but if you know many Great Books educators you will be struck by the fact that most of them regard scientists, on the whole, as a stupid lot. Stupid, that is, in contrast to liberal arts professors, and particularly professors active in Great Books work.

The science "classics" reprinted in the Hutchins-Adler fifty-four volume set of *Great Books of the Western World,* 1952, are so dated and often so technical that they have almost no value for any reader except a specialist in the history of science. As I. Bernard Cohen, assistant professor of the history of science, Harvard, phrased it in a review of the set (*Saturday Review,* Sept. 20, 1952), "The 'great books' of science in this collection have only a kind of archeological value: not only are whole areas such as geology omitted, but almost all the major currents of scientific thought of the last two and a half centuries are not represented."

At St. John's College, Annapolis, where Robert Hutchins' educational views have been most successfully practiced, they make, it is true, a great hubbub about science. The school's catalog boasts that more mathematics and laboratory work are required than at any other college, and there is even a pretentious listing of all pieces of apparatus used by the student, down to such items as compass, calipers, and ruler. But so heavy is the emphasis on highlights in the past history of science, that little time is left for acquiring a solid grasp of current scientific opinion.

Anthony Standen, a British chemist (now a U.S. citizen), took a much-publicized swipe at "scientism" in 1950 with a book called *Science Is a Sacred Cow.* Standen had taught at St. John's from 1942 to 1946, an experience, which according to the *Catholic World* (Feb., 1950), "led eventually to his conversion to the Church."

As a group, Standen feels, modern scientists are cocksure, arrogant, conceited, and not nearly as smart as they think they are. One presumes that educators fired with classical zeal—like Mortimer Adler and Robert Hutchins—are modest, unassuming fellows. John Dewey is chided for suggesting that the future of civilization depends on the spread of the scientific outlook. Has not Hilaire Belloc told us that the more widespread science becomes, the worse the world gets? (We will take a look at Belloc's knowledge of science in Chapter 11.)

There is the familiar beating of drums for Aristotle. The Greek philosopher was right after all when he said a heavy body falls faster than a light one, because air resistance has less effect on the heavy one, so why give Galileo so much credit? Besides, Galileo didn't drop his two weights from the Leaning Tower of Pisa at all. It was some other tower. Standen doesn't tell us that Aristotle used the falling body example as part of a completely cock-eyed proof that there couldn't be a vacuum.

"The first purpose of science," Standen informs us, "is to learn about God, and admire Him, through His handiwork." Social scientists are foolish enough to think they can develop a science of ethics without a theology. Biological scientists try to make us think evolution proceeds by slow stages, whereas in fact, there are equally good grounds for thinking it might have gone by jumps. (Standen doesn't reveal his ulterior motive here. If evolution jumps, then man can have a soul sharply separated in history from the animal soul.) When a biologist spouts a piece of nonsense, such as the view that the basic aim of animal life is comfort, "the necessary and sufficient answer to this is, 'Flapdoodle.' "

"We must watch scientists carefully," he concludes, "to see that they do not put anything over on us." Charles Fort expressed the same thought this way: ". . . if nobody looks up, or checks up, what the astronomers tell us, they are free to tell us anything that they want to tell us."

Science "is the great Sacred Cow of our time," writes Standen. If scientists only had a sense of humor about it, they would realize how laughable it is for them to bow low before it. One suspects, however, that scientists would be more amused by a cow mentioned by Fort. On May 25, 1899, the *Toronto Globe* carried a story about a cow that gave birth to two lambs and a calf.

"I don't know how that will strike all minds," commented Fort, "but to the mind of a standardized biologist, I'd not be much more preposterous if I should tell of an elephant that had produced two bicycles and a baby elephant."

Good old Fort!

Skyward ho!

CHAPTER 5

FLYING SAUCERS

CHARLES FORT died in 1932, fifteen years before the flying-saucer craze began. It is a pity he did not live to witness this mass mania, because in many ways, it was a triumph of pure Forteanism. Mysterious objects are seen in the sky. They elude all "official" and "scientific" explanation. Wild Fortean hypotheses are invented to explain them, and discussed seriously by the man in the street as well as by seemingly intelligent authors and editors. As we shall see later, Fort himself collected hundreds of press clippings about mysterious lights and objects in the sky, and speculated at length on their extra-terrestrial origin. But first, let us chronicle briefly the major events in the history of the flying-saucer delusion.

It all began on Tuesday, June 24, 1947. Kenneth Arnold, owner of a fire control supply company in Boise, Idaho, was flying his private plane above the Cascade Mountains of Washington. Arnold is a handsome, athletic chap (former North Dakota all-state football end) in his middle thirties, who uses his plane for distributing his firefighting equipment. As he neared Mt. Rainier, nine circular objects, in diagonal chain formation and moving at high speed, passed within twenty-five miles of his plane. He estimated their size as slightly smaller than a DC-4 which also happened to be in the sky. They flew, he later wrote, "as if they were linked together," swerving in and out of the high mountain peaks with "flipping, erratic movements."

At Pendleton, Oregon, Arnold told a reporter that the objects "flew like a saucer would if you skipped it across the water." Next day, the wire services blanketed the nation with the story, using the word "saucer" to describe the objects. Actually, Arnold's original statement did not say the objects were saucer-shaped. But the word caught on, and the mania was underway. Newspapers all over the country were swamped with phone calls from excited people who had seen "saucers" over their farms, towns, and cities. Most of these stories were printed and put on the wires with little or no checking. If the observer did not use the word "saucer," the local paper or press-service stringer put it in, and if the wire story failed to mention "saucer," papers receiving it were likely to use the word in their headlines. In a few weeks, saucers had been reported from every state in the union, as well as Canada, Australia, England, and Iran.

Occasionally, sky objects of other shapes broke into the news. There were balls of fire, ice cream cones, flying hub-caps, doughnuts, and one wingless, cigar-shaped craft with rows of lighted windows, long orange-red exhaust, and blue flames dancing along the underbelly. David Lawrence, in his *U.S. News,* disclosed that the saucers were secret U.S. aircraft, "a combination of helicopter and a fast jet plane." Walter Winchell had inside information that the strange platters were from Russia. Andrei Gromyko, in a rare burst of confidence, revealed that possibly the saucers were coming from a Soviet discus thrower who didn't know his own strength.

Three military men lost their lives investigating the saucers. The first tragedy occurred shortly after the original sighting. A report reached Arnold that a weird doughnut-shaped craft had spewed forth large quantities of lava-like rock on Maury Island, a few miles off the coast near Tacoma, Washington. Arnold flew to Tacoma to investigate. On the way, incidentally, he spotted another cluster of about twenty-five small (two or three feet across), amber-colored saucers.

The entire Maury Island episode later proved to be a hoax elaborately planned by two Tacoma men who hoped to sell the phony yarn to an adventure magazine. Both men eventually made a full confession. Arnold, however, was completely taken in by the hoax, and his phone call to Air Force Intelligence, at Hamilton Field, California, brought two officers to the scene. On their way back, the left engine of the B-25 bomber they were flying burst into flames. Two enlisted men also in the plane parachuted to safety after being ordered to jump. Eleven minutes later the plane crashed and both officers were killed.

According to Arnold's several thud-and-blunder accounts of all this, the plane was carrying a corn-flakes box filled with samples of the mysterious lava. No trace of the box was reported found in the wreckage. "Were both of these men dead long before their plane actually crashed and is that the reason their plane was under little or no control?" Arnold asks. In all his writings about the saucers he betrays this suspicion of mysterious forces and conspiracies thwarting his efforts to get at the real truth.

The second tragedy was perhaps the most dramatic event in the history of the saucer mania. It occurred in January, 1948, at the Air Force base near Fort Knox, Kentucky. A round, white object spotted in the sky was chased by Captain Thomas F. Mantell, Jr., in a P-51 fighter plane. The object rose rapidly. Mantell followed it to 18,000 feet, then radioed to the ground, "Going to 20,000 feet. If no closer, will abandon chase." That was the last message from him. Apparently he blacked out in the high altitude, and after reaching about 30,000 feet the plane went into a fatal dive.

At first the military forces brushed aside the flying-saucer mania as mass delusion, but after the reports grew to vast proportions, the Air Force set up a "Project Saucer" to make a careful investigation. After fifteen months they reported they had found no evidence which could not be explained as hoaxes, illusions, or misinterpretations of balloons and other familiar sky objects. Later, President Truman also issued an official denial that the military were working on any type of airborne craft which corresponded to saucer descriptions.

In February, 1951, the Office of Naval Research distributed a ten-page report on the Navy's huge skyhook balloons, used for cosmic-ray research. The report pointed out in detail the ease with which these giant plastic bags—a hundred feet in diameter—could be mistaken for flying disks. The balloons reach a height of 100,000 feet, and are often borne by jetstream winds at speeds of more than 200 miles per hour. If the observer guesses the balloon to be farther away than it is, then, of course, estimates of speed can be incredibly high.

At a distance, a balloon loses entirely its three-dimensional spherical aspect. It takes on the appearance of a disk. From beneath, instruments hanging below the balloon's center can easily be mistaken for a "hole," giving the disk the shape of a doughnut. If viewed from the side, the disk seems to be flying on edge, like a rolling wheel.

Through a telescope or binoculars, the flatness of a balloon is greatly magnified because of a curious optical illusion. A telescope does not present an image of the object as it would appear if you were closer to it. Instead, it takes the image, exactly as it appears in the distance, and enlarges it for the eye. You *seem* to be closer to it, but the perspective it would normally have if you *were* that close is not present at all. As a result, a globular object viewed through a telescope looks remarkably like a plate. If you have ever looked through binoculars at cars coming toward you on a long highway, you may recall how odd and flat the cars appear.

In addition, the plastic composition of a skyhook balloon offers a surface that seems highly metallic in reflected sunlight. Most of the saucer reports describe the disks as silvery in color. At sunset the balloons may shine in the sky for thirty minutes after the earth has become dark. "If your imagination soars," the Navy release said, "the light reflection from one side may impress you as the glow of an atomic engine. The wisp of the balloon's instrument-filled tail may impress you as the exhaust. The sun's rays may suffuse the plastic bag to a fiery glow."

The first skyhooks were sent up in 1947, the year flying saucers were first reported. Arnold's original description of what he saw above the Cascade Mountains tallies remarkably well with what he would be expected to see had he flown near a group of smaller plastic balloons often used in place of the single large one. He estimated their size as smaller than a plane and the distance as about twenty-five miles—or twice the length of Manhattan Island. At this distance they would have been mere specks in the sky, and since Arnold was seeing them with unaided eyes, we cannot trust his guesses as to their actual size, shape, distance, or speed. Estimates of speed presuppose accurate knowledge of distance, and this in turn cannot be gauged unless the exact size is known.

Similarly, all the details of Captain Mantell's unfortunate death suggest he was chasing a skyhook. Moreover, it is known that such a balloon was in the area on the day he made his fatal climb. Even the descriptions of several hundred small "saucers" which sailed over Farmington, New Mexico, on March 17, 1950, read like descriptions of balloons—though of course they could not have been skyhooks. They were white and round. They "fluttered." They seemed to "play tag" with each other in the sky.

One of the few points on which all observers of flying saucers agree is that there is no noise. This excludes, of course, any known type of propulsion, but is precisely the way a balloon behaves. Observers have sometimes insisted that what they saw could not be a balloon because it was moving against the wind. They forget that wind directions in the stratosphere may be quite different from wind directions on the ground. At the time of the Navy's report, 270 skyhooks had been released from various spots in the United States, often remaining in the sky more than thirty hours. Frequently, lost balloons were actually traced by following press reports of flying saucer sightings!

After the Navy's release on skyhooks, reports of flying saucers decreased markedly, and green fireballs, streaking across southwestern skies, caught the public fancy. In the spring and summer of 1952, however, there was a new wave of saucer sightings, as well as a dramatic saucer scare in the nation's capital when mysterious blips of light kept appearing and vanishing on radar screens.

Many factors seem to be involved in the saucer mania. Although skyhook balloons, singly or in clusters, may account for most of the reliable reports, one must not forget that many other types of balloons are riding the skies. Weather balloons often carry steady or blinking lights and various shaped metal gadgets. Radar-target balloons trail large targets of aluminum foil. Guided missiles, and experimental aircraft of unusual design, may also account for some of the saucer sightings.

In addition, one must consider a score of possible illusions arising from faulty observations of planes, flying birds, the planet Venus, reflections of lights on clouds, and similar phenomena. The theory that the disks are mirages produced by unusual weather conditions has been advanced by Donald H. Menzel, a Harvard professor of astrophysics, and will be defended by him in a forthcoming book, *The Truth About Flying Saucers*. Normally such illusions would be rare, but under the pressure of mild mass hysteria, they greatly increase in number and, of course, are more likely to be reported. Even delusions without external cause can be induced in the minds of neurotics If there is a strong public belief with which the delusions may be identified.

Lastly, there are the lies and semi-lies. A book could be written about flying saucer hoaxes perpetrated in the past few years by pranksters, publicity seekers, and psychotics. Unfortunately, exposure of the hoax seldom catches up with the original story.

Even more difficult to expose are the semi-lies—accounts which have a basis in fact, but may be grossly exaggerated. For example, an observer sees a balloon but is convinced it is a saucer. Others are skeptical and this irritates him. So to convince them, he adds details, or exaggerates what he has seen. He may do this without being aware of it, and later recall the episode not as he saw it, but as he has added to it in his desire to convince himself and others. This is a well-known human failing and there is no reason to suppose it could not be involved in hundreds of so-called saucer sightings.

It is possible, of course, there may be some type of experimental aircraft resembling a disk and flying without sound that is still officially top-secret. But this seems extremely unlikely. Information now available on the cosmic ray balloons, together with the factors mentioned above, are sufficient to account for all that has happened. Official denials by the military and by the President have the earmarks of authenticity. Naturally, it will always be impossible to prove there *never was* a flying saucer. Believers in the elusive platters are likely to be around for decades. But there is every reason now to expect that the saucer mania will go down in history as merely one more example of a mass delusion.

At this point, the reader may well ask, "What has all this to do with pseudo-science?" The answer is: very little if it were not for the fact that a widespread belief has developed that flying saucers are not only real, but are spaceships from another world. This view has been exploited in numerous magazine articles and three hardcover books published by reputable presses.

The first magazine to promote the extra-planetary theory was *Fate,* a pocket-size pulp specializing in articles on telepathy, spiritualism, and other occult subjects. The publisher is Raymond Palmer who formerly edited a science-fiction magazine called *Amazing Stories.* It was as editor of this magazine that Palmer was responsible for the greatest of all science-fiction hoaxes. It is known as the Great Shaver Mystery,[1] and involves a series of stories which first appeared in Palmer's magazine in 1945. The tales were Palmer's expansions of briefer drafts by a Pennsylvania welder named Richard Shaver. Drawing on his "racial memories," Shaver described in great detail the activities of a midget race of degenerates called "deros" who live in huge caverns beneath the surface of the earth. By means of telepathy and secret rays, the deros are responsible for most of earth's catastrophes—wars, fires, airplane crashes, shipwrecks, and nervous breakdowns. What happened to Judge Crater? He was kidnapped by the deros! The monsters even stole copy from Palmer's desk!

Shaver's stories were presented as solid fact, and so convincingly that thousands of naive readers were, and perhaps still are, taken in by them. More adult science fiction fans, who objected to Palmer's ethics in running this series, finally kicked up such a protest that the publisher of *Amazing* ordered the stories killed. Shaver's latest work, distributed by Palmer, hangs the saucers on a race of Titans (the original masters of the deros) who fled into outer space two hundred centuries ago and are now returning.

It is not, therefore, to Arnold's credit that his first article on flying saucers—titled "I *Did* See the Flying Disks"—appeared in the first issue of *Fate,* Spring, 1948. In addition, Palmer recently announced a privately printed book on the saucers, soon to be on sale for five dollars, which he has written in collaboration with Arnold.

Arnold has contributed several other pieces to *Fate.* His "Phantom Lights in Nevada" (Fall, 1948) is about strange disks of pale red or yellow light seen hugging the ground at night in the Oregon Canyon Ranch, near McDermott, Nevada. "More than fifty of the shepherds of the area have seen the mysterious lights," he wrote, "and it has been noted that dogs bark at them, proving they are visible to animals as well as humans." Another article by Arnold, "Are Space Visitors Here?" (Summer, 1948), describes globes of a blue-green-purple color. They were seen by a fisherman in Ontario, and Arnold suspects they are spacecraft from another planet. His latest piece, "The *Real* Flying Saucer,'' appeared in the January 1952 issue of *Other Worlds*—another Palmer magazine.

Arnold himself published, and sells for fifty cents, a pamphlet of fifteen pages titled *The Flying Saucer as I Saw It*. This pamphlet, like the saucer, must be seen to be believed. The Maury Island hoax is taken seriously and there is even a drawing of the giant doughnut belching forth lava rock. Arnold reveals that a Tacoma reporter who wrote about the episode died suddenly of unknown causes (presumably killed by saucer men). A plane crash on Mt. Rainier, in which thirty-two Marines perished, is likewise linked to the disks on the ground that the crash occurred shortly before Arnold's first saucer sighting. Furthermore, in 1947 a suspension bridge near Riggins, Idaho, was mysteriously ignited by ''something'' of such intense heat that the steel cables burned like wood!

Arnold accepts the report that one saucer, chased by a pilot, made evasive maneuvers in response to the pilot's *thoughts*. Two pages are devoted to pictures of ''radar angels''—white splotches of light which sometimes appear on radar screens, though what connection they have with saucers is not made clear. There is a photograph of a mummified man fourteen inches tall, discovered in 1932 in the Rockies and now owned by a man in Caspar, Wyoming. Arnold thinks this lends credence to reports that tiny men have been found in saucers that have crashed *(Fate,* Sept., 1950, ran an article by Ray Palmer on this mummy). He also thinks there is some sinister connection between the saucers and mystery submarines reported off the coasts.

A press clipping is reproduced in which Arnold says to a reporter, ''I realize it's the 'data of the damned' to make a report on these things. . . . Who's to determine what is and what isn't a fact?'' The quoted phrase is, of course, Fort's. Arnold prints a number of Fort's ''data'' on mysterious sky objects, and adds that he himself has a collection of many similar reports.[2]

The second magazine to publicize the spaceship theory was *True*. In its January 1950 issue, an article by Donald Keyhoe opened as follows:

> After eight months of intensive investigation, the following conclusions have been reached by *True* magazine:
>
> 1. For the past 175 years, the planet Earth has been under systematic close-range examination by living intelligent observers from another planet.
>
> 2. The intensity of this observation, and the frequency of the visits . . . have increased markedly during the past two years.

Two months later, *True* ran another piece on saucers. Written by Commander Robert B. McLaughlin, on active duty in the Navy, the article developed the theory that the platters were piloted by Martians small enough to fit into twenty-inch disks. "It is staggering to imagine intelligent beings that small," McLaughlin confessed, "but we must not disregard any possibilities." The Commander speculated at some length on the craft's mode of propulsion, coming to the conclusion that it probably has three sets of motors, using light radiation (from an atomic source) as pressure "against a heavily shielded curved reflector."

The most recent article arguing the extra-planetary theory appeared in *Life,* April 7, 1952, at a time when other magazines and newspapers had almost relegated the disks to limbo. This article may have been a major cause of the revival of saucer reports in the months which followed. Einstein was prompted by the revival to issue the statement that he had no curiosity about the saucers, and Father Francis J. Connell, dean of Catholic University's School of Sacred Theology, pointed out that Catholic belief does not exclude the possibility of intelligent life on other planets. "If these supposed rational beings should possess the immortality of body once enjoyed by Adam and Eve," Father Connell said, "it would be foolish for our superjet or rocket pilots to try to shoot them. They would be unkillable."[3]

Donald Keyhoe's book, *The Flying Saucers Are Real,* was issued in 1950 by Fawcett Publications, publishers of *True.* It gives the impression of a sincere, though scientifically naive, effort by a journalist and former Marine pilot to round up all the information he can on the topic. He tells the story of his research chronologically, almost like a work of fiction. As it progresses, you see the author's growing suspicion that military officials are not playing square with him. Gradually, he comes to the conclusion that the Air Force's "Project Saucer" was set up not only to investigate, but also to conceal—to conceal from the public the fact that the saucers are from another planet.

"As I waited for a taxi," he writes, "I looked up at the sky. It was a clear summer night, without a single cloud. Beyond the low hill to the west I could see the stars. I can still remember thinking. *If it's true, then the stars will never again seem the same.*"

Keyhoe is convinced that the earth has been under periodic observation by another planet, or planets, for at least two centuries. He thinks this observation increased in 1947 as a result of our series of atom bomb explosions which aroused the curiosity of the space men. Their visits are "part of a long range survey, and will continue indefinitely. No immediate attempt to contact the earth seems evident. There may be some unknown block to making contact, but it is more probable that the space men's plans are not complete."

Keyhoe believes that Mantell was not only chasing a space ship, but that the space men—by some unknown power—killed Mantell before he could reach them. In an earlier chapter, he writes, "The secret of the spaceship's power is more important than even the hydrogen bomb. It may someday be the key to the fate of the world.''

Romantic and preposterous as these speculations certainly are, they seem like the remarks of a cautious scientist when compared to the second of the three books which have been written about the topic. Frank Scully's *Behind the Flying Saucers,* published in 1951 by Henry Holt, is filled with so many scientific howlers, and such wild imaginings that when *True* magazine, Sept., 1952, revealed it to be a hoax, there was little cause for surprise. Scully is the Hollywood columnist for *Variety.* His previous book, *Fun in Bed,* suggests his status as a scientist and thinker.[4]

The major theme of Scully's book is that the saucers are flown here by ''magnetic propulsion'' (whatever that is) from Venus. They travel with the speed of light (or faster) and are piloted by Venusians who are exact duplicates of earthlings except they are three feet high and have teeth completely free of cavities. A mysterious ''magnetic specialist'' whom he calls ''Dr. Gee'' is Scully's chief source of information.[5] Four saucers have landed. Three crashed, but the fourth took off again. The crashed ships, including several dozen Venusian bodies, are now being studied at undisclosed government laboratories.

According to Scully, the saucers are made of a hard but extremely light metal completely unknown to our chemists. All the dimensions of the ships are divisible by nine. Their cabins revolve on an unfamiliar gear ratio. One ship ''defied all efforts to get inside of it, despite the use of $35,000 worth of diamond drills.'' The Venusians carried ''heavy water'' for drinking, and concentrated food wafers. A tiny radio operated on unknown principles. Booklets were found written in pictorial script which our experts are now trying to decipher.

All this is thoroughly mixed with a continual sniping at the Truman administration for its ''hush-hush policy'' on the saucers, its bungling bureaucracy, and its cowardly failure to take the people into confidence. Debunkers of the saucers are charged with following ''Party lines.''

The third and most recent book on the topic—Gerald Heard's *Is Another World Watching?*, Harper's, 1951—is the most terrifying of the three. Terrifying because there is the very real possibility that Heard (a sincere and devout mystic) actually believes everything he writes. He is the author of many learned, scholarly works in the fields of religion, psychology, and anthropology, as well as a number of mystery novels. As might be expected, he writes with considerably more polish than either Keyhoe or Scully, quoting occasionally from Shakespeare and John Stuart Mill, and making a great pother about fairness in all his reasoning.

Heard thinks the saucers come from Mars. Considering their small size, plus the speed with which they maneuver, he concludes that only an insect would be tiny enough, and have a hide sufficiently tough, to withstand the crushing inertial effect of sudden turns. These and other lines of thought convince him that the ships are piloted by Martian "super-bees" about two inches long and possessing an intelligence much higher than man's. Here is Heard's description of what these bees may look like:

> A creature with eyes like brilliant cut diamonds, with a head of sapphire, a thorax of emerald, an abdomen of ruby, wings like opal, legs like topaz—such a body would be worthy of this super-mind. I am sure that toward it our reaction would be: "What a diadem of living jewels!" It is we who would feel shabby and ashamed and maybe, with our clammy, putty-colored bodies, repulsive!
>
> Of course . . . we must allow that we should find it hard to make friends with anything that had more than two legs. . . .

Like Keyhoe, Heard believes the space men are here as scouts to investigate atomic explosions. Our sun is a Cepheid, or pulsing star, Heard states (which of course it isn't), and might possibly explode if our atomic blasts destroy its delicate chemical balance. Already our atom bombs have increased the size of spots on the sun, and sunspots "may be warnings of indigestive troubles—as spots on our own face sometimes tell about our deep interior conflicts. . . . Is it not possible that the Martians, who have so much to fear from sun trouble, may have read these signs?"

The Martian scouts, Heard believes, are seeking information about us but are careful to avoid direct contacts. A huge "mother ship" has been established as an earth satellite. From this ship the smaller, saucer-like craft go "ashore" to do the scouting. Mars' two satellites, he suspects, are not natural moons at all. They are artificial launching jetties for Martian spaceships!

A half dozen photographs of saucers are reproduced in the front of the book. The only two pictures clear enough to show anything were taken by Paul Trent, a farmer in McMinville, Oregon. They were printed in *Life's* June 26, 1950 issue. Trent's saucer bears a striking resemblance to the top of a garbage can tossed into the air.

Both Heard and Keyhoe take for granted that science has established the high probability of intelligent life on Mars. Actually, this is not the case. The fact is there is no evidence one way or the other. At the most, some obscure color changes on the planet may be interpreted as vegetation varying with the seasons. The so-called canals of Mars have had a highly dubious history. They were first reported in 1877 by an Italian astronomer, and later defended by the American astronomer, Percival Lowell. Unfortunately, later observers, with much better telescopes and

great visual acuity, have been unable to see them. The consensus among
modern astronomers is that the "canals" were subjective interpretations
with no reality outside the minds of Lowell and others who fancied they
could see them.[6]

The quickness with which the public will accept evidence of life on
nearby worlds is astonishing. One of the best examples was the famous
Moon Hoax perpetrated by the *New York Sun* in 1835. This was a series
of articles reporting what the great British astronomer Sir John Herschel
had seen through a new telescope at Cape Town, Africa. The articles
described life on the moon, with accompanying drawings of apelike
creatures who "averaged four feet in height, were covered, except on the
face, with short and glossy copper-colored hair, and had wings composed
of thin membranes. . . . Our further observation of the habits of these
creatures, who were of both sexes, led to results so very remarkable, that
I prefer they should first be laid before the public in Dr. Herschel's own
work . . . they are doubtless innocent and happy creatures, notwithstand-
ing some of their amusements would but ill comport with our terrestrial
notions of decorum."

The hoax was intended as a satire, but it was accepted as fact by about
half of New York City, and many believers remained unconvinced even
after the reporter, Richard Locke, publicly admitted the deception!

A somewhat similar, but more upsetting, prank was played on the
American public the night of Halloween, 1938, when Orson Welles
presented a radio version of H. G. Wells' *War of the Worlds*. The
broadcast opened with dance music, which was then interrupted by a
series of news flashes. The first flash reported a gas explosion on Mars.
The second, an earthquake tremor in New Jersey. Finally, there was a
spot broadcast from Grovers Mill, New Jersey, describing a huge, cylin-
drical spaceship out of which came bug-eyed monsters armed with death
rays. Other cylinders land and the monsters manage to destroy most of
New York before they die of disease germs for which they had no
resistance.

Six million people heard the broadcast, and it is estimated that approxi-
mately one million took it seriously enough to be in some degree frightened. Thousands wept, prayed, closed their windows to shut out poison
gas, or fled from their homes expecting the world to end. Phone lines
were tied up for hours. The panic was from coast to coast, but the
greatest hysteria was in the southern states among the poorly educated. If
the Moon Hoax could be believed in 1835, and an invasion from Mars
taken seriously in 1938, perhaps it is not so hard to understand a
widespread acceptance of the spaceship theory of flying saucers in a
decade that has split the atom and bounced radar off the moon.

Charles Fort, in one of his books, devotes a chapter to observations of cigar-shaped objects in the skies. He concludes: "Some of the accounts are not very detailed, but out of the bits of description my own acceptance is that super-geographical routes are traversed by torpedo-shaped super-constructions that have occasionally visited, or that have occasionally been driven into this earth's atmosphere."

Many of the strange sky objects discussed by Fort were in the shape of saucers, though of course Fort did not use that term. For example—a gray disk observed in 1870 by a sea captain; or the dark, circular object "with a structure of some kind upon the side of it, travelling at a great pace." The latter was observed one moonlit night in 1908 by employees of the Norwich Transportation Company, at Mousehead, England. "It seemed too large for a kite," they said, "and, besides, its movements seemed under control for it was traveling against the wind."

Fort had many explanations for these sightings. "Perhaps," he wrote, "there are inhabitants of Mars, who are secretly sending reports upon the ways of this world to their governments." On another page he speculated, ". . . I conceive of other worlds and vast structures that pass us by, within a few miles, without the slightest desire to communicate quite as tramp vessels pass many islands. . . ."

Fort's most daring hypothesis was that humanity was owned—owned by higher intelligences who visited earth occasionally to check on their charges. ". . . something now has a legal right to us, by force, or by having paid out analogues of beads for us to former, more primitive, owners . . . that all this had been known, perhaps for ages, to certain ones upon this earth, a cult or order, members of which function like bellwethers to the rest of us, or as superior slaves or overseers, directing us in accordance with instructions received—from Somewhere else. . . ."

"I think we're property," wrote Fort. This casual sentence inspired one of the best-known science fiction novels of recent years—*Sinister Barrier*, by Eric Frank Russell. Russell is chief British correspondent for the Fortean Society, and a half-believer in many of Fort's speculations.

The Society, incidentally, was not impressed at all by the flying saucer mania. In 1947 Thayer devoted issue No. 19 of *Doubt* to the saucers, but after that reported on them reluctantly. "Forteans have a legitimate gripe," he wrote, "at the usurpation of our long-time franchise upon lights and objects in the sky by the military and its lackey freeprez [free press]."

One suspects, however, that if Fort had lived until the flying-saucer era he would have thoroughly enjoyed reading the flood of Fortean speculations about the celestial platters. And how he would have chuckled at the frantic efforts of the military to persuade a dazed public that nothing sinister or extraordinary was taking place above their heads!

CHAPTER 6

ZIG-ZAG-AND-SWIRL

ALFRED WILLIAM LAWSON, Supreme Head and First Knowlegian of the University of Lawsonomy, at Des Moines, Iowa, is in his own opinion the greatest scientific genius living today. It's regrettable there isn't some sort of Nobel Prize which would recognize his fantastic career and incredible literary output. As proof, here are two testimonials from Lawson himself.

"His [Lawson usually writes of himself in third person] mind responds to every question and the problems that stagger the so-called wise men are as kindergarten stuff to him."

"When I look into the vastness of space and see the marvelous workings of its contents . . . I sometimes think that I was born ten or twenty thousand years ahead of time."

The publishers of Lawson's book, *Manlife*, have this to say about him: "In comparison to Lawson's Law of Penetrability and Zig-Zag-and-Swirl movement, Newton's law of gravitation is but a primer lesson, and the lessons of Copernicus and Galileo are but infinitesimal grains of knowledge." (Note: *Manlife* was published by Lawson.)

A preface to the same book, by someone with the improbable name of Cy Q. Faunce,[1] states: "To try to write a sketch of the life and works of Alfred W. Lawson in a few pages is like trying to restrict space itself. It cannot be done . . . Who is there among us mortals today who can understand Lawson when he goes below a certain level? There seems to be no limit to the depth of his mental activities . . . countless human minds will be strengthened and kept busy for thousands of years developing the limitless branches that emanate from the trunk and roots of the greatest tree of wisdom ever nurtured by the human race.''

Before hazarding a sketch of the life of this remarkable thinker, let us first survey briefly the basic principles of Lawsonomy. Although Lawson has written more than fifty books and pamphlets, the most inportant sources of his views are *Lawsonomy* (in three volumes, 1935-39), *Manlife,* 1923, and *Penetrability,* 1939. It is from these books that most of the quotations will be taken.

Lawsonomy is defined modestly by Lawson as ''The knowledge of Life and everything pertaining thereto.'' He has little use for the theories of ''so-called wise men and self-styled scholars . . . everything must be provable or reasonable, or it is not Lawsonomy. . . . If it isn't real; if it isn't truth; if it isn't knowledge; if it isn't intelligence; then it isn't Lawsonomy.''

At the base of Lawsonomy, underlying the entire structure, is a theory of physics so novel that Lawson was forced to invent new terms to describe it. In fact, Lawson himself has declared, ''The basic principles of physics were unknown until established by Lawson.'' Many of his books open with lengthy glossaries defining these new and revolutionary terms.

The concept of ''energy'' is completely discarded. Instead, Lawson conceives of a cosmos in which there is neither energy nor empty space, but only substances of varying density. Substances of heavy density tend to move toward substances of lesser density through the operation of two basic Lawsonian principles—*Suction* and *Pressure.* The law governing this movement is called *Penetrability.* ''This law was far too reaching for the superannuated professors of physics,'' Lawson writes, ''. . . but little by little the rising generations of advancing scholars have begun to grasp its tremendous value . . ..''

When a balance is reached between greater and lesser density, Lawson calls it *Equaeverpois.* An even more important principle is that of *Zig-Zag-and-Swirl.* Lawson defines this as ''movement in which any formation moves in a multiple direction according to the movements of many increasingly greater formations, each depending upon the greater formation for direction and upon varying changes caused by counteracting influences of Suction and Pressure of different proportions.''

This makes more sense than one might think, and can be explained as follows. No object in the universe moves in a simple straight line or curve because it partakes of many different motions which ultimately give it a zig-zag path, ''neither coming nor going.'' Lawson illustrates it colorfully by considering a germ moving across a blood corpuscle in the body of a man who is walking down the aisle of a flying airplane. The germ thinks he is moving in a simple straight line. Actually, the corpuscle is moving in the blood, the blood is circulating through the body, the body is walking down the aisle, the airplane is moving relative to the earth, the earth is both rotating and going around the sun, and the sun, with its system of planets, is rushing through space. Thus the path of the germ is a highly zig-zag one which "continues without direction or end." Lawson suggests that a "Supreme Mathematics" will have to be devised for computing such complicated paths.

The concepts of Suction and Pressure recur again and again in all of Lawson's thinking. "Currents"—such as rain, heat, blood, etc.—are due to these two forces. Light is a "substance drawn into the eye by Suction." Sound is another substance similarly drawn into the ear. Gravity? It is simply the "pull of the earth's Suction." In fact, Lawson candidly admits, "When one studies . . . Lawsonomy . . . all problems theoreticaily concocted in connection with Physics will fade away. . . ."

The human body, as might be expected, operates by means of thousands of little Suction and Pressure pumps. Air is sucked into the lungs, food into the stomach, and blood around the body. Each cell contains minute pumps. Waste matter is, of course, eliminated by Pressure. This "internal swirl goes on as long as the Suction and Pressure terminals . . . are properly maintained." When they cease to draw and push, the man dies.

The earth is a huge organism operating by Suction and Pressure. Although it swims in a sea of "Ether," a material of extremely rare density, it contains within its body a substance of even rarer density which Lawson calls "Lesether." This creates a Suction which draws into the earth, through an opening at the North Pole, various substances supplied by the sun and by gases from meteors. Some of this material is also sucked into the earth through surface "pores." Through the center of the earth, from pole to pole, extends a central tube. From it branch the arteries which carry life-giving substances to all parts of the earth, and veins that flush away waste matter. The South Pole is the earth's anus. Through it, by Pressure, are expelled the "discharged gases," although some of this waste matter also is eliminated through volcanic pores. The Aurora lights, at both poles, are caused by these gaseous movements in and out. On page 32 of *Penetrability* is a beautiful color plate showing the earth in violent action at both ends.

Sex, as might be expected, is simply Suction and Pressure. "Suction is the female of movement. Pressure is the male. . . . Female movement draws in from without, and male movement pushes out from within. . . . The attraction of one sex for the other is merely the attraction of Suction for Pressure."

Magnetism, which still baffles the "professors," is a form of low-grade sexual activity. If a magnet "has more female particles than male particles then it will have the power of Suction. . . . If it has more male particles than female particles, then it will have the power of Pressure and push matter away from it.''

Within the human brain, according to Lawson, two types of tiny creatures are living which he calls the *Menorgs* and the *Disorgs*. The Menorgs (from "mental organizers") are "microscopic thinking creatures that build and operate the mental instruments within the cells of the mental system." They are responsible for everything good and creative. "To move your arm requires the concentrated efforts of billions of Menorgs working together under orders from one little Menorg."

Unfortunately, the Menorgs have opposed to them the destructive, evil activities of the Disorgs ("disorganizers"), "microscopic vermin that infect the cells of the mental system and destroy the mental instruments constructed and operated by the Menorgs." As Lawson expresses it, "a Menorg will sacrifice himself for the benefit of the body, but a Disorg will sacrifice the body for the benefit of himself."

Of course much more could be written about the extraordinary principles of Lawsonomy, but this should be enough to give a picture of its great depth and scope. Let us turn now to a sketch of the life and career of its discoverer and some of the institutions he has founded.

"The birth of Lawson," according to Cy Q. Faunce, "was the most momentous occurrence since the birth of mankind." It took place in London, March 24, 1869. Lawson's parents (a mixture of Scotch and Scandinavian descent) left England shortly after his birth, going first to Canada, then to Detroit. His first study of Natural Law began, he writes, "at the age of three years while working on a small farm near Detroit. There he gained knowledge of insect life while picking potato bugs from the vines."

At the age of four, young Lawson "noticed that when he used the Pressure from his lungs to blow the dust within his bedroom, that it moved away from him and that when he used the Suction of his lungs by drawing in his breath that the dust was moved toward him." This was Lawson's first great discovery in physics.

After selling papers on the streets of Detroit, running a bobbin machine for his father, who had become a rug weaver, and working as a bootblack, Lawson ran away from home. For several years, he made his way around the country riding freight cars. In most of his books, which devote almost half of their pages to a pictorial description of his life, there is a drawing of himself standing on the cowcatcher of a speeding locomotive, much buffeted by the wind. The caption beneath reads: *Alfred Lawson studying atmospheric resistance to moving bodies.*

At the age of nineteen, Lawson became a pitcher for the Goshen, Indiana, ballteam. For the next nineteen years (until 1907), he was in professional baseball, both as player and manager. In many of his books, he reproduces photographs of himself wearing the uniforms of various ball clubs. They reveal a handsome, finely chiseled face, with dark curly hair, high forehead, and dreamy eyes.

It was when he began his baseball career that Lawson became corrupted by his friends. He began to earn money for money's sake. Worse than that, he took to tobacco and liquor, and the eating of meat. His health failed. His teeth decayed. At the age of twenty-eight, by a superhuman effort of will, he abandoned all these vices. His ailments stoped immediately, and from that day to this (he is now eighty-three), he has enjoyed perfect health. His first book, a novel called *Born Again,* 1904, was written about this experience. It is certainly one of the worst works of fiction ever printed, but Lawson says that "many people consider it the greatest novel ever written by man." If his claims are correct, the novel was also published in England, Germany, France, Switzerland, Italy, and Japan.

Collectors of early science fiction will be interested to know that *Born Again* is a Utopian fantasy in which the author predicts radio (transmitted by Suction and Pressure), poison gas, and several other features of the modern world. The hero, John Convert, tries to persuade the world to live by the principles of Natural Law which he learned in a mystical experience from a sleeping beauty named Arletta. After his kiss has awakened her, she initiates him into the organization of her home, Sageland—a Utopia destroyed by the flood of Noah. Convert later falls in love with a reincarnation of Arletta, a wealthy Chicago society woman named Arletta Wright. The plot is complicated by the fact that Arletta has a double, named Arletta Fogg, and Convert also has a double with the same last name as his. The evil Convert murders Arletta Fogg, and the good Convert is arrested for the crime. He is electrocuted one minute before Arletta Wright arrives with the evil Convert's confession. But John Convert's little band of followers have remained loyal through it all, and Arletta Wright dedicates her life and wealth to carrying on the great cause. John Convert dies wishing he could have another body in which to continue his work. If God could grant him this he would be willing to give his soul in exchange, and "take upon himself everlastingly, all of the misery, suffering, and torture now inflicted upon the rest of mankind."

Soon after he published this novel, Lawson began an astonishing career in aviation. In 1908 he established and edited in Philadelphia the first popular aeronautical magazine, *Fly.* From 1910 to 1914 he edited, in New York City, another magazine called *Aircraft.* It was a word he coined in 1908. He himself introduced it into the dictionary as editor of the aviation section of a revised Webster's. In 1917 and 1918 he designed and built war training planes for the Army. He claims he was the first to propose the idea of a flat-top airplane carrier, which he called to the attention of Congress and the Navy in 1917 by a series of weekly bulletins.

In 1919 he invented, designed, and built the world's first passenger airliner. It carried eighteen people, and although there was considerable doubt as to whether it would fly, Lawson himself piloted it from Milwaukee to Washington and back. The Lawson Aircraft Corporation was established in Green Bay, Wisconsin. In 1920, he built a twenty-six-passenger plane, and made a handsome profit flying it around the United States. It was the first plane to have sleeping berths. A year later one of his planes crashed, and soon the company followed suit.

From aviation, Lawson turned his attention to the social sciences, establishing the Humanity Benefactor Foundation, in Detroit. It published his second book, *Manlife*. The public paid little attention to his views, however, until the great depression came. Then suddenly he zoomed into prominence as the leader of an economic reform cult called the Direct Credits Society.

In his books *Direct Credits for Everybody,* 1931, and *Know Business,* 1937, are the basic tenets of the "Lawson Money System." The gold standard must be abolished. "Valueless money" is to be issued, not redeemable for anything. Also to be abolished are all interests on debts. Only such drastic measures will rid the world of its principal source of evil, the "pig-like maniacs known as financiers." The financiers, according to Lawson, milk undeserved interest from everybody. They control the country's press, schools, churches, and "every influential organization in the United States of America except the Direct Credits Society."

The Society published (and still does) a four-page tabloid called *Benefactor,* which at one time claimed a circulation of seven million, and was issued in ten different languages. It bears across the top the slogan, JUSTICE FOR EVERYBODY HARMS NOBODY. During the depression, most issues carried gigantic, single-word scare headlines—such as ISMS, THINK, JUMP, and WHICH—followed by one of Lawson's articles or speeches.

The most disturbing aspect of the Direct Credits Society is that it actually did attract tens of thousands of ardent followers. There is no more terrifying proof of the mass following that a worthless economic theory can achieve in the United States, in time of economic stress, than the pictures which appear in Lawson's book, *Fifty Speeches,* 1941. There are several hundred photographs of mass meetings, parades, lecture halls, office fronts, bands, and groups of DCS officers wearing a special white uniform and cap, and a diagonal red sash.

Parades and mass meetings were held in dozens of midwestern cities, but the largest was in Detroit on October 1, 1933. The floats, carrying plump and elaborately costumed women, were so preposterous that unless there were photographs you wouldn't believe them. They carried such banners as "All nations need direct credits for little children and feeble old folks." After the mammoth parade, Lawson spoke for two hours to 16,000 people assembled in the Olympia auditorium. When he made his entrance to the stage, amid hysterical flag-waving and the music of *Hail to the Chief,* he received an ovation that lasted fifteen minutes.

Special songs were written for these meetings, and one of Lawson's books, *Short Speeches,* 1942, reproduces the words and music of fifteen of them. They include such titles as *Hark to Lawson* by Ella Heft, and *God's Gift to Man* by Marie Pluks. Each stanza of the latter song ends with "Alfred William Lawson is God's great eternal gift to Man." And there is a stirring hymn of six verses called *Mighty Menorgs,* the second stanza of which runs:

> Menorgs are wondrous builders all,
>   Builders of the great and small.
> All of life they permeate,
>   All formations they create.
> Disorgs tear down eternally
>   While menorgs build faithfully.

In 1942, Lawson purchased the University of Des Moines. The school, which included fourteen acres, six buildings, and dormitories for about four hundred students, had been closed since 1929. It is now called the Des Moines University of Lawsonomy.

Lawson's opinion of American education couldn't be lower. "You don't begin to get bald on the inside of your heads until you start to go to high school," he once declared, "and you don't get entirely bald until you pass through college." His own educational views are summed up powerfully as follows:

> Education is the science of knowing TRUTH.
> Miseducation is the art of absorbing FALSITY.
> TRUTH is that which is, not that which ain't.
> FALSITY is that which ain't, not that which is.

The University of Lawsonomy is teaching, of course, the TRUTH. Only Lawson's own writings are used as texts, and they must be read by a student before he is eligible to attend. A basketball rule book was once banned because Lawson hadn't written it. Accredited teachers of Lawsonomy are called "Knowlegians," and top level Knowlegians are Generals. Lawson is Supreme Head and First Knowlegian.

There are no fees for enrollment. Board and room are furnished without charge, although students work part-time in the machine shop and on similar projects in agriculture, engineering, and other fields. Like all of Lawson's organizations, the University supposedly operates on a non-profit basis, without stocks, and is managed by trustees who pay for his "meagre living expenses" out of the sale of his books. Lawson insists that in 1931 he promised God he would never again accumulate wealth for personal use. He likes to describe himself as "moneyless and property-less." To prove it, he often turns his pockets inside out at meetings. His board members are secret, however, and Lawson's place of residence is equally vague. He seems to live somewhere "near Ann Arbor."

In March 1952, the Senate's Small Business Committee summoned Lawson to Washington to ask him how it came about that his school had purchased sixty-two war-surplus machine tools "for educational purposes," then resold forty-five of them for a handsome profit. Lawson professed ignorance of the details. "I don't know," he said. "I never go in for figures at all." His attempts to explain Lawsonomy to the Committee, and how it included mechanics, proved somewhat confusing. Lawson left the conference snorting, "The damnedest thing I've ever heard of in all my life," which moved Senator Blair Moody, of Michigan, to reply, "I don't know whether we're talking about the same thing, but I'm inclined to agree with you."[2]

At first, the University of Lawsonomy was coeducational but after one father sued to get his daughter out of the place,[3] Lawson decided to admit only men. They are accepted on a ten-year basis, and at present, about twenty such students are enrolled. They can be seen occasionally through the high picket fence which surrounds the campus. The faculty has an even lower degree of visibility, and there is a widespread theory among Des Moines newsmen that Lawson is the only member.

The use of liquor and tobacco is strictly forbidden on the campus. Such poisons overpower the Menorgs and let the Disorgs take over. "One may hunt the world over," Lawson once wrote with incontestable accuracy, "but can find no other animal strutting around with a lighted pipe or cigarette stuck in its face and using its mouth to suck in and blow out smoke, using at times the nostrils as human smoke-stacks. . . ." Lawson's antipathy to smoke is further indicated by his invention in 1946 of the Lawson Smoke Evaporator. Patent rights for this device (which eliminates factory soot by means of Suction and Pressure) have been turned over to the University.[4]

Lawson places a high premium on bodily vigor, and recommends to his students an elaborate set of health rules. He believes in a diet without meat, consisting mostly of raw fruits and vegetables eaten "from covering to core," including the seeds. "All salads," he once wrote, "should contain a sprinkling of fresh cut grass." The head should be dunked in cold water upon arising and before going to bed. He also believes in drinking lots of warm water, sleeping nude, and changing bed sheets daily.

He is against kissing. "Can you think of anything filthier than . . . a man and woman with their faces stuck together and spitting disease microbes into each other's mouths?"

Lawson has never married.

From the University, he believes, will come forth the salt of the earth. As the principles of Lawsonomy spread, from generation to generation, eventually a new species will be created—a super race capable of communicating by telepathy (operating by Suction and Pressure) and with great longevity of life. (See Lawson's book, *A New Species,* 1944.)

Today Lawson is a gaunt, lonely, silver-haired old man with shaggy white eyebrows overhanging steel-gray eyes. According to Oliver Rauch, treasurer of the University of Lawsonomy (as expressed in a personal letter), Lawson's eyes possess a "kaleidoscopic effect that appears to change as he thinks and talks." He feels himself surrounded by treacherous enemies. They are waiting for him to die so they can seize the holdings of his organizations. In recent years the conviction that he is a prophet of the Lord has increased ominously. One thousand Lawsonian Churches are currently being planned for cities in the Midwest. Since 1949, a Lawsonian Church in Detroit has been holding Sunday services, and a similar church has been built in Des Moines. His latest book, *Lawsonian Religion,* 1949, explains his religious views.[5] They are little more than a misty blend of transmigration, Lawsonomy, and Christianity without Christ.

Key concepts in the Lawsonian decalogue are love and unselfishness. "Alfred Lawson never hated nor harmed a man, woman, or child in his life," writes Lawson. "In days gone by when anybody struck harmfully at this writer, he merely took hold of the offender and threw him to the ground to show his superior strength and ingenuity, and then rose with a friendly smile to show there was no hatred in his system whatsoever. . . ."

Although the future of the earth looks black, as the Disorgs seem to gain the upper hand in the minds of men, Lawson is convinced the Menorgs will win the day. By the year 2,000, he predicts, all the races of the world will have accepted his principles. To usher in the Lawsonian Dispensation, however, will require millions of loyal disciples. "Therefore," he writes, "professors of Lawsonomy for Lawsonian ecclesiastical colleges and teachers of Lawsonian Parochial Schools, Church Messengers, Secretarial Forces, Pulpit Sermoners, Foreign Missionaries and various high dignitaries will all have to be educated at the University of Lawsonomy in large numbers as quickly as possible."

I know of no more inspiring close for this chapter than the last verse of a poem written by Lawson himself:

So come on, folks, the past is dead,
The future is alrighty,
And by the will, we'll win the till,
With strength from the ALMIGHTY.

CHAPTER 7

DOWN WITH EINSTEIN!

IN PHYSICS and chemistry, like all other branches of science, there is never a sharp line separating pseudo-scientific speculation from the theories of competent men. One fades into the other and there are always borderline cases. The British physicist, Oliver Heaviside, for example, was a fascinating mixture of scientist and eccentric. He was the only physicist of eminence to denounce Einstein, when relativity was first announced, and many of his speculations were so absurd that no reputable journal would publish them. At the same time, he made a host of solid contributions to electrical theory. Nikola Tesla—inventor of alternating current motors, the transformer (Tesla coil), and many other electrical devices of great value—grew steadily more paranoid with advancing years. John H. O'Neill's amusing biography, *Prodigal Genius,* discloses how Tesla spent the later years of his life—a lonely, uncommunicative egotist, intensely jealous of Edison, unwilling to shake hands for fear of germ contamination, frightened by round surfaces (like billiard balls and pearl necklaces), loving no one but the Manhattan pigeons he fed daily, and dissipating his great talent by trying to invent death rays, or devices for photographing thoughts on the retina of the eye.

Often, a theory of physics or chemistry will be unanimously rejected by the experts, but defended with such intelligence and restraint, and on such a technical level, that a layman is in no position to reach a firm conclusion about it. In the field of nuclear chemistry, for example, Dr. Albert Cushing Crehore has spent most of his life defending what he calls the "Crehore atom." Briefly, Crehore rejects the accepted view that the electrons of an atom, in its stable state, have orbits about the nucleus. He

68

thinks instead that the electrons are part of the nucleus itself. In his opinion, gravity is produced by the rotation of positive charges within the nucleus. Crehore's books, the latest of which, *A New Electrodynamics*, appeared in 1950, are universally considered worthless by his colleagues. On the other hand, Crehore was formerly assistant professor of physics at Dartmouth, with a doctor's degree from Cornell, and a distinguished record as a teacher and inventor. So, one hesitates to be dogmatic.

However, we do not have to be concerned with debatable cases of this sort. At the extreme end of the scale there is an abundance of physical and chemical literature so obviously balderdash that the label of pseudo-science can be applied without hesitation. No one but a theosophist, for example, would imagine that *Occult Chemistry,* 1908 (revised 1919), by Annie Besant and Charles W. Leadbeater, contained anything of value. The work is a study, by clairvoyance, of the structure of atoms, including the atoms of several elements not yet discovered by orthodox chemists. Unfortunately, most of this type of literature is extremely dull. We shall, therefore, limit our attention to only that small segment which is richest in interest and humor.

Any revolutionary scientific theory, once it has won acceptance by the majority of scientists, always finds itself under fire from cranks rebelling against the father-image of established authority. In the eighteenth and nineteenth centuries, the great symbol of authority was Isaac Newton. As one might expect, the crackpot literature attacking him ran into many hundreds of weighty volumes. Even the great Goethe produced a two-volume work on color, containing violent polemics against Newton's theories of light. Since Goethe had no understanding of experimental methods, and even less of mathematics, his attack proved one of the most irrelevant in the history of physics.[1]

In America, vigorous opposition to Newton came from the pen of a Methodist minister in New York City. His name was Alexander Wilford Hall (1819-1902) and his major opus, which runs to 524 pages in the twentieth revised edition, is titled *The Problem of Human Life*. Most of the book is an assault on evolution, but Hall also defends at great length an original theory of physics called "substantialism." According to this view, all so-called forces, including the force of gravity, are "substances." Of course they are made of much smaller atoms than "material" substances, but nevertheless are composed of actual particles. Light, heat, electricity, magnetism, and even sounds, are "substantial"—like the particles which transmit odors. In the first edition of this book (1877) Hall had a curious propensity for arguing in a trochaic tetrameter similar to Longfellow's *Hiawatha*. The following excerpt is representative:

I assert, without a question
That the chirping of a cricket
Or the twitter of a swallow
Scatters through the air around it
And through every object near it
Atoms real and substantial—
Matter of as true a nature
As the odoriferous granules
Issuing from the cryptic chambers
Of the rose or honeysuckle—

Reverend Hall was fond of pointing out that the sound of a locust could be heard for more than a mile. If the wave theory of sound were correct, he argued, it meant that a gigantic mass of air, weighing thousands of tons, had to be kept in constant agitation by a tiny insect. No sane person could believe this, he said, although he did not explain how the tiny locust could fill the gigantic space with a substance. Hall was very pugnacious about it all. For eleven years, he edited a monthly magazine called *The Microcosm* (and for two years another magazine, *The Scientific Arena*) in which he tried to prod contemporary scientists into debating with him. They refused. This of course convinced Hall that his theories were unanswerable.

The corpuscular theory of sound was also vigorously defended by another American, Joseph Battell (1839-1915), of Middlebury, Vermont. Battell was the owner of several farms and 20,000 acres of forest land in Vermont, a breeder of Morgan horses, and manager of the American Publishing Company. It was this firm which published his major work— three enormous volumes (each running more than 600 pages) titled *Ellen—or the Whisperings of an Old Pine*.

Few odder works than *Ellen* have ever appeared in the United States. All three volumes are in the form of a Platonic dialogue between a sixteen-year-old girl named Ellen and the narrator who happens to be an old Vermont Pine tree. It is not clear how either of the two managed to acquire their vast knowledge of science and mathematics. Like Hall, Battell is opposed to all wave theories, and especially the wave theory of sound which he brands a "monstrous lie." You might suppose that the vibration of a tuning fork produces the sound, but Battell quickly sets you straight. It is the sound emerging from the fork which causes the prongs to vibrate. The work is also filled with lengthy attacks on orthodox algebra and geometry, and is illustrated with about 200 photographs of Vermont mountain scenery. Ellen herself appears in many of these scenes.

After Einstein, Newtonian theories of gravity and sound ceased to be the principal bugaboo of the crackpot. Since Einstein was responsible for the greatest revolution in physics since Newton, it is not surprising that the literature attacking him is as large and violent as the earlier diatribes against his eminent predecessor. Often the attacks are made in the name of Newton, who had become the symbol of an abandoned, and therefore heretical, point of view.

Naturally, many of the early attacks on Einstein (most of them in French or German) cannot be considered pseudo-scientific. Frequently they were by professional colleagues who had difficulty accepting the new and bizarre doctrines. For example, Charles L. Poor, professor of celestial mechanics at Columbia University, wrote a book titled *Gravitation versus Relativity* which was published by Putnam's in 1922. The book was critical of the experimental evidence on which Einstein's views rested at that time. It raised important objections and concludes undogmatically, "The relativity theory may be true, but no substantial proofs have yet been submitted." A much less restrained work, but still one which cannot be labeled worthless, is Arthur Lynch's *The Case Against Einstein,* 1932. Many other anti-relativity books of the twenties and thirties are less the work of cranks than of journalists or amateur scientists who rushed into print before they had a sound comprehension of what they were opposing. Even today, it should be pointed out, many reputable physicists hold views which depart radically from Einstein's. The outstanding example is the "kinematic relativity" of the great British mathematician, Edward A. Milne.

Quite apart from all this critical literature, however, is another type of anti-relativity writing of entirely different tone and quality. Like similar literature opposing Newton and Darwin, it is the product of peevish, ignorant minds. It betrays no understanding of the views opposed, although the authors have had every opportunity for acquiring such knowledge. Even should Einstein later be found wrong in his major assertions, it would not elevate this literature into the realm of acceptable scientific controversy. At this stage in the history of physics, Einstein's theories have introduced enormous simplifications (notably the reduction of gravity and inertia to an identical phenomenon), and are slowly being confirmed by mounting experimental evidence. Unless an opponent of Einstein takes these huge achievements into full account, his objections are as irrelevant as the attacks of Hall and Battell on the wave theory of sound.

An outstanding example in recent times of such a worthless attack is *Back to Newton* by the self-styled French physicist, Georges de Bothezat. It was published in the United States in 1936, a few years before the author's death.

Although Bothezat considers his book a "rigorous refutation" of relativity, it consists less of relevant argument and evidence than of invective. Not only is Einstein accused of being "just unable to understand the great conception of Newton,'' but Bothezat seriously wonders whether Einstein even understands himself. Scientists who accept relativity ''by their general lack of any knowledge of physics are utterly unable to acquaint themselves with the subject.'' To explain the widespread acceptance of Einstein's views by the scientific world, the author suggests it may be due to ''the weakening of the critical spirit in science produced by the Great War.''

It would be a tedious task to go into the details of this muddled, poorly written work, but one aspect deserves, perhaps, to be mentioned because it is common to so many attacks on the theory of relativity. This is Bothezat's use of Dr. Dayton C. Miller's experiments as a prop for anti-relativity views. Miller was a responsible physicist who repeated, in the twenties, the famous Michelson-Morley experiment, obtaining results unfavorable to relativity.

The Michelson-Morley experiment is one of the major, if not *the* major, experimental foundations of relativity. It had been known for some time that the velocity of light, sent out from a moving object, is constant regardless of how fast the light source is moving. This meant that if measurements of light were made in different directions on the earth's surface, the speeds would be expected to vary because of the earth's motion relative to the paths of light. When Michelson and Morley first made such a test in 1887, they found to their great surprise that the velocity of light did not show the anticipated variations. It was one of the most unexpected developments in the entire history of science. Einstein's Special Theory of Relativity was, in a sense, simply a way of explaining the Michelson-Morley experiment's failure.

Forty years after this historic test, Dr. Miller decided to make the experiment once again. He did so, and obtained evidence of slight variations in the speed of light which he interpreted as a refutation of Einstein. He repeated the experiment many times, always with positive results. Dozens of articles were contributed by him to technical journals, and he remained convinced of the validity of his work until his death in 1941.

To this day, no physicist is sure just why Miller's experiments turned out as they did. But what Bothezat and other irresponsible critics of relativity do not tell the reader is that out of thousands of repetitions which have been made of the Michelson-Morley experiment, only Dr. Miller's gave positive results. The test has been performed in every scientifically advanced nation, at all altitudes, in all seasons, and with all types of equipment. In every case—except Dr. Miller's—the results were negative. Today the consensus among physicists is that some local disturbance affected Miller's apparatus, or perhaps he unconsciously made errors in recording his data.[2]

Frequently in the pseudo-scientific literature directed against Einstein, one meets with a violent prejudice against complex mathematical equations. The author, of course, does not understand them, so he rationalizes his ignorance by insisting that nature always obeys simple mathematical laws. A good example of this bias is the work of the American chemist, Thomas H. Graydon, of Santa Monica, California. In his *New Laws for Natural Phenomena*, published by a small press in 1938, he writes: "In that I go contrary to fixed methodical systems which seem to revel in the complex . . . my concepts may not be welcome by some intelligentsia whose positions of authority are sustained by the difficulties encountered in their orthodox methods. . . ." Graydon's method of simplifying astronomy is to abandon the notion of a gravitational "pull" altogether. Instead, he postulates a "push" that moves outward from the sun—a common theme, incidentally, in earlier attacks on Newton.

A mimeographed work by Graydon titled *Relativity's Failure,* issued by him in 1947, can be obtained from the Fortean Society. It finds Einstein's theory of gravitation less plausible than the "push" theory. If gravity did not exist, a planet would fall into the sun—not because it is "pulled," but because "it finds the least strain or the least time in which it can make a revolution. . . ." Fortunately, gravity exerts enough of a shove outward to keep the planets in place. Why do objects fall to the earth? ". . . because there is insufficient radiation emanating from the earth to hold them on orbits above the earth's surface."

As one would expect, Graydon considers himself a modern Galileo. He quotes a letter from a prominent astronomer to a publisher, advising against the acceptance of one of Graydon's manuscripts. Graydon's theory is so simple, the astronomer reasoned, that if it had any merit it would have been established long ago. "The ingenuity displayed in this letter," Graydon writes, ". . . is not a far cry from the rebuke Galileo received from Francesco Sizzi, a prominent astronomer of his day, when Galileo reported the seeing through his first telescope of four satellites revolving about the planet Jupiter."

More amusing than Graydon is another eccentric also much admired by Tiffany Thayer and the Forteans—George Francis Gillette and his "spiral universe." Biographical details on Gillette are scarce beyond the fact that he was born in 1875, attended the University of Michigan, and has held engineering posts with several large firms. But his four privately printed books are lasting monuments to his originality and ingenuity.

Relativity fares badly in all of Gillette's writing. "Einstein a scientist?" he asks. "It were difficult to imagine anyone more contrary and opposite to what a scientist should be. . . . As a rational physicist, Einstein is a fair violinist." Relativity is given such labels as "moronic brain child of mental colic," "cross-eyed physics," "utterly mad," "the nadir of pure drivel," and "voodoo nonsense." By 1940, he predicted (writing in 1929), "the relativity theory will be considered a joke." "Einstein is already dead and buried, alongside Andersen, Grimm, and the Mad Hatter."

Gillette has unbounded admiration for Newton, praising him as the greatest mental genius who ever lived. The "spiral universe" theory is, naturally, an improvement on Newton. As Gillette puts it, it "out-Newton's Newton."

Exactly what is the spiral universe? It is a little difficult to make out. The ultimate units—indivisible and unchanging—are called "unimotes." Our universe is a "supraunimote" and the entire cosmos is the "maximote." There is also an "ultimote" which is defined as the "Nth sub-universe plane." Here is a sample of Gillette's exposition:

> Each ultimote is *simultaneously* an integral part of zillions of otherplane units and only thus is its *infinite* allplane velocity and energy subdivided into zillions of *finite* planar quotas of velocity and energy.

"Bumping" is an important Gillette concept. "All motions ever strive to go straight—until they bump." In fact, everything in the cosmos finally reduces to motions bumping one another. "Nothing else ever happens at all. That's all there is." "In all the cosmos there is naught but straight-flying bumping, caroming and again straight flying. Phenomena are but lumps, jumps, and bumps. A mass unit's career is but lumping, jumping, bumping, rejumping, rebumping, and finally unlumping."

One of Gillette's greatest contributions to physics is his famous "backscrewing theory of gravity." It is difficult to do this concept justice, but perhaps these quotations will be helpful: "Gravitation is the kicked back nut of the screwing bolt of radiation." "Gravitation and backscrewing are synonymous. All mass units are solar systems . . . of interscrewed subunits." And finally, "Gravitation is naught but that reaction in the form of subplanar solar systems screwing through higher plane masses."

As might be anticipated, Gillette feels keenly the rejection of his views by what he calls the "orthodox oxen" of science. There is "no ox so dumb as the orthodox" he complains. They are the "would be scientists," the "built up favorites of publishers." They are "the reverse of true scientists. They are droll." It is all due to "their being cramped within Homoplania, ignorant of ultimotically related sub and supraplanias."

The fact that these "professors" with their "frozen beliefs" attack his theories he takes as a compliment. "The author would never have wasted his depleted resources," he admits bitterly, "in printing at his own expense *theories already granted.*" But he is aware that working against him are "all the mighty resources of mysticism which control the press, politics, publishers, colleges, public libraries, and all such direct avenues. . . ." Like Columbus, Galileo, and Copernicus, he is persecuted and misunderstood. Yet he bears up under all this with good humor. "The truth seeker is never a fanatic. He has no fantasies to be fanatic about. So he is serene, and humane, civilized."

Only one "professor," Gillette writes sadly in one of his books, has ever offered him encouragement—"a noble, brave-minded Russian. To him, Salute!" To the rest of the scientific world: "Pooh! . . . It will soon attain oblivion by its own efforts."

If the reader has failed to obtain a clear picture of Gillette's revolutionary cosmology, I refer him to Gillette's *Rational, Non-Mystical Cosmos,* revised third edition, 1933, in which he will find it carefully explained in 384 pages. If this proves too formidable, try the shorter work, *Orthodox Oxen,* 1929. According to the title page, it is entirely free of "Hi-de-hi mathematics," and is "bristling with new axioms." Moreover, there are innumerable diagrams of such impressive structures as "The all cosmos doughnut," and a "Laminated solid, solid, solid, solid." In some editions the pictures are hand-colored by the author.

A more dignified broadside against relativity was fired in 1931 by the Very Reverend Jeremiah J. Callahan. Rev. Callahan was at that time president of Duquesne University in Pittsburgh. To understand Father Callahan's indictment, however, it will be necessary first to make a few comments about the nature of non-Euclidian geometry.

In Euclid's classic *Elements,* all the proofs of his theorems rested on a set of assumptions called axioms and postulates. These assumptions were regarded as "self-evident" and impossible to prove. The Fifth Postulate, however, seemed more complicated than the others. It stated, in effect, that through a point outside a given line only one line can be drawn parallel to the given line. After Euclid, mathematicians regarded the postulate as a blot on Euclid's system which might be removed if it could be proved on the basis of the other assumptions. Thousands of such proofs were attempted, many extremely ingenious, but all of them were later found fallacious. Finally, in the nineteenth century, a Russian mathematician named Lobatchevsky and others demonstrated conclusively that the famous parallel postulate was completely independent of the other assumptions, and could not be proved by them.

Once this independence of the postulate was fully understood, mathematicians made an even more astonishing discovery. They found that they could replace the postulate by something which contradicted it—such as the assumption that through a point outside a line more than one parallel can be drawn. The new postulate could then be combined with Euclid's other assumptions to form a geometry that was logically consistent. These new geometries were called non-Euclidian. It was entirely out of mathematical curiosity and delight that the early work on non-Euclidian geometry was done, but when Einstein developed relativity, he discovered that a non-Euclidian approach to space led to enormous practical consequences. It was non-Euclidian geometry which provided the mathematical framework for his General Theory of Relativity.

One may say, therefore, that relativity and non-Euclidian geometry are inseparable. If non-Euclidian geometry can be found logically inconsistent, the framework of relativity collapses. And if one can prove the parallel postulate by means of Euclid's other assumptions, non-Euclidian geometry collapses. A simple means, then, of overturning relativity would be to prove the parallel postulate.

Thoughts similar to the above were running through the mind of Father Callahan one day when he was riding the New York subway. Not long after, he stated in a press interview, he interrupted a meal with a friend by shouting, "That can be proved!" By the close of 1931, Father Callahan had completed his proof and announced to the world that relativity had been overthrown.

In the nineteenth century the famous French mathematician, Lagrange, once appeared before a learned society to explain a proof he had worked out for a previously unsolved problem. No sooner had he started to read his paper than he suddenly stopped talking, frowned, then folded his papers and remarked, "Gentlemen, I must think further about this." Unfortunately, Father Callahan did not fold his papers. In 1931, he published his proof of the parallel postulate—a 310-page work titled *Euclid or Einstein*. Any competent geometer could have pointed out the error in this proof, but the amateur scientist is not noted for his willingness to seek helpful advice.

Einstein fares as badly under Father Callahan's rhetoric as he does under the turgid prose of Gillette. "We certainly cannot consider Einstein as one who shines as a scientific discoverer in the domain of physics," Callahan writes, "but rather as one who in a fuddled sort of way is merely trying to find some meaning for mathematical formulas in which he himself does not believe too strongly, but which he is hoping against hope somehow to establish. . . . Einstein has not a logical mind."

Reverend Callahan finds unintended humor in the "mental fog" and "huddle of meaningless words" which make up Einstein's geometry. "Sometimes one feels like laughing," he declares, "and sometimes one feels a little irritated, that such a hodgepodge could be seriously accepted anywhere for thought. . . . But there is no use expecting Einstein to reason."

Again: "His [Einstein's] thought is but odds and ends, unconnected bits, incongruous, undigested, and contradictory. . . . Whatever he is as a pure mathematician . . . he becomes the most out-and-out careless thinker the moment he gets beyond his symbols and his equations. . . . His thought staggers, and reels, and stumbles, and falls, like a blind man rushing into unknown territory.''

I have quoted these remarks at length because, like similar remarks previously cited, they are typical of pseudo-scientific criticism of great scientists. Even though Father Callahan is a man of the cloth, his attacks descend to a level almost indistinguishable from personal character assassination. Surely there must be psychological motives operating here of which the Reverend can scarcely be aware.

A good indication of Father Callahan's mathematical insight is the fact that in the same year he "proved" the parallel postulate, he also discovered a method of trisecting the angle![3] It was published as a pamphlet by Duquesne University. An announcement was made at the time that he was working on the duplication of the cube and the squaring of the circle, but apparently he has not succeeded in these two efforts. In 1940, at the age of 62, he retired as President of the University, and has since been living quietly in Isle Brevelle, Louisiana.

Among all the attackers of relativity, however, Fort—as so often is the case—seems to have the final word. In discussing the famed Michelson-Morley experiment, the cornerstone of relativity, he points out that the failure to find variations in the speed of light could lead to two conclusions. One—the conclusion of Einstein—that the velocity of light is an absolute regardless of the earth's motion. The other—a simpler, "more graspable" conclusion—that the earth is not moving at all!

And then he adds, "Unfortunately for my own expression, I have to ask a third question: Who, except someone who was out to boost a theory, ever has demonstrated that light has any velocity?''

CHAPTER 8

SIR ISAAC BABSON

THOMAS EDISON once remarked to Roger Babson, who is best known as a stock-market tipster, "Always remember, Babson, you don't know nothin' about nothin'. You've got to find something that isolates from gravity. I think it's coming about from some alloy."

Babson never forgot this remark, and in 1948, with an excess of capital on hand, he founded what is perhaps the most useless scientific project of the twentieth century. It is called the Gravity Research Foundation. Although the Foundation is interested in any and all types of work on gravity, its principal function is to stimulate a search for some type of "gravity screen"—a substance which will cut off gravity in the same way a sheet of steel cuts off a light beam.

This notion of a material "opaque" to gravity is a common one in early science fiction. In H. G. Wells' fantasy, *The First Men in the Moon*, a spaceship operates by means of such a substance—a complicated alloy of metals (with some helium thrown in) called "cavorite," after the name of its inventor. Since Einstein, however, the concept has become almost obsolete. The reason is that if relativity theory is correct, such a screen would be unimaginable. According to Einstein, gravity is not a "force" which pulls objects to earth, but rather a warping of the space-time continuum. The warping causes an apple to fall, but a "screen" between apple and earth would have no effect for the simple reason that there is no force to be screened off.

If Babson is aware of all this, he remains blithely undismayed. "I'm no scientist," he told the press, "but I do know what I'm trying to find out and how I'm going about it . . . few people realize that Edison experimented with more than 8,000 materials before he finally hit on the right one that gave him a filament for his electric light bulb.''

Since there are millions of possible alloys and no conceivable reason why one should be tried before another, the process of testing all of them becomes a tedious one. Evidently Babson soon gave up the idea, because at present, the Foundation is conducting no chemical experiments of its own. Housed in a large brick building in New Boston, New Hampshire, its chief purpose is to serve as a clearing house and information bureau for all scientists doing research on gravity. New Boston was picked because it is a non-industrial, self-sustaining community a safe distance from Boston in case Boston is bombed in World War III.

The response of professional scientists to the Foundation's work has been disappointing. "The mention of gravity too often brings a smile," writes William Esson, one of the Foundation's trustees, "as if the inquiry is not taken seriously. This has discouraged a frank discussion of the subject. Much experimenting which has been going on has been under cover. Hence a . . . purpose of the Foundation has been to get in touch with these people scattered throughout the world and let them know that they have at least one sympathetic friend." Every person who writes to the Foundation is classified in its files according to the specific type of gravity research he is undertaking. When the Foundation finds several people working along the same lines, each is notified of the others. In this way they can exchange ideas and report their co-operative progress.

One scientist, Harold V. McNair of Middletown, Pa., died in 1950 after willing to the Foundation all his files and apparatus. He had been working for forty years on the possibility of harnessing gravity. "Unfortunately," reads the Foundation's second annual report, "his formula was clear only to himself. We have shown it to everyone we can think of and will continue to do so in the hope that some day his work may be carried on."

In 1949, the Foundation ran a small ad in two magazines, *Popular Mechanics* and *Popular Science,* which read: "GRAVITY. If you are interested in gravity, write us. No expense to you." This proved a dubious mode of public relations, and now the organization's chief means of arousing interest in gravity is an annual prize essay contest. The essay must be limited to 1,500 words and deal with one of three topics—the possibilities of discovering (1) "some partial insulator, reflector or absorber of gravity," (2) "some alloy, or other substance, the atoms of which can be agitated or rearranged by gravity to throw off heat," (3) "some other reasonable method of harnessing the power of gravity."

The awards are generous—$1,000 for the first prize, and five additional awards of $100 each. The 1949 contest brought a total of 88 essays, which Babson read with huge delight. "It was just like opening

Christmas presents," he said. The first prize went to David B. Wittry, a University of Wisconsin student, for a paper which surveyed some historical failures by previous gravity researchers. In 1950, a Princeton graduate student took first prize, and in 1951, top honors went to Dr. Myron J. Lover, of Ozone Park, N. Y. Dr. Lover wrote on *Thermodynamic Aspects of Gravithermels*. *Time* magazine devoted considerable space to a story of the 1950 awards and a Foundation report boasts, "Einstein's new theory was announced on the same page, but we had a better story than they gave him."

In the fall of 1951, the Foundation held its first Summer Conference in New Boston. The audience heard lectures on gravity as well as a few speeches on business conditions by staff members from the famed Babson Institute. Special "gravity chairs" designed to aid blood circulation were available, and guests who complained of pains in their arms and legs were told about "Priscolene," a patent medicine which Babson is currently promoting as an "anti-gravity pill" to aid circulation. The conference also placed on display the original bed of Isaac Newton, recently acquired by the Foundation—presumably because Newton at one time rested on it by the force of gravity. "It is the hope," reads a bulletin from the Foundation, "that New Boston will gradually become the center where physicists, engineers, metallurgists and others especially interested in the causes and the possibilities of gravitation will come as a mecca in the summer."

One of the organization's trustees, Clarence Birdseye,[1] had the thought that perhaps some laboratory worker would stumble on a gravity insulator while he was working on something else. This prompted a mailing of literature to 2,500 laboratories in addition to usual mailings to colleges, secondary school science teachers, and scientific journals.

The Foundation maintains a library of books on gravity. It is growing steadily, although it is handicapped at the moment by not having anyone on the staff who reads French or German. Friends of the organization are being asked to "remember the Gravity Research Foundation in your will by leaving your gravity files or apparatus to us, where they will be in friendly hands and your work can be carried on."

An idea that has fascinated the Foundation since its inception is that variations in gravity caused by changing relations of the sun and moon have a measurable effect on human beings. Babson is convinced, for example, that it is easier to go upstairs during a high tide, apparently unaware that the difference in weight is so infinitesimal it would be much more effective to toss away a dollar bill before making the climb.

Nevertheless, the Foundation is busily engaged in compiling statistics about human affairs and seeking to correlate them with variations in the moon's pull. For several years, a number of mental hospitals have been collecting data for the Foundation on how mental conditions of patients are affected by phases of the moon. During full moon, the sun and moon are on opposite sides of the earth, working against each other, which may possibly disturb something in the brain or spinal fluid. Several hundred letters were mailed to police chiefs to find out if they had more calls during full moon. Replies indicated they did. Insurance companies were asked to report whether accident rates correlated with the moon's phases.

When a disastrous train wreck occurred in New Jersey on February 6, 1951, the Foundation noticed it was a day of no moon. Fifty-four railroad companies were immediately notified of this fact and requested to be on the alert in the future. President Truman decided to fire General McArthur on April 5, 1951, also in the dark phase of the moon. "This," according to the organization's third annual report to trustees, "dictated another mailing." The report also reveals that a twenty-year file of *Time* magazine has been purchased, and "we have just started to go through all the issues picking out important news events to see how they correlated with various phases of the moon." Each year the Foundation issues an almanac showing the moon's phases so that anyone interested can study the moon's effect on his own thoughts and actions.

Recently the Foundation has been mailing out questionnaires in which a person's weight (i.e., the degree gravity pulls on him) can be checked against various body types and temperaments. The Foundation suspects that weight (gravity) will be found to have a greater effect on temperament than one's body type has. If this is true, then a gravity screen would be able to change a person's weight and thereby alter his temperament! Until this is discovered, however, the Foundation points out that it is still possible to change one's personality traits slightly by altering one's posture.

"Certainly the gravity pull on a person's brain is *different* when lying down," a bulletin reads, "or when rocking, or when bowing. . . . anyone can test this by changing the position of his head when mad or unduly excited. The results from bowing in prayer are partly due to changing the direction of the gravity pull on the brain."

Moreover, the same bulletin states, "many thoughtful people believe that spiritual forces can modify the pull of gravity as illustrated by the story of certain Old Testament prophets having risen to the skies, and the Ascension of Jesus. The incident of Jesus walking on the water should not be ignored. People often ask why Angels are always shown defying gravity. . . ."

The ability of birds to defy gravity has also long fascinated Babson. On one occasion Thomas Edison pointed to a flying bird and said, "The bird can do what no man can do—namely, fly under its own power. I wish, Babson, you would take a greater interest in birds. . . ." This led Babson to make a collection of birds. It is now on the campus of the Babson Institute, but eventually will become the property of the Gravity Foundation. "Since starting the Foundation," writes Trustee Esson, "various cartoonists have portrayed the Foundation's trustees as flying through the air with wings. . . . Yet . . . it is very possible that a wing will someday be constructed of some anti-gravity alloy which will enable lightweight, muscular persons to fly under their own power. Animals such as flying squirrels are an illustration of what may be accomplished along these lines."

Many short essays on gravity by Babson and others are available from the Foundation for ten cents each. The vapidity of this literature is almost unbelievable. An essay on "Gravity and Posture" by Mary Moore points out that a properly fitted corset "prevents gravity from pulling us too far forward or too far backward, which in so doing, makes us old before our time." "Gravity and Sitting" by Babson is an attack on chairs. He recommends sitting cross-legged on the rug, or on a low bench "with the knees raised" and "balancing on the buttocks." He thinks a careful study of sitting should be made, and concludes with an unconscious pun by stating, "It is very possible that the balancing school of thought will win out in the end."

Babson also feels that one should sleep in a foetal position with one or both knees drawn up. "To prevent sleeping on one's back," he writes, "a rubber ball—two inches in diameter—may be buttoned in the rear center of the neck band of one's night clothes. This can best be accomplished by having a pocket back of the neck into which the ball would be kept during the night, and yet from which it could be removed when the night clothes go to the laundry. All of the above shows that the entire problem deserves much more study."

Another essay by Babson, "Gravity and Ventilation," discusses a topic in which he has long been interested. When a young man, Babson became ill with tuberculosis and was advised by his doctor to remain in the West where he had gone to convalesce. He chose instead to return to his home in Wellesley Hills, Massachusetts. To insure an abundance of fresh air, he refused to close any of his windows. During the freezing winter, he wore a coat with an electric heating pad in back and his valiant secretary did her typing by wearing mittens and hitting the keys with rubber hammers. Babson got well and has been a fresh-air fiend ever since. He thinks children should be trained to enjoy fresh air blowing against their face from a fan, and that air from pine woods has "chemical and/or electrical qualities" of great medicinal value. His essay suggests that gravity should be used to clear bad air from a building by giving a slight slope to all the floors and having air outlets at the lower end of the room. Apparently this drains off the bad air in the way a sloping roof drains water. Such a house has actually been built in New Boston, with floors sloping a half-inch to the foot.

In a bulletin titled "Weather Conditions and Political Victories," Babson develops the thesis that gravity affects crops, crops affect business, and business affects elections. He analyzes twenty-seven presidential elections, from 1844 through 1948, to prove that in 75 per cent of the cases, a party continued in power when weather and business were good, and was tossed out when weather and business were bad.

An essay by Trustee Esson deals with the possibility of finding an alloy which will give out heat when gravity acts on it. It would be formed into wires, with weights suspended on the ends. "When such an alloy . . . is discovered," Esson writes, "the next step is to attach various sized weights thereto and ascertain the comparative results of such experiments. . . . There are also interesting experimental possibilities such as placing these wires in a vacuum or in certain gases; but these factors can be studied after the original experiments of heating an element with tension (or compression) have been developed." One must applaud Esson for having the scientific insight to realize that the alloy should be discovered first before its properties are carefully explored.

The present president of the Foundation is George M. Rideout who is also vice-president of Babson Reports, Inc. Some conception of Rideout's scientific attainments may be gained by reading his essay "Is Free Power Possible?" He begins by outlining his private theory that "gravity is a form of stored up 'magnetic' waves which, for a billion years, have been thrown out from the sun and absorbed by every particle which now attracts any other particle." He is not sure this is true, but in any case we do not have to know the source of gravity in order to harness it. All we need, he writes, is a gravity insulator which we can place under one side of a rotating wheel. The result would be, of course, a simple perpetual motion machine. Rideout speculates at some length on the tremendous economic and social changes which would result from such a discovery of unlimited free power. He is careful to point out, however, that this need not harm established public utilities. "Central power plants will continue to be used; but their expansion could take place without increasing the consumption of coal, oil or gas. These valuable natural resources could be conserved for the chemical and other industries which would merely change their form instead of destroying them."

The Foundation resents the label of "perpetual motion machine" for its projected gravity device because so many crackpot attempts to create perpetual motion have been made. "These machines, as a rule, work by levers and weights," writes Trustee Esson. "None of these have been really satisfactory, although the one that has been most publicized was that known as Keely's Motor. Not having seen this motor in operation, the Foundation is not prepared to comment on it, but it soon hopes to make a study of this motor.'' Apparently Mr. Esson is unfamiliar with the well-authenticated story of John Keely's famous confidence swindle. If he is interested, he should read the concluding pages of Charles Fort's *Wild Talents* where he will find an interesting Fortean analysis of how the fraud operated.

Whether the Gravity Research Foundation will ever build a perpetual motion machine is problematical, but there is no doubt that Babson, now in his late seventies, is thoroughly enjoying the search. He is a tall, silver-haired man, with a white mustache, tiny goatee, humorous eyes, and a fondness for bow ties that look as though they might light up. He neither smokes nor drinks. In fact in 1940, he was the Prohibition Party candidate for president. Of late, gravity has been making it increasingly hard for him to climb stairways, an inconvenience that stimulates his interest in the Foundation.

There is little danger of Babson's running out of funds. He has a finger in hundreds of business operations, most of them fabulously successful. His chief enterprise, Babson Reports, Inc., which gives stock-market advice, brings in a net profit of more than $100,00 annually. It is a branch of his holding company, Business Statistics Organization. In addition, he is whole or part owner of such diversified enterprises as a company that supplies lobsters, a firm manufacturing fire-alarm boxes, a sand and gravel company, a dime store chain, an office building in Boston, the Babson sheep-ranch lands in New Mexico and Arizona, a cattle ranch in Florida, and a diamond company. He has written over fifty books, most of them on money and investment, and several hundred newspapers carry his weekly column on business trends.

In addition to the Babson Institute, a business college in Wellesley Hills, Massachusetts, he has also founded two other business schools—one for women in Florida, and Utopia College, at Eureka, Kansas. Babson picked Eureka as the site for Utopia in 1946 because it was the exact geographical center of the United States, and therefore least likely to be hit by an atom bomb. Buildings of the college are connected by underground tunnels. As another precaution against World War III, which he regards as inevitable, he has deposited $1,000 in each of one hundred different banks throughout the central states.

Babson attributes his interest in gravity to the drowning of his seventeen-year-old grandson in 1947. It reminded him of the similar death by drowning of his oldest sister when he was a small boy. To him, it seemed as though gravity were a "dragon" that had seized both of these loved ones and dragged them to the bottom. "Since Michael's death I have become more and more interested in the subject of gravity," he writes poignantly.

Long before this accident, however, Babson was a great admirer of Isaac Newton. He claims that his method of predicting stock-market changes is simply an adaptation of Newton's third law of motion—namely, that for every action there is an equal reaction. Apparently he interprets this in the image of a ball bouncing off a wall. After the stock market has gone down, Babson starts predicting it will come up again. After it has gone high, he predicts it will come down. The prediction which first brought him renown was in 1929 when he announced the great stock-market crash a few months before it happened. He does not like to remember, however, that a year later he was predicting the depression would soon be over, and that investors who followed his advice lost heavily.

Babson's autobiography has the Newtonian title of *Actions and Reactions,* and his wife owns one of the world's largest collections of books by and about Newton. As much as Babson admires Newton, however, there is one respect in which he has failed to emulate the great scientist. He has failed to acquire more than an elementary knowledge of physics.

And this, of course, is something of a tragedy. At the present stage of physical theory, no one has the foggiest idea of how to go about looking for a gravity screen. To establish an organization dedicated to this search is like setting up a foundation to study craters on the other side of the moon. Eventually, astronomers will map these craters, perhaps before the end of the sixties, but to devote an institution to it now would be premature and an absurd waste of funds.

Although Babson is an engineering graduate of the Massachusetts Institute of Technology, he confesses he hated most of his courses. Rideout, president of the Gravity Research Foundation, has had no training as a physicist. Nor are any scientists affiliated with the organization beyond the task of serving as judges for the essay contests. All this may explain why the Foundation has not, in the four years of its existence, issued a single scientific statement of any value.

There is no question that Babson, a kindly, devout Congregationalist, has the best of intentions in keeping the Foundation going. But surely there is a touch of the sin of pride in his refusal to accept advice from competent physicists on how money could best be spent for the good of science and humanity. Surely there is a touch of willful stubbornness in his inability to see the essential ludicrousness of his undertaking. He would do well to ponder Chesterton's remark that Satan fell by the force of gravity.

CHAPTER 9

DOWSING RODS AND DOODLEBUGS

IN RECENT DECADES, geophysicists have invented a number of delicate instruments for exploring underground geological features. In the oil industry, for example, a seismograph can map with a fair degree of accuracy the depths of various strata. It cannot detect the presence of oil, but it can discover a rock structure in which the probability of oil may be great enough to warrant drilling. As yet, science has found no means by which the actual presence of deeply buried minerals (such as oil or water) can be detected from the surface short of actual digging.

Pseudo-science operates under no such limitations. In this chapter, we shall discuss two curious instruments for sub-surface exploration, the dowsing rod and the pendulum, and glance briefly at the use of these and similar gadgets (known as "doodlebugs") in oil prospecting.

Dowsing is the art of finding underground water or other substances by means of a divining rod which is usually a forked twig. The dowser grips the twig firmly in each hand, the forked part pointing upward. As he walks over the ground, suddenly the stick twists in his hands as though moved by a powerful, invisible force. At times the turning is so violent that bark is peeled away by the fists. Where the rod points downward is, of course, the spot where water is to be found.

All types of wood have been used for dowsing, but the traditional favorites seem to be hazel, peach, and willow. Some dowsers are able to work with other substances—ivory, metal, wire, and so forth. A few diviners, especially in India where they are often on salaries from local water boards, use only their bare hands. Bare-handed dowsers describe a sensation like an electric shock on their palms when they are above underground water. The ability to dowse seems to be confined only to certain individuals. Some of the most famous dowsers have been illiterates, completely puzzled by their odd ability and offering no explanation for it.

The employment of various shaped rods for divination purposes goes all the way back to the ancient Greeks and Egyptians. Their use was restricted, however, to predicting future events, detecting guilt, and similar forms of magic. In the Middle Ages, it was associated with the power of Satan, although many churchmen made use of divination rods. The forked twig, for finding minerals, apparently did not appear until the fifteenth century when it was used by German prospectors in the Harz mining region. When German miners were imported to England in the century following, they brought the practice with them. It was in England that the use of the twig was transferred from minerals to the search for water.

By the close of the seventeenth century, dowsing had become widespread over England and all of Europe, and was a topic of violent controversy among scientists. Such formulas as "In the name of the Father, Son, and Holy Ghost, how many fathoms is it from here to the ore?" often were used by dowsers, but in general, the Church continued to frown on the practice. In 1659, a Jesuit father, Gaspard Schott, wrote a book denouncing dowsing as Satanic, though he later changed his mind in favor of a theory of unconscious muscular action. Baron de Beausoleil, a famous seventeenth century dowser, was charged with sorcery and died in prison. In 1701, the Holy Inquisition issued a decree against the use of divining rods in criminal trials.

By the eighteenth century, the "Age of Enlightenment," dowsing had become a common and respectable practice, no longer associated with the Devil's wiles. The first significant "scientific" study of the subject was made in 1891 by Sir William F. Barrett, professor of physics at the Royal College of Science, Ireland. *The Dowsing Rod,* by Barrett and Theodore Besterman, published in 1926, is one of the leading references on the subject. The book's thesis is that the turning of the rod is due to unconscious muscular action on the part of the dowser, who possesses a clairvoyant ability to sense the presence of water.

From 1909 to 1943 a French writer, Henri Mager, issued a series of publications on dowsing, developing the theory that the twig was controlled by electro-magnetic waves. An English translation of his *Water Diviners and their Methods* was published in 1931. In England, two members of the British Society of Dowsers, J. Cecil Maby and T. Bedford Franklin, conducted a series of experiments. Their 452-page book, *The Physics of the Dowsing Rod,* 1939, likewise defends the view that radio waves are responsible for the phenomena.

Although the literature on dowsing is so immense that one could compile a bibliography of bibliographies on the topic, one recent book stands out as the most significant ever written. That book is *Psychical Physics*, a 534-page treatise by Dr. Solcol W. Tromp, professor of geology at Fouad I University, Cairo. It was published in 1949 by the Elsevier Publishing Company, and although written in English, was printed in the Netherlands.

A publisher's leaflet states that Dr. Tromp was born in 1909, took his doctorate in geology in 1932 at Leyden University, Holland, and was appointed professor at Fouad I University in 1947. His experiments on dowsing were conducted at Leyden and later in Delft, Holland. This research convinced Tromp that dowsing was a real phenomenon, and due to electro-magnetic fields surrounding the underground substances. These fields, he believes, affect similar fields in the brain of the dowser. His book contains hundreds of impressive charts, tables, and diagrams, including twenty-five pages of electro-cardiograms obtained by attaching electrodes to the hands of dowsers.

The first half of Tromp's ponderous work is devoted to electromagnetic fields in the earth's atmosphere, under the ground, and surrounding living organisms. Somehow, it is the interaction of these fields which explains dowsing. Tromp is convinced that not only can a dowser locate underground substances, but he can also determine a person's sex. If the dowser stands on the left side of a reclining man, and lets the twig turn above the man's head, the dowser's right hand rises higher than the left. If the reclining person is a woman, the left band goes higher. For some dowsers, the hands are reversed. One might think this would worry Dr. Tromp. Not at all. It simply means that dowser's hands have "different polarities."

For many hours, or even days, after a person has risen from a chair or bed, a good dowser can obtain the proper sex reaction from the spot where the person has been. Dr. Tromp calls this a "shadow phenomenon."

The material of which the rod is composed does not matter. The rod serves, Tromp believes, merely as a muscular indicator. If a dowser washes his hands with a hot salt solution before working, his ability "increases enormously." When a copper wire is connected from the rod to the earth, he loses his ability entirely. He operates poorly if his forearms are exposed to bright sunlight. Wearing rubber soles also increases his sensitivity.

Dr. Tromp admits that dowsers frequently make a bad showing when they are being tested. But this does not bother Tromp. He lists several dozen factors which may cause failure—fatigue, lack of concentration, poor physical condition, worry, too much friction on soles of shoes, all sorts of atmospheric conditions, the presence of electric lines in the area, humidity of soil, and so on. "Trees and their roots are particularly likely to create disturbances that prevent accurate measurements," he writes. Although he has tried in his experiments to take all these considerations into account, it is quite clear they are so numerous and intangible that he has a ready excuse for every dowsing failure. Nothing remotely resembling a controlled experiment is reported in the entire volume.

A large section of Tromp's book is devoted to a field very similar to dowsing, and which he believes is likewise due to electro-magnetism. Commonly called radiesthesia, this phenomenon makes use of a small pendulum formed by a weight suspended on a chain or thread.[1] Usually the weight is a finger ring or bit of metal, although like in dowsing rods, a wide variety of substances have been used.

Radiesthesia seems to have developed first in Europe in the eighteenth century, when it was used for medical diagnosis. The pendulum was suspended above a patient, and without any conscious effort on the part of the operator, it soon began to swing mysteriously. Sometimes it rotated in circles. The nature of its swinging was supposed to indicate the person's ailment. Later, the device was widely used to locate underground minerals and buried treasures.

Dr. Tromp believes that every object in the world has a characteristic "aura," or electro-magnetic field. A person sensitive to these fields will unconsciously translate his impressions into muscular action and this causes the weight to swing. The pendulum can be used, Tromp believes, for distingulshing metals, paintings of different artists, different colored papers, drugs, plants, and sexes.

The pendulum, according to Tromp, rotates clockwise above females, counterclockwise above males. He does not mention that for many operators, a circle in either direction indicates female, and a a back and forth movement, male—while still others find a circular movement for male and an oval for female. The same "shadow phenomenon" obtains over spots where a person has been resting. Similar reactions occur above urine, or the womb of a woman with child.

Tromp's book closes with a discussion of the homing instinct of certain birds and animals, and other phenomena which he thinks have electromagnetic explanations. The stigmata of religious mystics, telepathy, firewalking, psychometry, and the influence of a mother's experiences on her unborn child are topics he feels should be investigated with electromagnetic fields in mind.

"Most scientists of the twentieth century seem to lack the courage and the romantic feeling to tackle problems which at first sight seem incredible. . . ." he concludes his book. "It is the unconventional scientist who enables the work to progress more rapidly." The last pages are devoted to a bibliography of 1,496 titles, of which 700 relate to dowsing and radiesthesia.

An indication of the popularity of radiesthesia in England and Europe today is the number of societies and periodicals devoted to the topic. In England, there is a monthly review called *The Pendulum*. Similar publications are being issued in France and Germany, and recently two new magazines on radiesthesia appeared in Italy. "Now there are so many scientists who believe in dowsing," wrote Charles Fort, "that the suspicion comes to me that it may be only a myth after all."

In 1936, the German and Italian war departments seriously considered the pendulum as a device for finding water, and it was actually used by Germany in the North African campaign of the last war. Radiesthesia fascinated the pseudo-scientists who flourished in Hitler's military forces. There is good evidence that sections of the German Navy actually used the pendulum above huge maps of the North Atlantic in an effort to locate enemy battleships!

This type of long-distance radiesthesia, incidentally, is much frowned upon by Dr. Tromp, as well as the use of the pendulum or dowsing rod above photographs. It doesn't fit into his electro-magnetic theory.

One of the favorite arguments of believers in dowsing, repeated by Tromp, is that anything which has lasted so long must have something to it. If you point out that astrology has an even longer history, the odds are high the believer will grant that astrology also may have a valid basis. This is, in fact, Tromp's position. In an earlier book, *The Religion of a Modern Scientist*, 1947, he argues that planets and stars may well influence the earth's electro-magnetic fields, which in turn may mold a person's character and future. He thinks, however, the astrologers made a grave mistake in using the date of birth instead of the date of conception! This error negates most of historical astrology, but Tromp feels it might be corrected and astrology given an empirical bases. There are accompanying photographs bearing such captions as "Capillary figures of a solution of silver nitrate, iron sulphate and lead nitrate taken two hours before the conjunction of Saturn and the Moon."

In comparison with the "scientific" trappings of Tromp's *Psychical Physics*, Kenneth Roberts' recent book, *Henry Gross and His Dowsing Rod*, 1951, reads like a work of fiction. Of course Roberts does not pretend to be a scientist, a fact amply borne out by his numerous geological howlers. He accepts, for example, a theory that underground water veins bear no relation to the ground-water table, but come from huge "domes" which are pushed up from great depths. In fact, one dome is supposed to come from 57,500 feet below ground! At this depth the earth's heat would have turned the water to steam, and geysers, not wells, would have resulted. Of course the whole concept of water being pushed up into domes is geologically absurd, but to make it worse, Roberts suggests that such domes are driven up ''by the same sort of pressure that drives up oil. . . .'' Roberts is apparently unaware of the elementary geological fact that oil floats on water, and is therefore flushed upward by water until it is trapped under inverted bowls of impervious rock.

Roberts, as everyone knows, is a writer of popular historical novels. Before his success as a novelist, he had been a roving editor of the *Saturday Evening Post*. Although long interested in dowsing, it did not become a major interest for him until recently when he discovered the dowsing abilities of Henry Gross. It is the story of Gross' rapidly growing prowess which forms the content of Roberts' book and leads him to declare that ''when the potentialities of the rod are more clearly understood and utilized, it may rank with electricity and atomic power.''

''Why . . . shouldn't scientists, in addition to spending time, energy and money on questionable laboratory experiments with dowsers,'' he asks, ''devote more of their energies to developing an invaluable, even though mysterious, phenomenon that, properly utilized, would prevent wars, move mountains, turn deserts into lands of plenty, feed the hungry, cure the sick and change the face of the world?''

Henry Gross, the man who introduced Roberts to this earthshaking power, is a stocky, gray-haired, bespectacled game warden in Biddeford, Maine. Roberts lives on a nearby farm at Kennebunkport. When forest fires in 1947 created a serious drought in the region, Roberts asked dowser Gross to find a new well for him. He was so impressed by Henry's accuracy that he obtained for his friend a supply of foreign books on dowsing and radiesthesia. Fascinated by what he read, Gross began experimenting with his talent, and in less than three years developed into one of the most fantastic super dowsers of all time.

Not only is Gross able to trace the winding course of underground water veins, but he also specifies the exact depth of the water, its direction and rate of flow, and whether it is good to drink. This information is obtained by asking the rod questions which can be answered by yes or no. The rod dips for yes, remains still for no. Although Gross prefers to use a stick of fresh-cut maple, he can dowse with any kind of material—even a long blade of grass. Wearing rubber gloves or rubber-soled shoes has no effect on his dowsing.

In 1949, Henry Gross discovered that he did not even have to be in an area to dowse it, but could be many miles away. Roberts recounts many of Henry's sucesses in long-distance dowsing, the most dramatic of which was in 1950 when he managed to locate three fresh water sources in Bermuda. Gross was in Kennebunkport at the time he did the dowsing—800 miles away! Mrs. Roberts had moved a pencil over a map of Bermuda, and Henry's rod dipped when the pencil touched the proper spots. Although Bermuda had not a single fresh water well at the time, and was suffering a severe drought, Henry's predictions later proved correct.

Once Henry dowsed a map of Africa and found a huge vein beneath the Sahara Desert. It ran all the way across Africa's western hump and emptied into the Gulf of Guinea! Unfortunately, this has not yet been verified. "There, by jingoes," Gross exclaimed after he finished plotting this underground river, "that's something I'd like to go and work on—something BIG!"

Not only can Henry find water, but he can also use the rod to locate people. If he's calling on friends, he doesn't need to push the doorbell to find out whether they are home. The dowsing rod tells him exactly who's there. He can find people who try to hide in the woods. He can locate lost objects. On one occasion he found an outboard motor that had dropped into sixteen feet of water.

On another occasion, Gross and Roberts experimented with drinks. When Henry touched the tip of his rod to rye whiskey, it would then dip only over rye, and not over other kinds of liquor. If he touched the stick to bourbon, it dipped only over bourbon. And so on. It was not necessary even to smear the tip with the beverage. Merely touching it to a sealed bottle of brandy would make the rod dip over other bottles of brandy. "It didn't seem reasonable to us," Roberts writes, "that emanations from scotch or brandy should be able to pass through a glass bottle—until we reflected that emanations from the North Magnetic Pole can pass through anything at all. . . ."

As a result of his reading the literature on radiesthesia, Gross began experimenting with a pendulum. He found it worked beautifully. Unlike most pendulums, however, it moved in circles over *male* anatomy, back and forth over *female*. Roberts reports that when Henry first swung a pendulum over his hand, "I felt a tingling sensation . . . that became as sharp as an electric shock, so that I snatched my hand away. . . ." In Henry's hands the pendulum correctly indicated the sex of animals, eggs, and even photographs of people. There is one curious exception, however. If a woman's blood contains an Rh-negative factor, the pendulum gives a male reaction!

What is one to make of all this? Of course it is difficult to assay evidence presented in a form resembling fiction by a man who is a passionate believer in the phenomena for which he is pleading. But even making generous allowance for considerable loading of evidence on Roberts' part (significant omissions, unconscious exaggerations, and so forth), it is not impossible to reach a few conclusions.

There is nothing to suggest, for instance, that Henry Gross is either a charlatan or hoaxer. He seems simple and sincere, completely baffled by his peculiar ability but at the same time intensely interested. On the other hand, there is everything to suggest that Gross, without conscious awareness, is transmitting his thoughts to the dowsing rod or pendulum by unconscious muscular movements.

The unwitting translation of thoughts into muscular action is one of the most firmly established facts of psychology. In individuals particularly prone to it, it is responsible for such "occult" phenomena as the movement of a Ouija board, table tipping, and automatic writing. It is the basis of a type of mind reading known in the magic profession as "muscle reading." Someone hides a pin in a room, and the performer finds it quickly by having a spectator take hold of his hand. The spectator thinks he is being led by the magician, but actually the performer permits the *spectator to lead him* by unconscious muscular tensions. Many famous muscle readers are able to dispense with bodily contact altogether, finding the hidden object merely by observing the reactions of spectators in the room.

Concentrate intently on the toes of your left foot. Do you find yourself wriggling them slightly? Or at least suppressing a strong desire to do so? Many people cannot read without slight movements of their lips. One of the most surprising demonstrations of unconscious muscular behavior can be performed with a home-made pendulum. A ring suspended from two feet of thread will serve admirably. Tell yourself it will swing in circles over a woman's hand, back and forth over a man's—then try it and see. For most people, this works so successfully that such pendulums have been sold in novelty stores in America for decades as "sex indicators." The explanation is, of course, that unconscious and invisible movements of your hand are sufficient to start the pendulum swinging in whatever manner you *expect* it to behave. A dowsing rod operates on exactly the same principle. The stick is given a strong tension in the hands. Although it remains upright, it is so precariously balanced that the slightest muscular movement will send it suddenly downward. It is not surprising that good dowsers are almost always equally successful with the pendulum.

But what, it may be asked, about tests in which the dowser or pendulum operator does not know in advance how his instrument is supposed to behave? The answer is that whenever such tests are given under controlled conditions, the results are no better than might be expected from the law of averages. Even in Kenneth Roberts' violently partisan book, he records an abundance of failures by Henry Gross *whenever conditions approaching a scientific test were arranged.* For example, Henry was unable to distinguish mason jars containing water from jars containing sand when the jars were concealed inside paper sacks. He was unable to find envelopes containing coins when they were placed on the ground beside empty envelopes. If the tip of his dowsing rod was touched to a coin and the coin tossed on the floor, Henry's rod would point to it. But when someone held the coin behind his back, then brought his hands in front with the coin concealed in a fist, Henry was completely unable to determine which hand held the coin.

When coins were put in certain pill boxes and not in others, Henry was unable to tell one box from another. When gold watches and brooches were buried under sod and Henry tried to find them, the results were unsatisfactory. Once Henry thought his stick (in the tip of which he had inserted a piece of gold) had located goldbearing ore in a brook near his house. The ore was sent away for analysis. It proved not to be gold. Roberts explained this last failure by saying that water had flowed over gold ore and then over the rock, leaving just enough trace of gold to affect Henry's rod! Since Roberts offers no indication of how he knew all this, his explanation is a little less than convincing. On another occasion, Professor Joseph B. Rhine of Duke University conducted a series of experiments in which Henry tried to determine when water was or was not in motion through a pipe. Henry failed miserably.

For the water pipe failure Roberts has two rationalizations. In the first place, he thinks water does not really stop moving in a pipe when the faucet is turned off. It is left in a state of "agitation." Besides, Rhine was testing Gross in an artificial, unnatural situation. Later, he writes, he kicked himself for not having thought of building a twenty-foot square platform, covering it with earth, and running a hose under it. The hose could have been twisted into various positions beneath the platform, then Henry asked to trace its contours by walking across the platform's top. Under such conditions, Roberts is positive "Henry's rod would have pointed unerringly to the hose . . . a thousand times out of a thousand tries. Unfortunately, however, I *hadn't* thought of it."

Unfortunately indeed! Guessing the position of the hose with one hundred per cent accuracy in a thousand tries would have been precisely the sort of test which the book so glaringly lacks. How a working scientist would shudder at Roberts' assurance about the outcome of an experiment which has not yet been tried! Alas, one fears, this test will never be made. Nor is Henry likely to try the blindfold test with which he was once challenged by a wise professor at the University of Massachusetts. This test is even simpler. Let Henry find a spot where his rod dips strongly. Then let him be blindfolded securely and led about over the area to see if his stick dips repeatedly when he walks across the same spot. Could anything be fairer?

When Henry's rod failed in some of the tests mentioned above, Roberts' reaction was typical. Did it suggest that he should endeavor to set up other tests, such as his hose idea, which might yet place Henry's ability on some sort of scientific footing? It did not. Instead Roberts writes, "If the . . . experiments proved nothing to the scientists, they proved a good deal to me. They proved above all else that I should have as little as possible to do with dubious skeptics or geologists in any future dowsing experiments on which I might venture. . . ."

Tests of Henry's ability to operate a pendulum were equally dismal. In the obstetrics department of the Maine General Hospital, he held a pendulum above the abdomens of sixteen pregnant women. After the births, his predictions were checked for accuracy. Out of the sixteen cases, Henry was right in seven. In fact, Roberts' niece did better in a similar test with eleven women. She was right in nine cases, and one of the failures was later discovered to have an Rh-negative blood factor!

As for Henry's successes in finding water, many factors combine to make his work seem much more astonishing than it really is. In the first place, there often are many surface indications which, to a man thoroughly familiar with the terrain, are clues to underground water. Game warden Gross is certainly well acquainted with the area, and there is no question that he is a shrewd, intelligent man. On page 46 he traces a vein to a spot which had always "hampered . . . farm equipment because of its sogginess." On page 276 he picks a spot for drilling where the vein is so close to the surface "that the surrounding sod and brush, in spite of the extreme drought, were moist and green." There is every indication that he is completely familiar with the clays, shell deposits, sands, and other formations characteristic of the region. Although he may not be conscious of it, he could easily be picking up important geological clues from the surroundings he knows so well.

In many cases when Henry's dowsing confirmed knowledge already possessed by someone, it is not incredible to suppose he may have acted as an unconscious "muscle reader"—gathering clues from the reactions of those present and transmitting this information to his hands.

But the most important factor of all is the simple fact that if you dig deep enough, almost anywhere, you are bound to hit water. Water fairly close to the surface is far more plentiful than one might think, and the odds of finding it at shallow depth, at a spot picked at random, are in many regions very high. Actually, water seldom occurs in "veins" (except in rare cases where rock fractures or cavities permit it). There is merely a variable porosity of ground-water below a certain level, which varies from year to year and season to season. In most areas, it is impossible to *avoid* finding ground-water, though it is seldom in sufficient quantity to supply more than a local household.

Bermuda is such a region. As on similar limestone islands, the fresh rain water soaks down through the porous lime and floats above the salty ground-water. A well dug to this spot, anywhere on the island, will skim off fresh water, though only in small quantities. Roberts once expressed to reporters the opinion that the water Henry found in Bermuda came from underground streams that originated on the North American continent!

Even allowing for Roberts' loading of data, he records several notable failures on the part of Henry's water dowsing. Like Dr. Tromp, he has easy excuses. In one case, when Henry had predicted water at sixteen feet, it proved to be six feet, but this was because the "vein had run into an obstruction . . . and set up a pressure area." On another occasion, the use of dynamite had diverted the vein. At still another spot, the water was pushed out of its course by the concussion of a drill. Even the weight of a bulldozer moving across the surface, Roberts claims, will crush and divert veins. Since it is difficult to sink a well without some ground disturbances of these sorts, it is not hard to find excuses for failures. If the water is there—as in most cases it would be regardless of how the spot were selected—fine. If not, the vein was "diverted." And how does Roberts know the vein was diverted? Because it wasn't where Henry said it would be, and Henry is never wrong.

Estimates of depth, it should be pointed out, are extremely difficult to verify. Roberts often speaks of Henry's predictions being accurate to the inch, though anyone who has ever tried to measure the depth of the source of water in a well knows how preposterous such precision would be. In some cases, Henry was off many feet, as in the case of one of the Bermuda wells. From Kennebunkport he estimated the depth at fifteen feet. His on-the-spot check showed the depth to be thirty feet. The final drilling found water at seventy-three feet.

Measurements of rate of flow are even more difficult to make, and moreover, have wide seasonal variations. As Thomas Riddick, an engineer in the water-works field, wrote in his article "Dowsing Is Nonsense"[2] *(Harpers,* July, 1951): "There is no water-works engineer in America who would certify the yield of any well to the quarter-gallon per minute even after a forty-eight hour pumping test with the most precise measurements." Needless to say, such careful tests are seldom made after one of Henry's quart-per-minute predictions.

Water dowsing is far from a harmless and inexpensive superstition. At the moment, untold numbers of dowsers throughout the entire world are being paid handsomely for their services. Recently Roberts and Gross have formed a company called Water Unlimited, Inc., which charges a minimum of $100 (plus travel expenses) for dowsing a single well. For some jobs, the cost may run as high as a thousand.

An even greater waste of funds results from the use of pendulums and more complex devices for the finding of oil. From the early days of the oil industry until the present, geologists and oil producers have been plagued by charlatans and cranks who operate some type of fantastic device supposed to be an infallible indicator of oil. "Oil smellers" was the term originally applied to such contraptions, though in recent years the term "doodlebug" has become more common.

Usually it is the gullible farmer who pays.[3] A smooth-talking doodlebug operator will charge anywhere from ten dollars to a small fortune for a survey of the farm. Later, if an oil company drills on the property, it often meets stubborn resistance from the farmer, who is convinced the well is being sunk in the wrong spot. Ten years ago, he insists, he paid a "geologist" to go over the land and show him exactly where the oil was.

In many cases, operators of doodlebugs firmly believe in the device. Usually they are proud of their lack of technical knowledge, and bitter against orthodox geologists unwilling to open their minds to something new. Often they claim to have submitted their invention to a world renowned scientist (usually a personal friend), and the scientist admitted he was unable to understand how it worked. Most modern doodlebugs operate on a gravity principle, or they detect electro-magnetic radiations of obscure sorts which emanate from the oil. They are often elaborately designed and beautifully constructed, and sometimes resemble a piece of Rube Goldberg apparatus. Invariably the inventor is the only man capable of operating it correctly. Should a test well actually be drilled on the advice of the doodlebug, and the well prove a dry hole, the operator always has (if he can be found) a ready answer. Some type of "interference" in the area is responsible for a bad reading.[4]

A petroleum geologist in Tulsa recently sent me a copy of a circular letter received by oil operators in the city early in 1952. The letter is typical, and reads in part:

> Years of intensive geological study, both mining and petroleum . . . have given me a broad, general knowledge. . . . I have spent over $100,000 and years of research on a special method of magnetic logging . . . which method gives promise of being over 90 per cent accurate.
>
> . . . I have decided to offer my services and vast experience commercially to a selected few independent oil operators. . . .
>
> My fees are very reasonable in view of the many millions each operator stands to make through their use. . . .
>
> . . . My time is extremely limited, so if you are interested. . . .

His fee for a general report on an area is $1,000.

The chances of finding oil in a wildcat well are, of course, much lower than finding water in a wildcat water well. But as one would expect, successes do occur—and none has been more fantastic than the discovery of the West Edmond field in central Oklahoma in 1943.

The discovery well was drilled by a wildcatter named Ace Gutowski, with money from the Fox Brewing Company of Chicago. How did Ace locate the drilling site? A farmer named J.W. Young, who lived in the area, had a bottle covered with goatskin and filled with a mysterious substance the composition of which he refused to divulge. The bottle was suspended on a watch chain.

It would swing north and south over oil—east and west over salt water.

A geophysicist friend of mine saw Young demonstrate his doodlebug in 1944 in a restaurant in Edmond. The bottle was held over specimens of sand. It worked very well, but my friend was struck by the fact that Young suffered from a palsy which gave a noticeable tremor to his upraised hand.

Nevertheless, Gutowski's well opened up the largest oil pool discovered in Oklahoma for twenty years! Moreover, the pool was under a structure of a type not detectable by orthodox geological or geophysical methods.

But the story has a topper. Farmer Young, with unbounded faith in his doodlebug, had bought land in the area. The new field was gradually extended until it finally bordered Young's property. And there it stopped. On Young's land was drilled the first dry hole in the history of the field!

CHAPTER 10

UNDER THE MICROSCOPE

JUST AS SOME astronomers, peering through telescopes, are able on occasion to see things other astronomers are unable to see (e.g., Lowell and the Martian canals) so some biologists, peering into their microscopes, observe remarkable events which somehow elude their colleagues. In a later chapter, we shall learn how the eminent Austrian psychiatrist, Wilhelm Reich, has witnessed the formation of protozoa from "bion particles." Here we shall survey briefly a few of the more bizarre reports of similar microscopic marvels.

The long abandoned doctrine of abiogenesis, the spontaneous generation of living forms from non-living matter, has always been a promising field for the modern pseudo-biologist. In 1836, an English amateur scientist named Andrew Crosse, puttering around with electrical experiments on his estate in the Quantock Hills, was startled to find microscopic insects appearing as a by-product of his research. He had been trying to produce artificial minerals by sending an electrical current through certain chemical mixtures. Here is Crosse's description of what he observed through hls microscope:

> On the fourteenth day from the commencement of the experiment I observed through a lens a few small whitish excrescences, or nipples, projecting from the middle of the electrified stone, nearly under the dropping of the fluid above. On the eighteenth day these projections enlarged, and seven or eight filaments . . . made their appearance on each of the nipples. On the twenty-second day these appearances were more elevated and distinct, and on the twenty-

sixth day each figure assumed the form of a perfect insect, standing erect on a few bristles which formed its tail. Till this period I had no notion that these appearances were any other than an incipient mineral formation; but it was not until the twenty-eighth day, when I plainly perceived these little creatures move their legs, that I felt any surprise, and I must own that when this took place I was not a little astonished. . . .

About a hundred insects were born in this manner. Crosse reported that the smaller ones seemed to have six legs and the larger ones eight. He guessed them to be of the genus *Arcurus* (mites), "but of a species not hitherto observed." Eventually they left the liquid entirely, and flew about the laboratory, hiding in dark spots as though avoiding light. Later experiments produced the mites in great numbers. Another amateur by the name of Weeks, who lived in Sandwich, repeated Crosse's experiments, taking great precautions to keep air-borne eggs from entering the liquid. He not only found the same mites, but also noted that their numbers varied directly with the percentage of carbon in his solutions.

The details of Crosse's research may be found in the *Memorials of Andrew Crosse,* by Mrs. C.H.A. Crosse, 1857; *A History of the Thirty Years Peace,* by Harriet Martineau, 1849; and Lieutenant Commander Rupert T. Gould's *Oddities, a Book of Unexplained Facts,* 1928.

In 1872, another English scientist, Henry Charlton Bastian, created an academic tempest with his two-volume work, *The Beginning of Life.* Bastian used the term "heterogenesis" for the process of spontaneous creation he believed was taking place constantly in nature, and which he claimed to have witnessed in his microscope. When colleagues repeated his experiments, however, taking better precautions against air bacteria, the results were disappointing.

In 1906 John Butler Burke, also British, published a similar work titled *The Origin of Life.* Using radium, Burke was able to produce primitive artificial forms—half living and half crystalline—which he called "radiobes." This work also created a stir at the time, but eventually was forgotten after other scientists failed to confirm Burke's findings.

Much more astonishing than the work of any of the previously mentioned men were the discoveries of Morley Martin, an amateur biochemist who began his strange adventures with the microscope in 1927 at his privately financed laboratory in Andover, a village in Hampshire, England. Briefly, what he did was this. He took a piece of Azoic rock (the Azoic era is the earliest in geological history—so early that no fossil traces have yet been discovered in it), and heated it until it formed a light ash. From this ash, by a complex chemical process, he obtained what he called "primordial protoplasm." When this was subjected to X-rays (exercising great care to prevent air contamination), its crystals gradually developed into microscopic forms of living plants and animals—chiefly fishes. In an area one inch across, for example, he once found 15,000 of these miniature fishes. It was his belief that they had been preserved in a state of suspended animation during the millions of years since the Azoic Age.

When Morley Martin died in 1937, he had published only a pamphlet, now exceedingly rare, titled *The Reincarnation of Animal and Plant Life from Protoplasm Isolated from the Mineral Kingdom*, 1934. Unfortunately, his laboratory notes explaining the details of his research were kept in a code that has not been deciphered. The best reference to his work is a chapter on him in Maurice Maeterlinck's *La Grande Porte*, 1939. The following paragraphs are quoted from a translation of this chapter in a mimeographed booklet, *The Morley Martin Experiments*, issued in 1948 by the Borderland Science Research Associates of San Diego:

> Under the enlargement of the microscope globules were seen taking shape within the protoplasm and forming vertebrae which elongated into a spinal column in which the ribs were inserted; then came the outlines of the limbs or claws, the head, and the eyes. These transformations are normally slow and require several days, but at times they took place under the eyes of the observer. One crustacean, for example, having developed its legs, walked off the field of the microscope.
>
> These emergents therefore live, sometimes move, and develop as long as they find sufficient nourishment in the protoplasm in which they were born: after that, their growth is checked or else they devour one another. Morley Martin has, however, succeeded in feeding them by the help of a serum whose secret he kept. . . .

Bacteriologists have remained unimpressed by these sensational findings, but Theosophists are less narrow-minded. Numerous articles have appeared in their periodicals pointing out how Morley Martin's work supports the views of Madame Blavatsky concerning primordial archetypes of life which existed in the earths "fire-mist period," and which were the source of forms later exploited by the evolutionary process.

American scientists seem to have lagged behind the British in their ability to observe such striking phenomena through their microscopes, but at least one native scientist achieved results which deserve comparison. He was a homeopathic physician in Seattle named Charles Wentworth Littlefield.

Not much is known about Littlefield beyond the fact that he was born in 1859, of Scottish ancestry, grew up on a farm near Muncie, Indiana, and began to practice medicine in 1886 somewhere in Arkansas. In 1896, he graduated from the Kansas City Homeopathic (we will discuss home- opathy in a later chapter) Medical College. Later he settled in Seattle, where he practiced as a physician and surgeon until his death. Judging from a photograph in one of his many books, he was a tall, stern-faced, bespectacled man, with heavy black eyebrows, large ears, and a bushy beard that completely concealed his mouth.

Littlefield attributes the origin of his curious investigations to an event which occurred when he was working as a boy on the farm with his brother. Young Charles' foot was badly cut in an accident. To stop the profuse bleeding, his brother obtained the help of a man in the area who had the ability to check haemorrhages by reciting a charm based on a Biblical passage. The charm worked like a charm. Littlefield learned what to recite, and later when he became a doctor, made frequent use of the charm to stop the bleeding of patients after an accident or during an operation.

About 1905, Littlefield began asking himself why this charm worked so well. On what elements in the blood did it operate? He knew that certain organic salts played important roles in the clotting process. Ac- cordingly, he obtained one of these salt solutions, and put a drop of it on a glass slide. While it was drying beneath his microscope, he recited the incantation three times, simultaneously concentrating on the image of a chicken. When he looked through the eyepiece—behold, a miracle! The salt crystals had formed themselves into the unmistakable shape of a chicken! Hundreds of experiments followed—all with similar results. By concentrating on an image, Littlefield was able to obtain that pattern in the crystal formation. It was his theory that the process of evaporation created a ''subtle magnetism'' which made the salts susceptible to mental control.

These early experiments of Littlefield were widely reported in the nation's press, and were described by the doctor in several popular magazine articles. The definitive reference, however, is *The Beginning and Way of Life,* a 656-page work, one hundred thousand copies of which were privately printed in Seattle in 1919. In the preface, the author extends his thanks to St. Paul, St. John the Revelator, and the English physicist, Michael Faraday, for having dictated portions of the book from the spirit world.

The book contains over a hundred photomicrographs of crystal struc- tures. They fall into three broad categories. One group consists of plant and animal forms which developed spontaneously in droplets of salt solution left undisturbed for several days. These forms include shells, crabs, fish, snakes, octopuses, one ape, and several human forms. Unlike Morley Martin's creatures, the structures were not alive, although Littlefield seems to think that some of them, the octopuses in particular, were in a kind of semi-living state. A second class of photographs shows patterns formed by mind control—faces of St. Paul and St. John, Uncle Sam, John Bull, a fowl, and an overshot water wheel. The third class of photomicrographs are called "unconscious thought pictures." The doctor describes these as forms "fixed in the salts which I had no intention of fixing, but after finding them, remembered having entertained the image." Examples include the head of Mephistopheles, the word "token," and a woman carrying a dog under her arm on a windy day.

Dr. Littlefield was utterly unable to comprehend why orthodox biologists were not excited by this research. "Why do not these men try these experiments," he asked, "and thus demonstrate their truth or falsity, instead of hurling meaningless invective against the man who presents them as a possible solution of a question of vital importance in science?"

The final third of the doctor's opus is devoted to his methods of healing. In addition to usual homeopathic remedies, he also made use of a "Rainbow Lamp" which he invented and patented in 1918. This treated almost every variety of ailment by shining colored lights on the patient. Six photographs are devoted to the buttocks of a Seattle waitress who had been badly burned. When Dr. Littlefield was called on this case, he recognized the burn as so severe that in most instances it would have been fatal. Fortunately, he was able to install his light machine and bathe the waitress' backside with rainbow colors. The photographs show the burn's healing progress. After ten weeks, the skin was good as new. It was the doctor's hope that proceeds from the sale of his book could be used to establish a Rainbow Temple in Seattle equipped with light machines in individual rooms and a large rotunda for mass rainbow therapy.

The book closes with many touching letters from patients telling how they were cured of cancer, tuberculosis, epilepsy, Bright's disease, and other ailments after taking "Dr. Littlefield's vitalized tissue salts." The salts could be obtained from the doctor at $2.00 for a half-ounce bottle.

A more recent work, *Man, Minerals, and Masters,* was published by Littlefield in Los Angeles, 1937. It contains more photomicrographs of such thought-produced forms as the Great Pyramid, a phallic symbol, a head of the Sphinx, a woman in white, an Indian head, a right foot, a left foot, the eye of God, Mount Zion, the Rosy Cross, an open book showing a map of the United States, and "Three Tibetans sitting in mountains." It often takes considerable imagination to see the shapes you are supposed to see, but there is no denying a connection between the crystal patterns and the captions beneath. One cannot resist suspecting, however, that the captions came *after* the crystal formations rather than

before, or perhaps the doctor made several hundred tries until he succeeded with each thought.

In the same book, Littlefield describes a new technique he developed
for aiding the crystallizing process. It involves the formation of "mantras"
—invocations which he obtained by numerology. For example, the
doctor decides to produce a formation of the crucifixion scene. After
considerable effort he obtains the mantra, "Lamb Sacrificed for Man."
Why is this a mantra? If you take the letters in the word "Lamb," and
substitute numbers (using 1 for A, 2 for B, etc., up to 9 for I, then
beginning again with 1 for J and continuing thus through the alphabet)
you obtain the figures 3-1-4-2. These add up to 10. The two figures in
10—1 and 0—are added once more to give the final figure of 1. A similar
process applied to the next word, "Sacrificed," yields the number 3.
"For" also gives 3, and "Man" gives 1. The complete phrase is
represented, then, by the figures 1-3-3-1. If two of these figures are alike,
you have a mantra. If more than two are alike, you have an extra-
powerful mantra. By using the above mantra, Littlefield obtained a
crucifixion scene which is reproduced in his book. A similar scene of the
Resurrection was produced by concentrating on the mantra, "Mental
Mastery of Life Power."

The belief that microbes may be constantly entering the earth's atmosphere from interstellar space has almost as large a following among
eccentric microscopists as the synthetic creation of life. Lord Kelvin, the
British physicist, was the last eminent scientist to defend this view. He
pointed out that the intense cold of space would be ideal for preserving
bacteria in a state of suspended animation. The theory has more recently
been revived by two Swedish scientists, Svante A. Arrhenius, and Professor Louis Bachman.

Another Swedish professor, Knut Emil Lundmark, thought bacteria
might be transported here by meteors. This view received support
as late as the thirties when Professor Charles B. Lipman of the
University of California sterilized some meteorites, ground them to
powder, put them in culture dishes, and found microbes. Other scientists have been unable to duplicate this experiment and suspect Lipman had insufficient knowledge of difficult laboratory procedures
necessary to prevent air-borne bacteria from sneaking into culture
dishes.

An equally surprising meteoric discovery was made seventy-five
years ago by Dr. Otto Hahn, a German geologist. Hahn found microscopic shells, corals, sponges, and crinoids in meteorites. His large
book on the subject, *Die Meteorite (Chondrite) und ihre Organismen*,
published in Tübingen in 1880, contains dozens of photomicrographs,

beautifully printed in sepia, which he thought proved his findings. Unfortunately, all they prove is that Hahn photographed tiny crystal structures, then—like Littlefield—let his imagination gallop away with him.

CHAPTER 11

GEOLOGY VERSUS GENESIS

> Some drill and bore
> The solid earth, and from the strata there
> Extract a register, by which we learn
> That He who made it, and revealed its date
> To Moses, was mistaken in its age!
> —*William Cowper*

DURING THE Middle Ages and Renaissance, no one knew quite what to make of fossils. The most popular explanation was that they were relics of plants and animals destroyed by the flood of Noah, but many distinguished scholars believed they had grown in the bowels of the earth by some sort of natural or occult process. A few argued that Satan placed them there to mislead the faithful. Others suggested God himself created them—either to befuddle scientists, test believers, or make crude experimental models of living forms.

One of the last to defend the divine origin of fossils was Professor Johann Beringer of the University of Würzburg. In 1726, he published an extensive monograph describing some curious fossils he had discovered, many of them bearing replicas of the sun and moon, and various Hebraic words. Actually, they had all been carefully baked out of clay and planted in the earth by his students, but the poor professor fell completely for the hoax until one day he discovered a fossil with his own name on it! He spent the rest of his life buying up copies of the work, which of course made it much in demand as a collector's item. As a crowning irony, a descendent had the treatise reprinted after the professor's death, and reaped a handsome profit. It is really a sad tale—a scientist remembered today only for his gullibility.

Gradually, through the eighteenth and nineteenth centuries, naturalists became impressed by the fact that fossils in the lower and older beds of rock were relatively simple, and as one moved up through later strata, fossils became more complex. Was it possible life began in some simple form, millions of years ago, then slowly evolved through long geological ages into earth's present fauna and flora? Finally, in 1859, Darwin published his *Origin of Species*. It was not the first book on evolution, but it was a book which marshalled such an imposing array of facts that it was no longer possible to dismiss the theory as fanciful and impious speculation.

The blow which this book let fall on the back of Christendom is hard to overestimate. Certainly it was a major factor in the widening Protestant split between fundamentalists and so-called modernists. The modernists, of course, accepted the new theory. It was, they said, simply God's method of creation. If we do not take the *Genesis* story too literally—if we interpret the "days" as geological epochs—we can even read it as a rough description of evolutionary history.

The fundamentalists rejected evolution in toto. Most of them clung to the flood theory of fossils, which had been eloquently championed by Luther. Others, including British Prime Minister Gladstone, defended the view of a French naturalist, Baron Cuvier, that there had been a series of creations, at various intervals in geologic time, each following a cataclysm which buried earlier forms of life. Still another theory, which survives today in the notes of the popular *Scholfield Reference Bible,* was that fossils are remains of an earlier "Pre-Adamite" creation which flourished between the first and second sentences of the opening verse of *Genesis!*

Thousands of books were published in the nineteenth century, most of them in England, attempting to harmonize geology and *Genesis*. In this dreary and pathetic literature, one book stands out from all the others as so delightful and fantastic that it deserves special mention. It was called *Omphalos* (the Greek word for navel), and was written by zoologist Philip Gosse, father of the eminent British poet and critic, Edmund Gosse. Not the least of its remarkable virtues is that although it won not a single convert, it presented a theory so logically perfect, and so in accord with geological facts that no amount of scientific evidence will ever be able to refute it.

Gosse admitted geology had established beyond any doubt that the earth had a long geological history in which plants and animals flourished before the time of Adam. He was also convinced that the earth was created about 4,000 B.C., in six days, exactly as described in *Genesis*.

How did he reconcile these apparently contradictory opinions? Very simply. Just as Adam was created with a navel, the relic of a birth which never occurred, so the entire earth was created with all the fossil relics of a past which had no existence except in the mind of God!

This is not as ridiculous as it may seem at first. Consider, for example, the difficulties which face any believer in a six-day creation. Although it is possible to imagine Adam without a navel, it is difficult to imagine him without bones, hair, teeth, and fingernails. Yet all these features bear in them the evidence of past accretions of growth. In fact there is not an organ or tissue of the body which does not presuppose a previous growth history.

The same is true of every plant and animal. As Gosse points out, the tusks of an elephant exhibit past stages, the nautilus keeps adding chambers to its shell, the turtle adds laminae to its plates, trees bear the annual rings of growth produced by seasonal variations. "Every argument," he writes, "by which the physiologist can prove . . . that yonder cow was once a foetus . . . will apply with exactly the same power to show that the newly created cow was an embryo some years before creation." All this is developed by the author in learned detail, for several hundred pages, and illustrated with dozens of wood engravings.

In short—if God created the earth as described in the Bible, he must have created it a "going concern." Once this is seen as inevitable, there is little difficulty in extending the concept to the earth's geologic history. Evidence of the slow erosion of land by rivers, of the twisting and tilting of strata, mountains of limestone formed by remains of marine life, lava which flowed from long-extinct volcanoes, glacier scratchings upon rock, footprints of prehistoric animals, teeth marks on buried bones, and millions of fossils sprinkled through the earth—all these and many other features testify to past geological events which *never actually took place*.

"It may be objected," writes Gosse, "that to assume the world to have been created with fossil skeletons in its crust—skeletons of animals that never really existed—is to charge the Creator with forming objects whose sole purpose was to deceive us. The reply is obvious. Were the concentric timber-rings of a created tree formed merely to deceive? Were the growth lines of a created shell intended to deceive? Was the navel of the created Man intended to deceive him into the persuasion that he had a parent?"

This question of whether Adam had a navel is by no means a forgotten one. A few years ago North Carolina's Congressman Carl T. Durham and his House Military Affairs subcommittee objected to a cartoon of Adam and Eve in Public Affairs Pamphlet No. 85 *(The Races of Mankind* by Ruth Benedict and Gene Weltfish). The cartoon disclosed a pair of navels. The subcommittee thought this had something to do with communism. Their fears were somewhat allayed when it was pointed out that Michelangelo had painted a navel on Adam in his Sistine Chapel murals.

So thorough is Gosse in covering every aspect of this question that he even discusses the finding of coprolites, fossil excrement. Up until now, he writes, this "has been considered a more than ordinarily triumphant proof of real pre-existence." Yet, he points out, it offers no more difficulty than the fact that waste matter would certainly exist in the intestines of the newly-formed Adam. Blood must have flowed through his arteries, and blood presupposes chyle and chyme, which in turn presupposes an indigestible residuum in the intestines. "It may seem at first sight ridiculous," he confesses, ". . . but truth is truth."

Gosse's argument is, in fact, quite flawless. Not a single truth of geology need be abandoned, yet the harmony with *Genesis* is complete. As Gosse pointed out, we might even suppose that God created the earth a few minutes ago, complete with all its cities and records, and memories in the minds of men, and there is no logical way to refute this as a possible theory.

Nevertheless, *Omphalos* was not well received. "Never was a book cast upon the waters with greater anticipation of success than was this curious, this obstinate, this fanatical volume," writes the younger Gosse in his book *Father and Son*. ". . . He offered it, with a glowing gesture, to atheists and Christians alike. . . . But, alas! atheists and Christians alike looked at it and laughed, and threw it away . . . even Charles Kingsley, from whom my father had expected the most instant appreciation, wrote that he could not . . . 'believe that God has written on the rocks one enormous and superfluous lie.' . . . a gloom, cold and dismal, descended upon our morning tea cups."

Compared to Gosse's brilliant reconciliation of geology and the Bible, later attempts seem pale anticlimaxes. Yet they continued to be written, and are being written today. A bibliography of Protestant attacks on evolution, published in the United States alone since the turn of the century, would run into many thousands. Most of this literature is too frowsy to deserve even a passing glance, but occasionally a dignified, scholarly work finds its way into print. For example, Dr. Louis T. More, brother of the distinguished critic Paul Elmer More, and a professor of physics at the University of Cincinnati, gave a series of lectures in 1925 at Princeton which were published as a book titled *The Dogma of Evolution*. It is an infuriating book. Although Dr. More attacks evolution (using all the old and outworn arguments), he nowhere lets the reader know exactly what his own explanation is of the fossil record. One gathers he believes the different "species" came into being as a result of a series of independent creative acts. A somewhat similar blast at evolution was an article in the *Atlantic Monthly* (October, 1928) by another member of the Literary Humanist movement, Dr. Paul Shorey, head of the Greek Department at the University of Chicago. Titled "Evolution, a Conservative's Apology," the essay bulges with pleasant quotations from literary and philosophic greats, but otherwise reveals only the author's lack of scientific background.

Among twentieth century Protestant opponents of evolution, one man and one alone stands head and shoulders above all others. He is the "geologist" whom Bryan cited as his chief authority at the famous Scopes trial in Tennessee, and almost every fundamentalist attack on evolution in the last three decades has drawn its major ammunition from his writings. He is, in fact, the last and greatest of the anti-evolutionists.

The name of this remarkable man is George McCready Price. According to the current *Who's Who,* he was born in Canada in 1870. After obtaining several degrees at various Seventh Day Adventist schools and holding a number of teaching posts here and there, he became a professor of geology at a small Adventist college in Nebraska. At present, he is living in Loma Linda, California, having retired in 1938 from the faculty of another Adventist school, Walla Walla College in Washington.

Although Price has published some twenty books, the most important source of his views is *The New Geology,* a 726-page college text book issued in 1923. It is a classic of pseudo-science. So carefully reasoned are Price's speculations, so bolstered with impressive geological erudition, that thousands of Protestant fundamentalists today accept his work as the final word on the subject. Even the sceptical reader will find Price difficult to answer without considerable background in geology.

The heart of Price's approach can be stated in a few words. The great "sacred cow" of evolution, he points out, is the belief that fossils proceed from simple to more complex forms as you move from older to younger strata. Unfortunately, there is no adequate method of dating the ages of strata except by means of the fossils they contain. Thus a vicious circularity is involved—like a dog chasing its tail. The theory of evolution is assumed in order to classify fossils in evolutionary order. The fossils are used to date the beds. Then the succession of fossils from "old" to "young" strata is cited as "proof" of evolutionary development.

Price's own opinion is that the entire creation took place a few thousand years before Christ, in six literal days exactly as described in *Genesis.* The different beds do not indicate different geological ages. They were all deposited simultaneously by the Great Flood, in turn caused by an astronomical disturbance which sent huge tidal waves crashing around the earth, buckled up the present mountain ranges, and

destroyed the mild climate of Eden.* "The Grand Canyon of the Colorado," he wrote a few years ago, ". . . may not be very much older
than the Pyramids of Egypt." Fossils, according to Price, are simply the
records of antediluvian flora and fauna buried by the convulsion of the
Flood.

If all this is true, then in outcrops where several or more fossil-bearing
beds are found in one place, one would expect the fossils to be in the
reverse of the evolutionary order as often as conforming to it (though
Price makes allowance for the fact that marine life would tend to be
buried first, land life later, and birds last of all). This, Price declares, is
precisely the case, and much space in his books is devoted to pictures and
descriptions of such "upside down" areas. To explain away these embarrassing beds, Price asserts, traditional geologists invent imaginary faults
and folds. The following quotation on this point will introduce Price's
persuasive style:

> . . . there is scarcely an artificial geological section made within
> recent years that does not contain one or more of these "thrust
> faults," or "thrusts." But the really important thing to remember
> in this connection is that it is solely because the fossils are found
> occurring in the wrong order of sequence that any such devices are
> thought to be necessary—devices which, as has already been sug-
> gested of similar expedients to explain away evidence, deserve to
> rank with the famous "epicycles" of Ptolemy, and will do so some
> day.

To a reader unacquainted with geology, this has a plausible ring. How
is he to know, for example, that there are dozens of well established
criteria by which a geologist can determine whether a fault or fold has
occurred? In many cases, of course, the entire fold or fault is clearly
visible. When this is not the case, there are other indications for which
the trained geologist looks. An overturned fold, for instance, actually
turns beds upside down. This naturally inverts such fossil evidence as
wind ripple marks, mud cracks, rain prints, and foot tracks. Trilobites are
found on their backs. The center of gravity of large particles buried in
what was formerly mud will be high instead of low. And so on. In the
case of faults, there is usually a clear fault line of demarcation and often

---

*Many ingenious theories have been advanced by pseudo-scientists to account for the
waters of the Deluge. The Fortean Society is particularly fond of the view of Isaac Newton
Vail (1840-1912), of Barnesville, Ohio. His book, *The Waters Above the Firmament,* 1886
(later revised), is the best source for his theory that each planet passed through a phase in
which it had a ring like Saturn. In the earth's case, this ring was the source of the flood
waters. Vail's work is currently carried on by the Annular World Association, in Azusa,
California.

slickened sides of rock along which the fault moved, and other mechanical evidences of shifting.

When Price describes his upside-down areas and says there is no evidence for faulting or folding other than the reversed fossil order, it simply isn't so. All one need do is go to some of the original studies of the areas in question, and an abundance of technical evidence will be found for the faults and folds—evidence which has nothing whatever to do with the fossils. Price is fond, for example, of citing Chief Mountain, in the Alberta-Montana region of the Rockies, where older strata are found resting on younger. Although *The New Geology* reproduces seven photographs of this mountain (one is the book's frontispiece), Price neglects to tell the reader that at the base of this mountain the fault line of the overthrust can be seen clearly, with slickened faces of rock which testify to the faulting movement. The Hart Mountain, in Wyoming, is another of Price's upside-down spots. He fails to mention, however, that the fault line is easily traceable for some twenty-five miles.

Price also does not point out that the number of upside-down areas are exceedingly small in comparison with the tens of thousands of outcrops where fossils are always found in correct evolutionary order. In fact, the upside-down spots occur in the number which one would expect in view of the amount of folding and faulting that obviously has occurred. Actually the relative ages of major beds were fairly well worked out before the theory of evolution became current, and in recent years, dating by radio-active methods has added striking proof of the over-all correctness of dating by previous methods.

Concerning fossil remains of early man, Price's views follow a statement by Mrs. E. G. White, the inspired prophetess of the Seventh Day Adventist cult. In her book *Spiritual Gifts*, 1864, Mrs. White had written:

> If there was one sin above another which called for the destruction of the race by the flood, it was the base crime of amalgamation of man and beast, which defaced the image of God, and caused confusion everywhere. . . .
>
> Every species of animals which God had created was preserved in the ark. The confused species which God did not create, which were the result of amalgamation, were destroyed by the flood. Since the flood, there has been amalgamation of man and beast, as may be seen in the almost endless varieties of species of animals, and in certain races of men.

"I am sure," Price once wrote, "that Sister White's statements were given very providentially for our guidance. . . . I am confident that, if they had not been given us, we ourselves would now be in confusion and perplexity over this 'species' question. . . ."[1]

Sister White's statement about amalgamation was dropped from later editions of her book, just as Hitler's assertion that non-Aryan races were due to early Aryan-ape mating was omitted from the second edition of *Mein Kampf*. Nevertheless Price has remained loyal to her utterances. In his opinion, the men who lived before the flood were so completely destroyed that no fossil remains have been found. ". . . Since we are told that the Lord wished to destroy that ungodly race, He probably did a good job of it and buried them so deeply that we have not yet discovered their remains," he writes. The fossil human bones which have been uncovered are, he believes, those of men who lived after the flood.

Early Adventists frequently referred to certain primitive tribes—such as the African bushmen, Hottentots, and Digger Indians—as examples of degenerate hybrids, and on a few occasions, the entire Negro race. Price does not go quite this far. He thinks the Negro and Mongoloid races are degenerate types produced by amalgamation of the pure races God created at the Tower of Babel. Modern apes, however, are probably hybrid men. Here are Price's words on this matter:

There are no clear and positive evidences from paleontology which would prove that the existing anthropoid apes existed before the great world cataclysm, or the Deluge. These present-day anthropoid apes may be just as much a product of modern conditions as are the Negroid or the Mongolian types of mankind. And if I were compelled to choose between saying that the apes are degenerate or hybridized men and that man is a developed ape, I am sure it would not take me very long to decide which it would be. Nor do I think it ought to take any well-informed scientist long to make the choice.[2]

The notion that higher apes are not animals at all, but either primitive or retrograde humans, has been held by a number of post-Renaissance naturalists, from the Scottish anthropologist, Lord Monboddo, to the American writer of popular books on animals, Ivan T. Sanderson. Monboddo thought the orang-outang belonged to the human species—a thought which inspired the character of Sir Oran Haut-ton in Peacock's novel, *Melincourt*. Sanderson has similar views about gorillas. "Seeing these creatures in life, listening to their calls and talk . . . I can only regard them as a retrograde form of human . . . life," he writes in his book *Animal Treasure,* 1937. (Sanderson's latest contribution to zoology is an article in *True,* June, 1951, about fresh dinosaur tracks found on a beach in western Florida.)

Price is, of course, a devout Seventh Day Adventist. Like Velikovsky he has strong religious motives for establishing the truth of Old Testament records. But is this sufficient to force a man of his intelligence into the curious and lonely role he has played for almost half a century? Other drives creep out occasionally when he writes of "having to do this work of reforming the science of geology almost singlehanded," but on the whole, his books are singularly free of the intense egotism which characterizes so much pseudo-scientific writing. Price writes quietly, simply, clearly. Here for example—astonishingly free of the usual bitterness—is his description of his difficulties in getting a hearing for his views:

> Twenty-five years ago, when I first made some of my revolutionary discoveries in geology, I was confronted with this very problem of how these new ideas were to be presented to the public. And it was only after I found that the regular channels of publication were denied me, that I decided to use the many other doors which stood wide open. Perhaps I made a mistake. Perhaps I should have had more regard to the etiquette of scientific pedantry, and should have stood humbly hat in hand before the editorial doors which had been banged in my face more than once. But I decided otherwise, with a full realization of the consequences; and I have not yet seen any reason for thinking that I really made a mistake. Some day it may appear that the reigning clique of "reputable" scientists have never had a monopoly of the facts of nature.[3]

After publishing *The New Geology*, Price seems to have expected the theory of evolution to wither and blow away. In 1924, he wrote, "Organic evolution is dead. . . . This volume is merely a sort of funeral oration. *Requiescat in pace.*"[4] A year later, in a debate with Joseph McCabe in London, he predicted that in two years, public opinion about evolution would change in England as he felt it already had done in America. Of course no such change had occurred in the United States at all. In fact not a single geologist considered Price's books even worthy of refutation. Did this raise doubts in Price's mind? Of course not. "My previous larger treatise on this subject has not been answered," he wrote. "It will not be answered. But it has been ignored, and probably will still be ignored, because very few even among men of science, have the patience to follow carefully a completely new line of argument based on unfamiliar facts."

Although Seventh Day Adventists consider the Catholic Church the work of Satan, many Catholic writers on evolution have taken Price's geology seriously. The most scholarly example is *The Case Against Evolution*, published in 1925. The author, George Barry O'Toole, accepts completely Price's naive criticism of strata chronology, and his chapter on "Fossil Pedigrees" is little more than a verbose summary of Price. Similarly, Arnold Lunn, in the 1932 revised edition of *Flight from Reason*, describes Price as "professor of geology in an American University" and praises him for having "poured well-deserved ridicule on the arbitrary rearrangement of strata.'' In the early thirties, Price contributed several articles to *Catholic World,* one of them on "Cranks and Prophets"[5] in which he makes the inevitable comparison of himself with great scientists who were considered cranks by their contemporaries.

From the beginning until now, however, the Catholic Church has reacted to evolution with considerably less frenzy than the orthodox Protestant groups. In the early decades following Darwin's book, the Church took no position on the theory beyond making clear that no Catholic could accept the gradual evolution of the human soul. The general reaction among Catholics at that time was, of course, one of hostility, but compared to the Protestant crackpot literature, very few Catholic books were written on the topic. Perhaps the Church had learned a bitter lesson from her experiences with Galileo.

On the lower levels of the Church, however, Catholic laymen have written many books against Darwin and his theory. In America, a typical work of this sort is *God or Gorilla,* by Alfred W. McCann, 1922. The author makes much of a "Triassic shoe sole fossil" which he says is proof that men were walking around in shoes in the Triassic period! A photograph shows what is obviously a common type of rock concretion. McCann is incensed because orthodox geologists refuse to take it seriously.

O'Toole's *The Case Against Evolution* is a much more academic work than McCann's, but beneath its ponderous style lies nothing new or significant. In England, Arnold Lunn's *Flight from Reason* is equally banal. Lunn grants that each "species" may have been modified over the aeons by slight changes, but none are connected by a common family tree. A special creation is required for the origin of each.

The most amusing British Catholic attack on evolution was in Hilaire Belloc's *Companion to Mr. Wells' Outline of History,* 1926. The lacunae in Belloc's scientific knowledge are equaled only by the heights of his cocksureness. Most of his arguments are so ancient and flimsy that not even Price had the courage to exhume them. The book prodded Wells into a reply, published later in the same year under the title *Mr. Belloc Objects.* Written in a mood of amused anger, it is a little masterpiece of polemics. Few literary debates in history have been so decisively won. Belloc produced a rebuttal pamphlet, *Mr. Belloc Still Objects,* twice as cocksure as the former work, but it was the shouting of a man too angry to realize how badly he had been wounded.

Belloc's great and good friend Gilbert Chesterton seldom touched on evolution in his writing. When he did, he nearly always wrote nonsense. In *The Everlasting Man,* for example, he wastes many pages—as do so many anti-evolutionists—convincing the reader there is a great difference between the minds of men and animals. It is a waste because no evolutionist wishes to deny these differences. Men and monkeys are end points of quite separate branches of the evolutionary tree, and the transitional forms belong to the dim past. ''. . . No one is more convinced than I am,'' wrote Huxley, the great disciple of Darwin, ''of the vastness of the gulf between civilized man and the brute.'' Yet Chesterton could write: ''The higher animals did not draw better and better portraits; the dog did not paint better in his best period than in his early bad manner as a jackal; the wild horse was not an Impressionist and the race-horse a Post-Impressionist . . . a cow in a field seems to derive no lyrical impulse or instruction from her unrivalled opportunities for listening to the skylark.''

Chesterton's thesis, of course, is that there is such a huge difference between men and animals—men speak, create works of art, laugh, wear clothes, feel guilt, form governments, worship God, and so on—that one cannot conceive of a transitional stage. The simple answer is that the same vast difference exists between a man and a newborn baby. The reply that a baby grows into a man is irrelevant. The point is that if a baby and man can be the end points of a continuum, with no sharp line which the infant hurdles to acquire ''human'' traits, then there is at least no *theoretical* reason why man and an animal ancestor (much more ''human'' than a newborn baby) might not lie on a similar continuum.

In the same volume, Chesterton makes fun of the fact that beautiful cave paintings, the work of prehistoric men, had been found in southern France. He assumes that because the artists were prehistoric, they must therefore be considered ape-like, and the fact they painted so well seems to him a huge joke on the anthropologists. Unfortunately Chesterton did not trouble to learn that the cave paintings were the work of Cro-Magnon man, a fully developed human type with a brain capacity slightly larger than modern man. In an appendix to the book, he makes a lame apology for this oversight.

At the moment, Catholic opinion is swinging rapidly toward full acceptance of evolution, with the firm qualification that at some point in geologic time the human soul was infused into a body which had evolved to a point ready to receive it. In fact, this view was defended as early as 1871 by a Catholic biologist, St. George J. Mivart, in his book *Genesis of Species*. Later, for other reasons, Mivart was excommunicated, but now his book is regarded as highly prophetic.[6] In 1950, the Pope issued an encyclical in which he warned Catholics against acting ''as if the origin of the human body from preexisting and living matter were already completely certain,'' but gave permission for a Catholic to believe it if he wishes. The official attitude is that the evolution of plants and animals is probably true, but that evolution of man's body is a question not yet decided. A Catholic scientist may work toward making it a probable hypothesis, but until this occurs, it must not be taught in Catholic schools.

It is interesting to note that Dr. Mortimer J. Adler of the University of Chicago and Great Books fame, and one of the nation's leading neo-Thomists, has for some time been carrying on a one-man crusade against evolution. In *What Man Has Made of Man,* 1937, Adler brands evolution a "popular myth," insisting it is not an established fact "but at best a probable history, a history for which the evidence is insufficient and conflicting . . . facts establish only one historical probability: that types of animals which once existed no longer exist, and that types of animals now existing at one time did not exist. They do not establish the elaborate story which is the myth of evolution. . . ."

"I say 'myth,' " Adler continues, "in order to refer to the elaborate conjectural history, which vastly exceeds the scientific evidence. . . . This myth is the story of evolution which is told to school children and which they can almost visualize as if it were a moving picture. It is the concoction of such evolutionary 'philosophers' as Herbert Spencer, Ernst Haeckel and Henri Bergson, as well as the invention of popularizers of science."

Dr. Adler makes clear he is not denying an orderly succession of living forms through the ages. What he objects to is the view that they lie on a continuum in which one species fades into another by imperceptible changes. The evidence indicates, he argues, that "species" differ not in degree but in kind, with a radical "discontinuity" separating them.

In *Problems for Thomists,* 1940, Adler examines in more detail the question of how many "species" exist. In other words, how many creative acts of God are required to explain the evolutionary jumps? He opposes the view of Jacques Maritain, a leading Catholic philosopher, that the number is very large and unknowable. Adler's own view, which he considers "almost completely demonstrated," is that there is a small number of species—probably four (matter, plant, animal, and man), but certainly more than three and less than ten. Within a species, changes have occurred, but each species itself is a fixed type—immutable in its essence, and coming into being only by an act of God. Adler suspects that each species was created in several different types, underived from each other—for example, the separate creation of flowering and non-flowering plants. The scientific evidence for this is "indecisive," he admits, but there is a theological argument of "great suggestive force" based on a passage in *The Wisdom of Solomon* (11:21), one of the Old Testament books considered apocryphal by Protestants.

In the April 1941 issue of *Thomist,* Adler contributed an article called "Solution of the Problem of Species" in which he argues that Maritain's position can be positively disproved. His own view, after an error has been corrected, can now be established with certainty. He made the error out of "excessive zeal," he states. "I might almost say that what blinded me was the brightness of the new light."

Adler's latest blast at evolution was in a lecture before a student Catholic club at the University of Chicago, in 1951. Men and apes, he declared, are as different "as a square and a triangle. There can be no intermediate—no three and one-half-sided figure." Most of Adler's arguments were straight out of the arsenal of Bible Belt evangelism. "Sometimes the difference between a child and a pig," he said, "is not very noticeable. But the child grows up to be a man and the pig seldom does." If a scientist would only produce an ape that could speak "in simple declarative sentences," Adler said, he would admit a close bond between man and monkey.

(In passing, it is amusing to note that an American amateur zoologist, Richard Lynch Garner, devoted most of his life to recording and analyzing simian speech, and finaliy developed the ability, so he claimed, to converse with monkeys in their own tongue. See his *The Speech of Monkeys,* 1892; *Gorillas and Chimpanzees,* 1896; and *Apes and Monkeys,* 1900. His books are not, however, highly regarded by other authorities.)

Only two explanations will fit all the facts, Adler concluded his speech. Either man "emerged" from the brute by a sudden evolutionary leap, or he was created directly by God. One assumes Adler did not mean a creation of body and soul, but rather the increasingly popular Catholic view of an infusion of soul into a body which had bestial parents.

Many questions are raised by this view of course, and no doubt there will be considerable Catholic speculation in the future about them. For example: where is one to place the dozens of well-preserved skeletons which have been found of Neanderthal man—a creature with a low forehead like an ape, a head that hung forward, no chin, and non-opposable thumbs? It was a creature that made fires, and buried its dead with ornamental stones. In his reply to Belloc, Wells posed the problem as follows:

> When I heard that Mr. Belloc was going to explain and answer the *Outline of History,* my thought went at once to this creature. What would Mr. Belloc say of it? Would he put it before or after the Fall? Would he correct its anatomy by wonderful new science out of his safe? Would he treat it like a brother and say it held by the most exalted monotheism, or treat it as a monster made to mislead wicked men?
>
> He says nothing! He just walks away whenever it comes near him.
>
> But I am sure it does not leave him. In the night, if not by day, it must be asking him: "Have I a soul to save, Mr. Belloc? Is

that Heidelberg jawbone one of us, Mr. Belloc, or not? You've forgotten me, Mr. Belloc. For four-fifths of the Paleolithic age I was 'man.' There was no other. I shamble and I cannot walk erect and look up at heaven as you do, Mr. Belloc, but dare you cast me to the dogs?''

No reply.

Another important question facing the orthodox is whether the initial infusion of the soul took place only in a single pair, or were many humans created simultaneously? The latter view would enable Cain to marry someone other than his sister. In the Pope's encyclical of 1950, however, this view was wisely condemned on the grounds that it would conflict with the doctrine of Original Sin. Again: at what age of life did the infusion occur? If the first man and woman were adults before they received souls, then they would have lived the early part of their lives as animals, and the latter part as humans. On the other hand, if the soul was infused at conception (or at birth), it would mean that the first man and woman were brought up and suckled by a soulless mother or mothers. There is nothing illogical about either of these views, but they strike one at first thought as rather odd.

To date, the most complete discussion of these tangled problems is *Evolution and Theology,* a book published in 1932 by Father Ernest C. Messenger. Father Messenger courageously defends the evolution of Adam's body, but insists that the formation of Eve was a miraculous event. A portion of Adam's side (not necessarily a rib) would contain "virtually the perfection of the species," and the creation of Eve from it would be analogous to what biologists call "asexual generation." Father Messenger concludes: "The formation of Eve *ex Adamo* seems to be so clear in Scripture and Tradition that, at the very least, it cannot be prudently called into question. Further, there is no reason to doubt it, other than the difficulty of understanding how it could take place.''

Perhaps the church theologians and Dr. Adler should consider more carefully the revolutionary geology of George McCready Price. There is a grand simplicity about his reading of the rocks, and none of the distressing issues which arise once the time chronology of the beds is accepted. "One direct creation of a beautiful and perfect world," Price wrote recently, ". . . might be believed; but this long-drawn-out agony of interminable ages . . . does not look like . . . an intelligent method of creating a world. It may be a naturalistic process . . . but it is much more like a cosmic nightmare than a creation.''[7]

Or better still, let them ponder these wise words of St. Augustine. "It very often happens there is some question as to the earth or sky, or other elements of this world . . . respecting which, one who is not a Christian has knowledge . . . and it is very disgraceful and mischievous and of all things to be carefully avoided, that a Christian speaking of such matters as being according to the Christian Scriptures, should be heard by an unbeliever talking such nonsense that the unbeliever perceiving him to be as wide from the mark as east from west, can hardly restrain himself from laughing.''

CHAPTER 12

LYSENKOISM

SELDOM BEFORE in the history of modern science has a crackpot achieved the eminence, adulation, and power achieved by Trofim D. Lysenko, the Soviet Union's leading authority on evolution and heredity. Not only have his opinions been pronounced dogma by the Kremlin, but his Russian opponents (whose views are held everywhere but in the USSR) in recent years have been systematically eliminated from their posts. Some have died in prison camps. Some have simply vanished. A few remain at work—but it is work in other fields of biology.

What is behind this seeming insanity—this apparently planned destruction of one branch of science? Indefensible though it is, we can understand it better if we know something of the ideological issues involved, and the principal events which preceded Lysenko's fantastic rise.

The story begins in the eighteenth century with a French scientist, Jean Lamarck. Lamarck has been truly called the Father of Evolution. Although his books were published a half-century before Darwin's and lacked sufficient facts to convince his colleagues, he had a magnificent vision of the slow evolutionary development of plants and animals through the vast aeons of geologic time. The mechanism by which evolution operated, Lamarck believed, was through the inheritance of traits which organisms acquired in response to their surroundings.

The classic illustration of this "inheritance of acquired characters" is Lamarck's description of how giraffes developed elongated necks. The giraffes found themselves in an area in which they could live only by eating leaves which grew high on trees. They stretched their necks to reach these leaves. Somehow, this stretching, and the desire to stretch, got passed on to later generations. Giraffes with longer and longer necks were born.

Darwin was actually a "Lamarckian" in the sense that he also accepted the inheritance of acquired characters, but he made the role which it played in evolution a small one. More important, he argued, was the fact that some giraffes would be born with shorter necks and some with longer. Those with the short necks would be more likely to die. Those with longer ones would survive. In this way, "natural selection," or "survival of the fittest," would eventually result in longer necked giraffes.

Modern evolutionary theory has discarded the Lamarckian approach entirely in favor of natural selection, though its explanation of how selection operates differs considerably from Darwin's. It is now known that the units of heredity are submicroscopic bodies called "genes," carried within the sperm and eggs. Each new animal has a combination of genes acquired from both parents. These genes determine the animal's growth. Occasionally, however, a "mutation" occurs—a gene is changed. When this happens the animal carrying the new gene grows up with something in its structure altered, though the alteration is usually slight.

Mutations occur at random and are not connected in any way with an animal's experience. When a mutation is harmful, the animal's chances of survival are lessened. If helpful, the chances are increased. In individual cases this means little, but over long periods of time these tiny changes have an accumulated effect on the population of a group. Eventually, the helpful mutations establish themselves as part of the collective heredity of the species.

There are many reasons for thinking this the basic process by which evolution works. In ant, bee, and wasp colonies, for example, "worker" insects are sterile and do not leave descendants. Yet they are marvelously adapted to their tasks. A Lamarckian explanation of this is unthinkable, but the mutation theory explains it fairly well. Another strong argument is based on the complete lack of any conceivable means by which genes of the germ cells can be connected to other parts of the body. If Lamarck was correct, the stretching of a male giraffe's neck would have to be transmitted somehow to that giraffe's sperm. No such connection exists. In fact, many highly adapted structures are simply dead matter secreted from the body like fingernails—a butterfly's wings for instance. There is no imaginable way the use of such non-living structures could influence the insect's genes.

A host of experiments have been designed to test Lamarckianism. All that have been verified have proved negative. On the other hand, tens of thousands of experiments—reported in the journals and carefully checked and rechecked by geneticists throughout the world—have established the correctness of the gene-mutation theory beyond all reasonable doubt. The chromosomes, which carry the genes, have been studied in great detail. In recent years, the electron microscope has made visible what are probably the genes themselves.

In spite of the rapidly increasing evidence for natural selection, Lamarck has never ceased to have loyal followers. In Darwin's day, the English satirist, Samuel Butler, devoted a half-dozen books to a defense of Lamarck and bitter attacks on Darwin. Later, George Bernard Shaw took up Butler's cudgel. In France, where Lamarck's views survived longer than in England or Germany, the philosopher Henri Bergson found that Lamarckianism fitted neatly into his concept of "creative evolution." Both he and Shaw were "vitalists" who felt that back of evolution was a creative *élan vital,* or "life force," which found expression in the constant striving of organisms to better themselves. In America, around the turn of the century, there were a number of neo-Lamarckians, of whom the most eminent was the psychologist William McDougall.

All these men were idealists. They objected to Darwin because they felt his theory left no room for free will and individual effort. Natural selection seemed a blind, purposeless struggle in which progress came about almost like an accidental afterthought. Just as orthodox Christians objected to evolution because it impressed them as a dreary, roundabout, wasteful method of creation compared to the story told in *Genesis,* so neo-Lamarckians felt that natural selection was a dreary, roundabout, and wasteful method compared to the transmission of acquired traits. There is indeed a strong emotional appeal in the thought that every little effort an animal puts forth is somehow transmitted to his progeny. It permits every individual to share directly in the process of evolution. The harder a rabbit tries to run, the faster will his offspring be able to run. The more a man uses his brain, the better brains will his children have.

Just as Lamarckianism combines easily with an idealism in which the entire creation is fulfilling God's vast plan by constant upward striving, so also does it combine easily with political doctrines which emphasize the building of a better world. One of the most eloquent defenders of this aspect of Lamarckianism was a Viennese biologist and socialist, Paul Kammerer. His book *The Inheritance of Acquired Characteristics* was translated into English in 1924, and from it, we can draw an extract which conveys vividly the strong appeal Lamarckianism has for the man with a social conscience.

"If acquired characteristics cannot be passed on," Kammerer wrote, "as most of our contemporaneous naturalists contend, then no true organic progress is possible. Man lives and suffers in vain. Whatever he might have acquired in the course of a lifetime dies with him. His children and his children's children must ever and again start from the bottom. . . . If acquired characteristics are occasionally inherited, then it becomes evident that we are not exclusively slaves of the past—slaves helplessly endeavoring to free ourselves of our shackles—but also captains of our future, who in the course of time will be able to rid ourselves, to a certain extent, of our heavy burdens and to ascend into higher and ever higher strata of development. Education and civilization, hygiene and social endeavors are achievements which are not alone benefiting the single individual, for every action, every word, aye, even every thought may possibly leave an imprint on the generation.''

Kammerer was responsible for a series of sensational laboratory experiments which seemed to prove the Lamarckian view. On the strength of them, he was offered in 1925 a chair at the University of Moscow, where the Lamarckian views of a Russian horticulturist named Michurin were popular. Shortly after he accepted the post, however, it was discovered that a number of his animal specimens had been deliberately faked. Kammerer denied everything, blaming one of his assistants. Few believed him. He willed his valuable library to the University of Moscow, his body to an anatomical school in Vienna, then shot himself with a revolver. He was the last Lamarckian whose writings and experiments carried, at least for a time, a tone of authority.

As Lamarckianism became more and more discredited throughout the world, it began to grow in popularity within the Communist Party of the Soviet Union. A Russian film glorifying Kammerer was produced in which the faking of specimens was blamed on reactionary capitalist enemies. Nevertheless, many Russian biologists continued to do excellent work in the Mendelian gene-mutation theory (so called after the pioneer work of the Austrian monk, Gregor Mendel), and it was not until the late thirties that Mendelianism began to be branded by the Party as "bourgeois idealism." Henceforth events moved with grim rapidity.

H. J. Muller, Nobel Prize winner, who served as senior geneticist for four years (1933-37) at the Institute of Genetics in Moscow, wrote two authoritative articles on Lysenkoism for the *Saturday Review of Literature*, December 4, and 11, 1948. The following excerpts give a vivid picture of what took place.

". . . In 1933 or thereabouts," Muller wrote, "the geneticists Chetverikoff, Ferry, and Ephroimson were all, on separate occasions, banished to Siberia, and Levitsky to a labor camp in the European Arctic . . . in 1936, the Communist geneticist Agol was done away with, following rumors that he had been convicted of 'Menshevik idealism' in genetics . . . it is impossible to learn the real causes of the deaths of such distinguished geneticists as Karpechenko, Koltzoff, Serebrovsky, and Levitsky. Certain it is, however, that from 1936 on Soviet geneticists of all ranks lived a life of terror. Most of them who were not imprisoned, banished, or executed were forced to enter other lines of work. The great majority of those who were allowed to remain in their laboratories were obliged to redirect their researches in such a way as to make it appear that they were trying to prove the correctness of the officially approved anti-scientific views. During the chaotic period toward the close of the war, some escaped to the West. Through it all, however, a few have remained at work, retained as show pieces to prove that the USSR still has some working geneticists.''

''Ironically,'' Muller comments, ''the great majority of the geneticists who have been purged were thoroughly loyal politically; many were even ardent crusaders for the Soviet system and leadership, as the writer well knows through personal contact with them.''

In 1936, the Medico-genetical Institute, one of the finest in the world, was attacked by *Pravda,* then closed. Solomon Levit, the founder, made a confession of his Mendelian errors, and has not been heard from since. N. I. Vavilov, the most distinguished geneticist in Russia—internationally famous and respected—was relieved of his many posts and accused of being a British spy. He died in 1942 in a Siberian labor camp, although it was not until several years later that biologists outside of Russia were able to learn what had happened to him. ''Thus,'' writes Muller, ''ended the career of a man who had . . . undoubtedly done more for the genetic development of Soviet agriculture than has ever been done by any individual for any country in the world.''

Several conferences were held in Russia, supposedly to debate the issues involved in the dispute. At these meetings, the chief defender of the Michurin (Lamarckian) point of view was Trofim Lysenko, a former peasant and plant-breeder who had been rising steadily in party favor. The conference of 1948 marked Lysenko's decisive victory. A nervous, shy man in conversation, with unruly dark hair and the tanned face of a farmer, Lysenko becomes a Russian William Jennings Bryan when he lectures. His eyes blaze with hidden fire. At the 1948 conference, he delivered a passionate address of 12,000 words. Defenders of Mendelian thought were savagely attacked as reactionary and decadent, groveling before western capitalism. They were the enemies of the Soviet people. At the close of the conference, he mentioned casually that his address had been approved by the Central Committee of the Communist Party. According to *Pravda* this is how the audience reacted:

> This announcement by the President evoked the general enthusiasm of the members of the session. With one impulse all those present rose from their seats and engaged in a stormy and prolonged ovation in honor of the Central Committee of the Party of Lenin and Stalin, in honor of the wise leader and teacher of the Soviet people, the greatest scholar of our epoch, Comrade Stalin. . . .

Party approval of Lysenko's speech meant, of course, the final and complete victory of Michurinism. It had been established as the Party line in biology. From now on it would be impossible to present evidence against it, or even to express covert sympathy for Mendelian views. One by one, the few courageous scientists who had raised objections to Lysenko's views wrote their pathetic public letter of confession, praising the Party for its wise guidance and promising to correct their errors. Professor Muller writes that the Academy of Sciences, headed by the late Vavilov's brother, "toed the Party line by removing from their posts in utter disgrace the greatest Soviet physiologist, Orbeli, the greatest Soviet student of morphogenesis, Schmalhausen, and the best remaining Soviet geneticist, Dubinin. Dubinin's laboratory, long known for the admirable work done there by numerous careful investigators, was closed down."

The Academy unanimously approved a long letter to Stalin which was published in *Pravda,* August 10, 1948, thanking him for his great assistance. "Carrying on the work of V. I. Lenin, you have saved for progressive materialistic biology the teachings of the great remolder of nature, I. V. Michurin, and, in the presence of the entire world of science, you have raised the Michurinist tendency in biology to the position of the only correct and progressive tendency in all the branches of biological science. . . . Long live the forward-looking biological Michurinist science! Glory to the great Stalin, leader of the people and coryphaeus of forward-looking science!"

Honor upon honor has been heaped upon Lysenko. He replaced his enemy, the great Vavilov, in all of his important posts. On two occasions he received the Stalin Prize. He was given the Order of Lenin and made a Hero of the Soviet Union. At one time he was vice-president of the Supreme Soviet.

Almost all the chief characteristics of the paranoid crank are exhibited by Lysenko. He is egotistical, fanatical, filled with hate for his enemies, and profoundly ignorant of scientific method. "Lysenko can only be described as illiterate," writes Julian Huxley, in his excellent book, *Heredity East and West,* 1949. "I use the word as meaning that it is impossible to discuss matters with him on a scientific basis . . . . Sometimes he appears ignorant of the scientific facts and principles involved, sometimes he misunderstands them, sometimes he distorts them, sometimes he counters them with bare assertions of his own beliefs."

In the opinion of Professor Muller, "Lysenko's writings along theoretical lines are the merest drivel. He obviously fails to comprehend either what a controlled experiment is or the established principles of genetics taught in any elementary course in the subject."

Here is the similar view, quoted by Huxley, of another leading geneticist, Professor S. C. Harland. "In 1933 . . . I saw Lysenko in Odessa, catechized him for several hours and inspected his practical work. It was quite clear that Lysenko was blazingly ignorant of the elementary principles of both plant physiology and genetics. . . . You simply couldn't talk to Lysenko—it was like discussing the differential calculus with a man who did not know his 12-times tables. When I say that some of his assistants were using plant pots without drainage holes, you amateur gardening readers will understand.''

In Huxley's opinion, Lysenko's views are so vague they cannot even be called a theory. Throughout, they have strong analogies with Marxist political dogma. The existence of genes is simply denied. Mendelians are obviously ''idealists'' since they study something that doesn't exist. Heredity is transmitted by every particle of the body (just as every worker in Russia contributes toward the future of the state). When a plant is suddenly given new environmental conditions there is a ''shattering'' of its heredity (like a political revolution). This shattering is a kind of shock treatment. It makes the plant peculiarly plastic to change. The new environment then produces desirable changes in the plant which are transmitted permanently to all later generations.

Unfortunately, few of Lysenkos' experiments have worked out properly when scientists outside the Soviet Union try to repeat them. In many cases, he does not publish enough data to permit evaluation or repetition. Some of his experiments are undoubtedly successful, but in all such cases, there are simple Mendelian explanations. Lysenko probably does not understand the theory of heredity well enough to realize this. Almost no precautions are taken to insure a controlled experiment. For example, he will grow specimens of a plant under new conditions. But not having made sure that the plant strain is ''pure'' (i.e., free of a wide variety of recessive genes buried in the strain) the new plants naturally show a wide variety of differences. Lysenko imagines these differences to be the direct effect of the new environmnt. By picking out individual plants which seem to have made the best ''adaptations,'' he automatically and without realizing it, carries out an elementary process of selection of Mendelian characters in which genes present in the originally impure strain are bred into prominence.

The only way to prevent this Mendelian effect is to use strains genetically pure, but to get such strains requires a laborious process of inbreeding. Naturally Lysenko is not going to bother with such a time-consuming procedure since he doesn't believe in the mechanism by which it operates. And of course no other Soviet biologist is going to attempt experiments which might cast doubt on what the Party has declared true. He remembers too well what happened to Vavilov and other Mendelians.

One of the most important phases of modern genetics is the application of statistical methods to the different types resulting from breeding experiments. It is, in fact, indispensable. But Lysenko does not believe in "chance" and is therefore opposed to the use of statistical methods. Michurin did not need statistics, he shouts, so why should he? This refusal to use a powerful scientific tool is another reason why it is extremely difficult for geneticists outside of Russia to judge the results Lysenko claims to have achieved.

Many of Lysenko's promised results never materialize. On one occasion, Vavilov made the mistake of saying it would take at least five years to develop a certain improved type of wheat. He was immediately accused of national sabotage while Lysenko loudly proclaimed that his methods would produce the wheat in less than two years. "Needless to say," writes Muller, "Lysenko has not been able to make good his promises."

The question naturally arises—why has the Soviet Union been willing to turn its back on all the positive achievements of modern genetics? Why has it returned to a discarded Lamarckian point of view which, in Huxley's words, is "little more than a survival of sympathetic magic," and in the opinion of Muller, "as much a superstition as belief that the earth is flat"?

One can only guess at the reasons. The fact that Mendel was a Catholic, or that the Nazis made incorrect use of his ideas to support theories of Aryan superiority, are probably negligible factors. More important is the fact that Lysenkoism offers a convenient means of glorifying a purely Russian "science" at the expense of the "foreign" science of capitalist enemies. It may also be true that Stalin and his lieutenants are suspicious of Mendelian theory because it is too abstruse for them to understand. Perhaps also they feel that a simpler theory should be taught Soviet farmers as long as practical results in increasing crop yields can be obtained. Actually, the backwardness of Russian agriculture is such that many quick improvements can be made by means of simple cross-breeding, accompanied by elementary selection, and also by means of improved agricultural methods. As long as Lysenko keeps busy making crosses, he is likely to keep turning out useful variations. Back of his successes, of course, will be Mendelian laws, though they will be explained in Michurin terms.

Perhaps the most important reason of all is ideological. We have already seen how neatly Lamarckianism fits into the emotion of constructing a new society. Evolution, in Mendelian theory, is a slow process which operates by means of random, purposeless mutations. The over-all result is progress, but a progress in which an individual cannot feel that his own improvements are directly passed on to children. Lysenkoism offers a more immediately attractive vision. Humanity becomes plastic—capable of being molded quickly by new conditions and individual efforts. Russian children can be taught that the Revolution has "shattered" the hereditary structure of the Soviet people—that each new generation growing up in the new environment will be a finer stock than the last. Thus, a foundation is being prepared for a new type of racialism. Every Soviet citizen, regardless of his genetic background, will soon be able to feel himself superior in heredity to citizens of decadent, bourgeois environments.

This view offers something of a problem in connection with territories "liberated" by the Soviets. From the Michurin point of view, the masses of China—who for thousands of years have lived in poverty and misery—would have acquired far too degenerate a heredity for it to be removed by merely a few generations of new Soviet environment. On the other hand, as Huxley has pointed out, Mendelian theory "makes it clear that even after long-continued bad conditions, an enormous reserve of good genetic potentiality can still be ready to blossom into actuality as soon as improved conditions provide an opportunity." But all this is far too complex for the Soviet politicians to understand. Perhaps they will be able to convince their "liberated" peoples that the new environment will so "shatter" their heredity that their children, or at least their grandchildren, will take a sudden leap upward to the Soviet level. In fact, this has already been alleged in some of the new "Peoples' Democracies."

The charge that Mendelian genetics is a form of idealism could not be further from the truth. The mutation theory bases evolution squarely on a material basis—namely genes—and its laws are the outcome of careful experimental research for the past fifty years. It is the Soviet view which is riddled with metaphysics. As Professor Muller writes, it "implies a mystical Aristotelian 'perfecting principle,' a kind of foresight, in the basic make-up of living things, despite the fact that it claims in the same breath not to be 'idealistic' at all." One is reminded of the equally metaphysical botanical views of Goethe. The great German poet spent many weary months wandering over Sicily trying to find the *Urpflanze*—an "ideal" plant from which he believed all other plants had degenerated as a result of environmental influences.

Actually, the controversy over Mendel has nothing whatever to do with religious, philosophical, or political beliefs. Just as a theist can regard evolution as God's method of creation, so can a theist regard random mutations as one of the means by which God's evolutionary program is realized. The results are the same regardless of the mechanism by which evolution occurs. Why cannot God make use of any device He wishes? Random mutations, upon which environment has a molding effect, can be as adequate an instrument for divine will, or the manifestation of an *élan vital*, as any other instrument. Substitute "nature" or "Dialectical Materialism'' for God in the above sentences and the arguments remain unchanged.

Likewise, the emotion of building a new society can be combined with Mendelian thinking just as successfully—in fact more so—than with the outmoded Lamarckian view. While Muller was in Moscow, he wrote a book called *Out of the Night*. In it he pointed out that once a culture achieves an equality of environment for all its citizens, it is possible to use modern Mendelian methods for increasing rapidly the general health and intelligence. The book was not acceptable to the Soviets. Muller is now a professor at Indiana University, and in Huxley's opinion, "probably the ablest and certainly the most all-around geneticist that the world has yet seen.'' To the Soviet biologists he is one of the world's most misled scientists, having sold his services to the imperialist war-mongers.

The really frightening thing about Lysenkoism is, of course, the fact that a great culture has unequivocally made scientific truth subordinate to political control. The Nazis provided an earlier instance of this sort of policy when they raised the theories of crackpot anthropologists to official state doctrines. There is no difference in principle between either of these examples and the similar rejection of Galileo's discoveries because they conflicted with a state orthodoxy. In fact Galileo's well-known confession, extracted from a beaten, exhausted man who was trying desperately to save his life, parallels in every sad sentence the similar "confessions'' of Soviet Mendelians.

There remain, however, two hopeful aspects of these shocking events. It may be that the steady deterioration of Soviet biology will be followed by a similar deterioration in other sciences. We know now how greatly the Nazi efforts to make an atom bomb were bungled by the control of political Neanderthals.[1] There is reasonable ground for hope that a similar state of affairs may, to some degree, hamper Soviet war research.[2]

Secondly—the rise of Lysenkoism provides a dramatic object lesson for the free world. Fortunately, our own sins in this respect have not been very grave. True, the Scopes trial in Tennessee was a victory for the views of George McCready Price. True, there is control of research by the demands of government agencies and large corporations which alone can finance the necessary giant laboratories. True, the excessive zeal of poorly informed politicians for keeping security risks out of government research projects, and for "classifying'' certain basic lines of work, has weakened both our war efforts and our fundamental scientific research.

But on the whole, in contrast with other countries and other ages, our science is enjoying a relative freedom that (until very recently, at least) is perhaps its greatest in history. Fundamentalists in the Bible Belt continue to read their dismal literature denouncing Darwin, but you are not likely to find a fundamentalist in any position of scientific authority or eminence. Only a few southern states have laws against the teaching of evolution, and even in those states the laws are constantly evaded in schools of higher learning. Scientists in hundreds of universities and institutes are working with unrestricted vigor on projects of their own choosing. Even in top-secret war research it is unthinkable that the President or Congress would make a decision about a scientific *theory*, then proceed to purge from their posts all who disagreed.

Let us hope that Lysenko's success in Russia will serve for many generations to come as another reminder to the world of how quickly and easily a science can be corrupted when ignorant political leaders deem themselves competent to arbitrate scientific disputes.

CHAPTER 13

APOLOGIST FOR HATE

AS WE HAVE seen, the teachings of Lysenko offer a convenient background for a new and novel type of racism. Whether such a doctrine will develop in Russia, and if so, how much harm it will bring the world, only the future can decide. No such doubt can exist, however, regarding the enormity of crimes which stemmed from Nazi racial theories. Never before in history has evil of such magnitude been so intimately connected with crackpot science.

Antisemitism in Germany, as everyone knows, was much older than the Nazis. Like the antisemitism of other European countries, it has a long and infamous history reaching back into the early Middle Ages. Protestant readers who associate its origin with Medieval Catholicism and the Inquisition may be surprised to learn that in Germany, the first influential and passionate anti-Semite was Martin Luther. His solution of the German "Jewish problem" was a simple one. Drive them out of Germany. "Country and streets are open to them," he wrote. ". . . They are a heavy burden like a plague, pestilence, misfortune. . . ." For Jews who refuse to leave he recommended, "that into the hands of the young strong Jews and Jewesses be placed flails, axes, mattocks, travels, distaffs, and spindles, and they be made to earn their daily bread by the sweat of their noses as it is put upon the shoulders of the children of Adam." He further suggested, "that their synagogues or schools be set on fire . . . that their houses be broken up and destroyed . . . and they be put under a roof or stable, like the gypsies . . . in misery and captivity as they incessantly lament and complain to God about us."

Intense German nationalism, coupled with the doctrine of a Master

Race destined to rule the world, did not come until much later; in fact it had its first theoretical formulation in the philosopher Fichte's *Addresses to the German Nation*, 1807. Oddly enough, the earliest anthropological defense of Nordic superiority was not by a German but by the French nobleman, Comte Joseph de Gobineau. It ran to four volumes, published in 1853-5 (the first volume was translated into English in 1915 under the title, *The Inequality of Human Races*). Gobineau had little interest in nationalism. His main concern was to combat the notion of democracy which had been made popular by the French Revolution, and to defend the virtues of aristocratic rule. By stirring together a mass of anthropological facts and superstitions, he arrived at a hierarchy of races with the Nordic (tall, blonde, fair-skinned, blue-eyed) at the top and the Negro at bottom.

Gobineau's views greatly excited Richard Wagner, the composer, and Wagner's son-in-law, Houston Stewart Chamberlain. Chamberlain was an Englishman, but he had settled in Germany where he became imbued with the growing German sense of destiny. In 1899, he published his *Foundations of the Nineteenth Century*, the second great work in the history of the Nordic myth, and a book of tremendous influence on the thinking of German people. In essence, Chamberlain combined Gobineau's theory of Nordic superiority with German patriotism and vigorous antisemitism.

The third great name in the history of German racism is a full-fledged professional anthropologist, Professor Hans F. K. Günther, of the University of Jena. Throughout the Nazi movement, he was their most distinguished authority on racial questions. It would be foolish, of course, to suppose that men like Günther, or such colleagues in racist anthropology as Ludwig Woltmann and Ludwig Schemann, were responsible for the Nazi crimes against the Jews. Antisemitism had a much deeper root in the cultural paranoia of the Germans. But the books of men like Günther stand as striking testaments to the ease with which a science can be perverted by strong emotional prejudices which a scientist derives not from his subject matter but from cultural forces surrounding him.

Günther admitted Germany was a mixture of races, but he thought it contained a higher proportion of pure Nordics than any other nation. He went to great lengths in his books to explain how a Nordic differs from a member of an inferior race. He keeps himself cleaner, for example. According to Günther, both soap and hairbrush are Nordic inventions. He is more athletic (in the Olympic Games held in Germany in 1936, Hitler was so miffed when Negroes took prizes that he refused to shake hands with them). He prefers the colors blue and pale green. Freckles on Nordic women are acceptable. Günther suspects they are a Nordic trait. But to sing about a "Nut-Brown Maid" is much to be condemned because dark-skinned maids are not Nordic. Nordic women are modest. In contrast to women of other races they keep their legs together when they sit in streetcars. As Wallace Deuel writes in *People Under Hitler,* from which the above samples are drawn, "Dr. Günther . . . apparently has devoted considerable time to this [streetcar-sitting] problem."

All racial mixtures are bad, according to Professor Günther. The hope of Germany is to prevent them and to increase the purity of the Nordic strain. Unfortunately, the very virtues of the Nordic are causing his decline. For example, he is very brave, hence more likely to become a soldier and get killed. Being adventurous, he emigrates. Being virtuous, he marries late and has fewer children. And being innocent of guile, he is easily trapped into marrying the "diabolically alluring" women of other races.

Under Hitler, the rising antisemitism of the German people—greatly stimulated of course by party propaganda, and given "scientific" backing by Professor Günther and other anthropologists—reached a crescendo. Never before in history has a nation set about so coldly and methodically to exterminate a people. There is no need to retell here in more than brief summary the now familiar story of this nightmare horror. A caste system was set up, depending on the amount of one's Jewish blood, with rigid and complex laws determining who could marry whom. Sex relations between Jews and Aryans were punishable by imprisonment. Even prostitutes were protected by law from "race defilement" by Jewish patrons. Only business contacts between Jews and Aryans were not considered a form of "race shame." Familiarities such as dancing and playing games were discouraged. Jews were sentenced to prison for kissing Aryan girls even when the girls desired it. "By kissing an Aryan girl, the Jew has insulted not only the girl herself, but also the entire German nation," the court declared in one such case.

Eventually, as the world knows, Jews were denied the right to work in professions and in business, their property was confiscated, their citizenship removed. In the final hysterical culmination of hate, came the concentration camps, the gas chambers, and the unbelievably sadistic medical experiments.

In this mounting madness, German pseudo-anthropology reached its greatest heights. "The non-Nordic man takes up an intermediate position between the Nordic man and the . . . ape," wrote Herman Gauch, in his *New Elements of Race Investigation,* 1934. Here is another typical declaration of "scientific" fact—from an address by Julius Streicher in 1935: "The blood particles of a Jew are completely different from those of a Nordic man. Hitherto one has prevented this fact being proved by microscopic investigation." Streicher, you recall, edited *Der Sturmer,* a violent, pornographic hate sheet that specialized in obscene stories about the Jews.

The word "Aryan" as used by Hitler finally lost all rational meaning. The Japanese, for example, had to be declared Aryan when they became allies. A Nazi journalist happened to have a grandmother who was an American Sioux Indian. In 1938, the Chamber of the Press, after much anthropological deliberation, actually ruled that Sioux Indians were officially Aryan! (No rulings are on record for other Indian tribes).

The outstanding literary expression of the Nazi mythology was, of course, Hitler's *Mein Kampf*. Ranking second was *The Myth of the Twentieth Century,* a 700-page work for educated Germans by the leading philosopher of Nazism, Alfred Rosenberg. All the usual psychotic elements of the work of an extreme crackpot are in this book. Orthodox anthropology and history are simply brushed aside. The "myth" is the German Nordic's awareness of his racial purity and his destiny to rule the world. Everything good is labeled "Aryan." Everything bad stems from Jews, Catholics, Russians, and orthodox Protestants. Even Christ was an Aryan, though his views were quickly corrupted by Jewish influence. It is worth noting that reputable German scholars did not take the book seriously when it appeared. They ignored it. But the book's influence was enormous.

No nation has been free of the poisons of racism and the United States is no exception. Our greatest sin, of course, is our treatment of our colored people—a sin inextricably bound up with racial pseudo-science. Even in the northern states, today the average white probably thinks of the Negro as slightly inferior by heredity, and in the South it is almost universally believed. In the light of modern anthropology, these views are on a level with the view that the earth is flat, but with one important difference—they are capable of causing infinitely more suffering. When Negroes are given environmental opportunities equal to that of whites, they make the same creative and intellectual achievements. There have been hundreds of carefully controlled research attempts to find a genetic difference between the two races which would justify ranking one above the other. All have failed. George Bernard Shaw summed up the American scene perfectly when he wrote (in *Man and Superman*) " . . .the haughty American nation . . . makes the Negro clean its boots and then proves the inferiority of the Negro by the fact that he is a bootblack."

In the early history of the United States, when racial feelings were the greatest, many books and pamphlets were written to prove the Negro inferior. It is difficult, however, to find a single work of this type that purports to be written by a professional anthropologist. Most of it is religious in character, relying chiefly on the Bible for support. There are many variations, but the basic themes of these shabby works are that God created different races which He did not intend to intermarry, and that the Negroes were ordained a race of servants. This view was held throughout the entire South. "Show me a nigger who can do a problem in Euclid or parse a Greek verb," John Calhoun once declared, "and I'll admit he's a human being."

It seems impossible now that any intelligent person could have regarded the Negro as a sub-human species, yet this view was by no means uncommon in the South even as late as the early years of this century. In 1867 the Reverend Buckner H. Payne, writing under the pseudonym of "Ariel," published a booklet (later expanded to a larger work) titled *The Negro: What is His Ethnological Status?* Payne's conclusion was that the Negro is an animal without a soul. It remained, however, for Charles Carroll, a resident of St. Louis, to give this demented theory definitive formulation. His two books on the subject—*The Negro a Beast,* 1900, and *The Tempter of Eve,* 1902—set a record in racial literature that probably will never be surpassed.

Carroll held the view that the Negro was created along with the animals as a higher ape, and for the purpose of providing Adam and his descendents with servants to perform tasks of manual labor around Eden. He possessed a mind, in common with other mammals, but not a soul. The "Serpent" who tempted Eve was in reality a Negro maidservant. The age-old problem of where Cain got his wife is solved neatly. Cain married a Negro—the first example in history of the heinous crime of amalgamation of man and beast. All the races except the white are hybrid products of mixtures between the race of Adam and Negroid animals.*

Do these hybrid offspring have souls? There is no indication Carroll even considered this a perplexing question. They do not, he declares. "Man cannot transmit to his offspring by the Negro," he writes, "the least vestige of the soul creation. Hence, *no mixed blood has a soul.*" (Italics his.) Brilliant intellectual achievements by mulattos do not bother Carroll. "The mere fact that Alexandre Dumas possessed a fine mind is no evidence that he possessed a soul."

If the red, yellow, and brown races, and all individuals who have a red, yellow, or brown ancestor do not have souls, then why bother sending missionaries to preach the Gospel to them? Like many crackpot

---

*How Carroll would have fumed if he had lived to read a three-volume work by the American Negro writer, J. A. Rogers! It is titled *Sex and Race,* and was published in 1941-44. Rogers defends the highly questionable thesis, which has a literature all its own, that the original human stock was a light-skinned, Negroid type. A portion of this stock evolved into the white strains, and another portion, in Africa, became darker. All present races are descendants of these two groups, in varying degrees of intermixture. Rogers is not a racist, however, since he does not believe in genetic inferiorities, and his books are good sources for obscure references on extreme racist views.

scholars, Carroll maintains a striking consistency with his premises. The Lord never intended the Gospel to be preached to these halfbreeds, he argues. That it is being done only indicates how sinful and corrupted and "negroized" modern churches have become. In fact almost all the ills which beset mankind since the Fall can be traced to a failure to recognize the bestial character of all peoples except the pure white descendants of Adam.

Carroll writes well. He quotes from eminent anthropologists. He reproduces tables of brain weights, and so on, designed to give "scientific" proof of his views. Ten illustrations in his earlier book are vicious caricatures of Negroes, contrasting them with pictures of Jesus, the Virgin Mary, and white men and women. There is nothing to indicate Carroll considers himself in any way "prejudiced." He is merely a humble "worker for the Lord," in the great cause of expelling the "Negro from his present unnatural position in the family of men, and the resumption of his proper place among the apes." He speaks of his opponents as "little narrow-minded bigots," ignorant alike of both Scripture and modern science. A portrait of the author, in one of his books, shows him to be an intelligent looking man with dark hair and eyes, and a black mustache and beard. Apparently his first book sold fairly well because a minister named William G. Schell felt obliged to publish a reply in 1901 titled *Is the Negro a Beast?*

Of course no one in the South today, let us hope, believes the Negro lacks a soul. But many of Carroll's anthropological arguments are still used in religious writings to prove the Negro's inferiority. The most common one is the fact that the Negro brain is, on the average, slightly smaller than the white's. But so is the white's slightly smaller than the Eskimo, Polynesian, American Indian, and Japanese brain, not to mention the brain of two African tribes, the Kaffir and the Amaxosa. Even Neanderthal man had a larger brain than the modern white!

These variations in physical anthropology mean, of course, absolutely nothing as far as mental capacities are concerned. Yet an incredible amount of supposedly scientific works have been devoted to nonsense of this sort. In 1871, for example, Paul Broca, a French anthropologist with a broad skull, wrote five volumes to prove that the broader the head the better the brain, and that the French have particularly broad heads.

The widespread folk belief among ignorant whites that colored people are anatomically closer to the monkey than other races is totally without foundation. Every race has certain characteristics similar to those of apes, and it is always easier to see these traits in some other race than your own. The Negro's thick lips, kinky hair, and small amount of body hair all mark him as *further* removed from the ape then a Caucasian. Apes have thin lips, straight hair, and lots of body hair.

The resemblance of whites to chimps forms an important part of a book which had considerable vogue in England and the United States when it first appeared in 1924. It was called *The Mongol in our Midst,* and was written by a physician named Francis G. Crookshank. The last edition, published in 1931, runs to 539 pages, with many amusing photographs. There are three major branches of the human race, Crookshank argues— White, Oriental, and Negro. The White branch is closely related to the chimpanzee, the Oriental to the orangutan, and the Negro to the gorilla. Because of breeding among races, there are occasional atavistic throw-backs—the Mongolian idiot indicates "orangoid" ancestors the microcephalic and the schizophrenic are throw-backs to "chimpanzoid" relatives, and the Ethiopian idiot is a "gorilloid" type. Needless to say, all this is pseudoscience of the purest sort.

By far the most persuasive of the racist writers in the United States were two lawyers—Madison Grant and Lothrop Stoddard. Their works came closer than any others to the writings of Nazi anthropologists. Both men accepted the myth of Nordic superiority, and warned against the degeneration of our Nordic strain by racial mixing with alien types. It is significant that in one of Günther's books, pictures of Grant and Stoddard appear with those of Gobineau and Chamberlain as the four great defenders of the Nordic ideal. The picture of Grant is a photograph of a bronze bust—probably to conceal the fact that he had a swarthy (i.e., non-Nordic) complexion.

Grant, who died in 1937, was a New York attorney, amateur zoologist, and trustee of the American Museum of Natural History. His two chief works on race problems are *The Passing of the Great Race,* 1916, and *The Conquest of a Continent,* 1933. Both books have introductions by the distinguished geologist, Henry Fairfield Osborn, to the everlasting embarrassment of Osborn's admirers.

The best way to characterize Grant's two books is by saying that they are eloquent attacks on the lines graven at the base of the Statue of Liberty:

> . . . Give me your tired, your poor,
> Your huddled masses yearning to be free,
> The wretched refuse of your teeming shore,
> Send these, the homeless, tempest-tossed to me.

Grant believes that America was originally settled by a racially pure strain of Protestant Nordics (This has been questioned by Dr. Ales Hrdlicka in his study, *The Old Americans,* 1925). Unfortunately, Grant continues, this superior stock, capable of carrying America to great heights, is rapidly being "debased" by a flood of immigrant "aliens." As a result, we are rapidly becoming a nation of hybrids—a "racial chaos" such as existed in Rome before it fell. Fortunately we are still 70 percent Nordic and 80 percent Protestant. If we can only put a stop to these "alien invasions," Grant writes, we may still grow in vigor and fulfill our destiny.

All racial inter-breeding, Grant thinks, is bad. "It must be borne in mind, that the specializations which characterize the higher races are of relatively recent developments, are highly unstable and when mixed with generalized or primitive characters tend to disappear. Whether we like to admit it or not, the result of the mixture of two races, in the long run, gives us a race reverting to the more ancient, generalized and lower type. The cross between a white man and a Negro is a Negro; the cross between a white man and a Hindu is a Hindu; and the cross between any of three European races and a Jew is a Jew."

The above statements have, of course, not the slightest scientific foundation. They are in the realm of folk beliefs, as mythological as the superstition that a "black baby" may be born to a white couple if one of the parents has a slight trace of Negro blood.

That the Negro is the lowest in the racial hierarchy is a fact Grant takes for granted. ". . . Evidence does exist," he writes, "to show that the intelligence and ability of a colored person are in pretty direct proportion to the amount of white blood he has. . . ." Even the mulatto who has enough white blood to "pass," still has "intellectual and emotional traits" which "may insidiously go back to his black ancestry, and may be brought into the White race in this way."

"The Southerners understand how to treat the Negro," Grant declares, "with firmness and with kindness—and the Negroes are liked below the Mason and Dixon lines so long as they keep to their proper relations to the Whites. . . ." Grant's solution of the "Negro problem" is to enact rigid state laws against intermarriage (we now have them in thirty states, by the way, many of which also prohibit White-Mongolian marriages), do everything possible to prevent social mixing, and finally, give the Negro birth control information so he will stop breeding so rapidly.

Lothrop Stoddard, another American attorney (with a Ph.D. from Harvard) holds substantially the same views as Grant. His best-known work, *The Rising Tide of Color*, was published in 1920 by Scribners. As the title suggests, the colored races (black, yellow, brown, and red) are rising like a great tide which threatens to engulf the superior whites. Like King Canute, Stoddard writes, the white man "seats himself upon the tidal sands and bids the waves be stayed. He will be lucky if he escapes with wet shoes."

Race, according to Stoddard, is the very "soul" of a culture. Civilization is merely the body. "Let the soul vanish," he says, "and the body moulders. . . ." What causes the race soul to vanish? Interbreeding. All racial mixtures are bad, and particularly bad when inferior races are involved. The more primitive the race, Stoddard contends, the more dominant its genes. "That is why crossings with the Negro are uniformly fatal."

"If the present drift be not changed," he continues, "we whites are all ultimately doomed. . . . And that would mean that the race obviously endowed with the greatest creative ability, the race which had achieved most in the past and which gave the richer promise for the future, had passed away, carrying with it to the grave those potencies upon which the realization of man's highest hopes depend. A million years of human evolution might go uncrowned, and earth's supreme life-product, man, might never fulfill his potential destiny."

It is astonishing how often that word "destiny" appears in the writings of men who assume their own race to be superior to all others. If the United States is to fulfill its destiny, Stoddard argues, it must somehow build "dikes" against the rising color tide. This can be done only if whites acquire a "true race-consciousness" and recognize the supreme importance of preserving superior hereditary strains.

One element must be fundamental in this new attitude, Stoddard insists, and "that element is *blood*. It is clean, virile, genius-bearing blood, streaming down the ages through the unerring action of heredity, which, in anything like a favorable environment, will multiply itself, solve our problems, and sweep us on to higher and nobler destinies."

In 1935, Scribners published another book by Stoddard on the same theme, *Clashing Tides of Color*. The "rising tide" has now become a "rip-tide—a confused welter of swirling eddies and choppy waves dashing against one another." He finds the new racial doctrines of the Nazis "hard to evaluate," but quotes favorably from the Nazi anthropologist, Günther. Stoddard's latest book, published in 1940 by Duell, Sloan, and Pearce, is a description of his trip to Germany shortly before the United States entered the war. Titled *Into the Darkness*, it is mildly critical of Hitler's racial and eugenics program, but on the whole gives the impression that the author is not unsympathetic with many of Hitler's racial aims.

Stoddard is not, of course, a trained anthropologist. Like Grant, he is a lawyer writing of topics about which he has only a novice's understanding. This lack of scientific background is even more characteristic of the authors of less sophisticated racist works—Charles W. Gould's *America a Family Matter*, 1920; Earnest S. Cox's *White America*, 1923; James D. Sayers' *Can the White Race Survive?*, 1929; *The Right to be Well Born*, 1917, by a millionaire horsebreeder in Lexington, Ky., William E. D. Stokes; and numerous British works of similar theme such as James H. Curle's *Our Testing Time*, 1926.

That none of these writers has status as a "scientist" is a point in the nation's favor. It indicates that as intense and pervasive as our racial feelings are among the general population—and especially among the poorly educated—we are still far from the Nazi state on such matters. No recognized professors of anthropology are giving scientific sanction to racial hatreds, and the acceptance of racist books by leading publishers has virtually ceased since 1935.

The most violent antisemitic propaganda in America has always come from our scientific illiterates. A major source for the racial lies which these people spread is the well-known forgery, *The Protocols of Zion.* The *Protocols* are a crude, incoherent work purporting to be the minutes of a series of secret meetings at which Jewish leaders plot to control the world. Actually, they are copied from a French work by Maurice Joly, published in 1865, which attacked the rule of Napoleon III. Joly's book had nothing whatever to do with Jews. Yet in spite of the fact that this forgery was completely exposed as early as 1921, the *Protocols* continue to turn up wherever anti-Jewish movements get underway. The Nazi philosopher, Alfred Rosenberg, published them in 1923 with an elaborate commentary. At the moment, they are still being circulated in the United States by professional antisemites.

One curious work by an American, Samuel Roth (a New York City publisher of semi-erotica), is worth mentioning briefly. Titled *Jews Must Live,* and sub-titled, "An Account of the Persecution of the World by Israel on all the Frontiers of Civilization," it has the questionable distinction of being the most antisemitic work ever penned by a native Jew. An appendix is headed "Do the Jews emit a peculiar odor?" The author's answer is yes. Roth published the book himself (all of his books are published by himself) in 1934. It had few sales in the United States, but was widely reprinted in Germany and by Nazi organizations in other countries. (Roth's latest book, *The Peep-Hole of the Present,* 1945, outlines a new system of philosophy and physics. The work is so obscure, confused, and badly proofread, however, that it was impossible to understand it well enough to include it in the chapter on Einstein's opponents, where it properly belongs.)

If the reader is quick to smile at the folly of anyone who takes such nonsense seriously, let him pause to reflect a moment on how widespread in the United States are racial superstitions not one whit less absurd. During World War II, the Red Cross was forced by public pressure to maintain separate banks of blood plasma, one donated by whites and one by Negroes, although the plasma was identical. Every pair of drinking fountains in the South—one labeled "White" and the other "Colored" —stands as a shameful monument to the persistence of crackpot racial views.

Everyone remembers the parable of the Good Samaritan. Few realize, however, that the reason Jesus chose the Samaritan as an example of the true "neighbor" to be loved, was that in ancient Jerusalem Samaritans were the despised minority. Substitute "Negro" for "Samaritan" and you begin to understand the parable in the way it was understood by Christ's listeners.

"Be beautiful, be natural, and be like God," the Nazi Streicher once declared. "National Socialism restitutes the divine order and therefore works at the order of our Lord," said Dr. Walter Gross, leader of Hitler's Office for Race Politics. It would be difficult to find a greater blasphemy in the history of Christendom than the coupling of racial pseudo-science with the name of the Nazarene who taught a doctrine of universal love and compassion.

CHAPTER 14

ATLANTIS AND LEMURIA

MOST ANTHROPOLOGISTS share with orthodox Christians the belief that the races of mankind originally came from a single parent stock. If this is true, the question of where this stock flourished becomes an intriguing one, and a volume could be written about the off-trail answers of pseudo-anthropologists and religious cranks. Russian ethnologists recently announced that the human race made its first appearance in Russia. A number of Oriental anthropologists, however, have long favored the Orient. Scarcely a spot on the globe has escaped being defended by at least one writer as the cradle of the human species.

The more unlikely the region, the more attractive it is to a certain type of thinker. It would be hard to imagine an area less likely than the North Pole, yet William F. Warren, a Methodist minister who was president of Boston University for thirty years, was firmly convinced that the Pole was the site of ancient Eden. His book *Paradise Found,* 1885, is a scholarly work running to more than 500 pages. In it, Warren draws on the sciences of geology, climatology, botany, zoology, anthropology, and mythology to prove that the climate at the North Pole was once exceedingly warm and pleasant. Here Adam and Eve were created. Later, the Deluge of Noah submerged the land on which Eden was situated, and changed the climate to its present frigidity.

A much more ingenious theory, however, had been published in America three years before Warren's book. It was written by Ignatius Donnelly, the Minnesota reformer whom we encountered in Chapter Three as a predecessor of Velikovsky. His book, *Atlantis,* had an enormous vogue when it was published by Harper & Brothers in 1882.

Subsequently it was translated into all major languages, and only a few years ago, issued in newly-annotated American and British editions. No book on the topic has surpassed it in popularity and influence.

Donnelly's thesis, in essence, is that the Biblical Paradise was on a vast continent which at one time existed in the Atlantic Ocean. It was there mankind arose from barbarism, after the Glacial Epoch, and developed the world's first civilization. The culture was a superior one, with a religion of sun worship, and advanced scientific knowledge. Colonizers from Atlantis spread in all directions. They were the first to populate the Americas, Europe, and Asia. Kings and queens and heroes of Atlantis became the gods and goddesses of the ancient religions. About 13,000 years ago a volcanic cataclysm shook the earth and the entire continent of Atlantis was submerged. Flood legends, like the story of Noah, are memories of this great catastrophe.

To support these contentions, Donnelly marshals a great mass of questionable geological, archeological, and legendary material—chiefly evidence of similarities between ancient Egypt and the Mexican-Indian cultures of early South America. In both regions, for example, one finds a knowledge of embalming, use of a 365-day calendar, the building of pyramids, myths of a flood, and so on. To provide a connecting bridge between the Mediterranean world and South America, Donnelly argues, one must assume the existence of an earlier culture on a continent situated between the two areas. Few writers on the subject have excelled Donnelly in reasoning, as one critic has put it, "from a molehill of fact to a mountain of surmise."

Actually, Donnelly's book is no more than an elaborate modern defense of the ancient Greek myth of Atlantis, first recorded by Plato. Plato described Atlantis as a barbarian culture, full of pomp and luxury, which at one time existed near the mouth of the Mediterranean. The gods became displeased with the island's decadence. As punishment they sent a great earthquake which caused Atlantis to sink in a single day and night. Perhaps it was a half-submerged Atlantean city that Poe described in his *City in the Sea*.

> No rays from the Holy Heaven come down
> On the long night-time of that town;
> But light from out the lurid sea
> Streams up the turrents silently—
> Up domes—up spires—up kingly halls—
> Up fanes—up Babylon-like walls—

The existence of Atlantis was widely accepted by the scholars of the Middle Ages. There are many references to it in medieval writings, but nothing of importance was added to Plato's story. Throughout the Renaissance there was much speculation about the myth, and in the nineteenth century a few individual works were devoted exclusively to it. However, it was not until Donnelly wrote his book that anyone suceeded in giving the Platonic legend a coherent, scholarly, and seemingly scientific defense. A no less eminent person than William Gladstone, Prime Minister of England, was so impressed by Donnelly's book that he asked the Cabinet to approve funds for sending a ship into the Atlantic to trace the outlines of the sunken continent. (Gladstone was something of a pseudo-scientist in his own right, having written a book about Homer in which he argued that the ancient Greeks were color-blind because of the paucity of color-words in the *Iliad* and *Odyssey*.)

Since Donnelly's book, an unbelievable number of similar works have appeared, though none has yet surpassed Donnelly's in ingenuity and eloquence. It would be a conservative guess that a list of all the titles on Atlantis, in all languages, which have been published in the present century would run to several thousand.[1] Most of them, of course, are the disordered products of eccentrics, without even literary merit to recommend them. The more colorful ones are the works of occultists, of one variety or another, who have access to secret sources of information not available to materialists. In the theosophical, Rosicrucian, and anthroposophical cults, the writers on Atlantis draw on knowledge possessed by initiates to whom it has been handed down through the ages from former initiates. In many cases, the authors have direct clairvoyant insight into matters Atlantean. A few books—such as Joseph B. Leslie's 807-page *Submerged Atlantis Restored,* published in Rochester, New York, 1911—were obtained entirely from departed Atlantean souls through the agencies of spirit mediums. Such writers are naturally in a position to obtain a vast, intimate knowledge of all the details of life in ancient Atlantis, and although their books are somewhat afield from what might be called pseudo-*science,* nevertheless many of them are amusing enough to deserve mention.

The sanest of the occult Atlantis scholars is a Scottish Presbyterian named Lewis Spence.[2] He is the author of forty learned books on folklore, more than half a dozen of which deal with Atlantis. Unlike other occultists, Spence relies almost entirely on geology, biology, mythology, and archeology for his views, though heavily salted of course with his imagination. He accepts Plato's story as substantially true. He thinks the Atlanteans were a composite race, with large brains, and that their first emigrants into Europe were of the race anthropologists call Cro-Magnon. "If a patriotic Scotsman may be pardoned the boast," he writes in *The Problem of Atlantis,* 1924, "I may say that I devoutly believe that Scotland's admitted superiority in the mental and spiritual spheres springs

almost entirely from the preponderant degree of Cro-Magnon blood which
certainly runs in the veins of her people. . . .''

Ten years ago, when World War II was beginning, Spence became
obsessed with what he thought was a parallel between the decadence of
ancient Atlantis and the decadence of modern Europe. In a book called
Will Europe Follow Atlantis?, 1942, he argues that all of Europe,
Germany in particular, is suffering from great moral decay. Just as occult
arts, such as astrology (in which Spence is a firm believer), were corrupted by the ancient Atlanteans, so the Nazis have transformed the
occult arts into a Satanic form of "black magic." Spence is specially
irritated by the view of Rosenberg and other German writers that the
Nordic race had its origin in Atlantis, since the evidence clearly shows
the Anglo-Saxons to be the true descendants. Just as surely as Atlantis
met with divine wrath, so will Germany suffer a great cataclysm—and all
Europe likewise—unless she repents and returns to true Christianity.
"Will that punishment take the selfsame form as the awful doom meted
out to the people of Atlantis?" he asks. "That is not for man born of
woman to say." Spence's earlier books on Atlantis had a great influence
on German Atlantean speculations, but there is no evidence that the Nazis
were impressed by this suggestion that Germany might lose the war by
being suddenly submerged. In 1944, Spence's *The Occult Causes of the
Present War* went into even greater details about the Satanic origin of
German occultism and the impending German doom.

Theosophists have always taken Atlantis for granted, and to the myth
have added a second one—the myth of Lemuria. This name was originally proposed by a nineteenth-century zoologist for a land mass he
thought must have existed in the Indian Ocean, and which would account
for the geographical distribution of the lemur. Madame Blavatsky, the
high priestess of theosophy, adopted the name and wrote in some detail
about the "Third Root Race" that she believed flourished on the island.

According to Blavatsky, five root races have so far appeared on the
planet, with two more yet to come. Each root race has seven "sub-races,"
and each sub-race has seven "branch races." (Seven is a mystic number
for theosophists.) The first root race, which lived somewhere around the
North Pole, was a race of "fire mist" people—ethereal and invisible.
The Second Root Race inhabited northern Asia. They had astral bodies
on the borderline of visibility. At first, they propagated by a kind of
fission, but eventually this evolved into sexual reproduction after passing
through a stage in which both sexes were united in each individual. The
Third Root Race lived on Lemuria. They were ape-like giants with
corporeal bodies that slowly developed into forms much like modern
man. Lemuria was submerged in a great convulsion, but not before a
sub-race had migrated to Atlantis to begin the Fourth Root Race.

The Fifth Root Race, the Aryan, sprang from the fifth sub-race of the Atlanteans. At the present time, according to theosophists, the Sixth Root Race is slowly emerging from the sixth sub-race of Aryans. This is happening in Southern California where, in Annie Besant's words, the "climate approaches most nearly to our ideal of Paradise." Eventually, the American Continent will sink, and Lemuria will rise from the Pacific to be the home of the Sixth Root Race. After the Seventh Root Race (which will develop from the seventh sub-race of the sixth root race) has risen and fallen, the earth cycle will have ended and a new one will start on the planet Mercury.

Later occultists have accepted the leads of Blavatsky, Besant, and other early theosophical leaders and expanded them in fascinating detail. A theosophical work by W. Scott-Elliott, *The Story of Atlantis,* 1914, is the richest source of information about the seven sub-races which succeeded each other on the Atlantean isle. The first sub-race, coming originally from Lemuria, was the Rmoahal. Members were ten to twelve feet high with mahogany black skin. The second sub-race, the Tlavatli, were copper-colored and worshipped their ancestors. Then came the Toltec, the highest culture of Atlantis, which lasted some 10,000 years. They were even more copper-colored, tall, and with Grecian features. Their science was very advanced. There were Toltec airships which operated by a cosmic force unknown today. Sometimes the Toltecs drank the hot blood of animals. (Annie Besant had written earlier that the Toltecs were twenty-seven feet high, with bodies so hard that "one of our knives would not cut their flesh, any more than it would cut a piece of present-day rock.")

After the Toltecs came the Turanians—irresponsible individualists but great colonizers. Then the Semites appeared, with a highly developed ability to reason and an inner conscience. They were a turbulent, discontented race, always warring with their neighbors. The sixth sub-race, the Akkadian, were the first legislators. Finally, the Mongolians. They immigrated to Asia and were the first Atlanteans to develop their culture off the island.

The late Rudolf Steiner, founder of the Anthroposophical Society, the fastest growing cult in post-war Germany, accepted all of Scott-Elliott, then added new details of his own from a source he said he was not permitted to place on record. In his *Atlantis and Lemuria,* 1913, he reveals that the Lemurians were unable to reason or calculate, living chiefly by instinct. They had no speech, but communicated with each other by telepathy. They lived in caves, and possessed a highly developed will power which enabled them to lift enormous weights. The atmosphere was denser then than it is now, the water was more fluid, and the earth was in a plastic, unconsolidated state.

The most indefatigable publicist for Lemuria, or "Mu" as he called it, was Colonel James Churchward, a Britisher who served in India with the Bengal Lancers. As a young officer assigned to famine relief, he formed a close friendship with the high priest of a temple-school monastery in India. According to the Colonel, the priest permitted him to see a collection of tablets written in ancient Mu script, and with the priest's aid, the tablets were finally deciphered. At the age of seventy, living as a retired officer in Mount Vernon, N.Y., Churchward began his series of Mu books, based on the monastery's tablets. *The Lost Continent of Mu*, 1926, was followed by *The Children of Mu*, 1931; *The Sacred Symbols of Mu*, 1933; and *Cosmic Forces as They Were Taught in Mu*, 1934. The Colonel died in 1936, at the age of 86.

Mu (pronounced "moo") was, according to Churchward, the original Eden where man was created 200 million years ago. It was a special creation, not the product of evolution. The Lemurians reached an advanced civilized state with a science far surpassing ours. Among other things, they had mastered "cosmic force" which enabled them to nullify gravity, a fact which should interest the Babson Gravity Research Foundation. It was the same force Jesus used when He walked on the water. In fact, Jesus studied the sacred religion of Mu from sages in India and Tibet, so his teachings are all Mu-derived. About 12,000 years ago a subterranean gas belt near the equator exploded. In the convulsion which followed, the entire continent of Mu sank beneath the sea and its 64 million inhabitants perished. The Pacific islands are the remaining "pathetic fingers of that great land." It was the same watery fate which later engulfed the continent of Atlantis.

The Colonel accepts all the apparatus of occultism—reincarnation, telepathy, spirit photography, astral bodies, and so on. One of his Mu books closes with the story of how a Rishi—the same high priest in India who showed him the Mu tablets—put him into a trance and the two of them visited the scene of their previous incarnations. The Rishi had continually called Churchward "my son." In this visit to the past, the author sees himself as a dead soldier, and beside him the weeping Rishi crying, "My only son and fallen in battle!"

The Mu books are uniformly crude in writing, and such a mishmash of geological and archeological errors that they are widely regarded, even by other Atlantean and Mu scholars, as a deliberate hoax. It is significant that no one ever saw the tablets which were the chief source of the Colonel's knowledge, nor did he anywhere identify the monastery where he found them.

A refreshingly new angle on the sinking of Atlantis has been provided by Hans S. Bellamy, whom we discussed in an early chapter as England's chief publicist for the Hörbiger cosmic-ice theory. In his book, *The Atlantis Myth*, 1948, Bellamy argues that the sinking of Atlantis was caused by the quakes which attended the earth's capture of its present moon. Bellamy's American competitor, Dr. Immanuel Velikovsky, disagrees. In *Worlds in Collision* the doctor informs us that Plato made a slight error when he set the date for the island's submergence as 9,000 years before the time of Solon. "There is one zero too many here," Velikovsky writes, and blandly adds, "The most probable date of the sinking of Atlantis would be . . . 900 years before Solon." This coincides, of course, with the convulsions produced by the first visit of Velikovsky's erratic comet.

How is one to explain the fascination which the myths of sunken continents have exercised over the minds of so many recent scholars? Although geologists agree that in remote ages the arrangements of land and sea were quite different from what they are now, there is unanimous agreement that no great continental sinkings have occured within the relatively brief time man has been on the scene. There is, in fact, not a shred of reliable evidence, geological or archeological, to support the twin myths of Atlantis and Lemuria. Yet the literature continues to proliferate. Since 1900, dozens of little periodicals in various nations have been devoted to Atlantean research. For a time, Spence edited such a magazine in England, and at the moment, Bellamy is on the staff of a similar British publication. Innumerable Atlantis societies have been formed. Twenty years ago, a Danish group actually printed Atlantean stamps and currency, and designed an Atlantean flag. In Chicago, in 1936, a Lemurian Fellowship was organized to promote the ancient wisdom of the Motherland and usher in the new Lemurian order. Its literature was written by a reincarnated Lemurian and published by the Lemurian Book Industries in Milwaukee. The group had plans for building "super cities" somewhere in Southern California, but apparently their Lemurian wisdom failed to provide them with sufficient funds for this project.

The psychological factors behind all this are not hard to understand. There is, of course, an obvious element of escape in dreaming about a vast, mythical land of wonder—like the Land of Oz—but it is more than this. There is a strong element of being one of the chosen—one who is rich in Atlantean blood, or a reincarnation of an Atlantean, or at least a person initiated into great mysteries hidden from the eyes of the common multitude. H. G. Wells, in his novel *Christina Alberta's Father*, penned a shrewd portrait of the lonely, imaginative, neurotic type of personality—a Mr. Preemby—who finds the Atlantis myth so compelling. One hears the voice of Preemby in these words of Bellamy:

There is magic in names . . . and the mightiest among these words of magic is *Atlantis*. When we have pronounced this word, nothing definite is revealed, but it is as if a sudden shaft of sunlight smote through the darkness of the past, allowing us a glimpse of cloud-capped towers, and gorgeous palaces, and solemn temples; and it is as if this vision of a lost culture touched the most hidden part of our soul.

Fantasy and science fiction have not escaped the magic spell. In Jules Verne's *Twenty Thousand Leagues Under the Sea* Captain Nemo's submarine visits the ruined towers of Atlantis. In another science-fiction story, *The Maracot Deep* by Conan Doyle, a group of scientists descend to Atlantis in a steel sphere and flnd there a flourishing city, covered with a watertight roof, manufacturing its own air, lit by fluorescent lights, and populated by a happy people. An Atlantean scientist, foreseeing the coming doom, had managed to save a portion of a city. The citizens have mastered atomic energy, and can project their thoughts on screens. In the chapter on flying saucers we mentioned the Great Shaver Mystery which ran in *Amazing Stories*. The evil deros, who figure in the Shaver tales, came originally from the lost Lemuria.

It is safe to say that speculations about Atlantis and the older Motherland of Mu will occupy the minds of pseudo-archeologists for many decades to come. Until the last square mile of the oceans' depths is fully explored, those who are so inclined may continue to believe, in the words of John Masefield, that:

> In some green island of the sea,
> Where now the shadowy coral grows
> In pride and pomp and empery
> The courts of old Atlantis rose.

CHAPTER 15

THE GREAT PYRAMID

The Great, the Mighty God, the Lord of Hosts . . . which hast set signs and wonders in the land of Egypt, even unto this day. . . .

Jeremiah 32: 18-20

THE LITERATURE OF Biblical archeology presents a bewildering panorama. It ranges all the way from competent, objective studies by men who took great pains not to draw unwarranted inferences from their artifacts, to the work of men who have twisted their material in every conceivable way to make it conform to Biblical records. Thousands of books and pamphlets have been written in the past hundred years to show that the "latest findings" of archeology confirm all the details of scriptural history—especially the miracle stories at which unbelievers scoff. In some cases, it is hard to believe the distortion has been unconscious. Professor Hubert Grimme, for example, of the University of Munich, published in the twenties a "translation" of a stone tablet which told how the infant Moses had been rescued from the bullrushes by Pharaoh's daughter. It later came to light that the professor had made free use of cracks and weather marks on the stone, combining them with the hieroglyphics to make the translation come out right.

In English, the most dignified books in this pseudo-archeological literature are by Sir Charles Marston—*The Bible is True,* 1934; *New Bible Evidence,* 1934; and *The Bible Comes Alive,* 1937. Among the less scholarly, John O. Kinnaman's *Diggers for Facts,* 1940, is a good recent example. From it you will learn that Abraham's home and even his signature have been unearthed, as well as evidence that St. Paul once preached in England. Kinnaman is cautious, however, about identifying the salty remains of Lot's wife. "There are many actual pillars of salt in that region," he writes, "but which may be the remains of the unfortunate woman, no one can tell."

Expeditions to Mount Ararat, to find Noah's Ark, take place every few years. Egerton Sykes, head of the Hörbiger Institute in England and editor of an Atlantis magazine, planned such an expedition a few years ago, but Russian authorities put pressure on the Ottoman officials at Ankara, and he was refused a visa. According to the Russians, his expedition was part of an "Anglo-American military plot to spy on the Russian borders in sight of Ararat." However, in 1949 another expedition, led by Dr. Aaron Smith, of Greensborough, N.C., did manage to climb 12,000 feet up the side of Ararat. Unfortunately, they failed to find the Ark. "We can't say if the Ark may have landed at a lower level," Smith reported, "or if it was completely buried by the debris of earthquakes, violent in this region. Again, it may exist on the north side of the range, under ice and snow. We have not found it; but we sure have cleared the way for others who may have better luck than we had."

It would be impossible, of course, to survey even briefly the literature of eccentric Biblical archeology. One aspect, however, known as Pyramidology, is sufficiently curious and colorful to warrant special attention. With this topic (which rivals Atlantis in the number of books devoted to it) the remainder of the chapter will be concerned.

The Great Pyramid of Egypt was involved in many medieval and Renaissance cults, especially in the Rosicrucian and other occult traditions, but it was not until 1859 that modern Pyramidology was born. This was the year that John Taylor, an eccentric partner in a London publishing firm, issued his *The Great Pyramid: Why was it Built? And Who Built it?*

Taylor never visited the Pyramid, but the more he studied its structure, the more he became convinced that its architect was not an Egyptian, but an Israelite acting under divine orders. Perhaps it was Noah himself. "He who built the ark was, of all men, the most competent to direct the building of the Great Pyramid," Taylor wrote. The picture is rather amusing, of poor old Noah, after his Herculean task of building the Ark and surviving the Deluge, being sent to Egypt to direct the even more Herculean labor of building the Pyramid!

Taylor's chief reason for thinking the Pyramid part of God's plan was the fact that he found in its structure all kinds of mathematical truths which far surpassed the knowledge of ancient Egypt. For example, if you divide the monument's height into twice the side of its base, you obtain a fairly close approximation of *pi* (the ratio of diameter to circumference of a circle). In addition, Taylor found elaborate reasons for thinking that the measuring unit used by the Pyramid's architect was none other than the Biblical "cubit" employed by Noah in the construction of the Ark, by Abraham in building the Tabernacle, and by Solomon in the architecture of his temples. The "sacred cubit" was, Taylor thought, about twenty-five inches, and based on the length of the earth's axis. Since the earth's diameter varies considerably, because of the flattening at the poles, what could be a more natural basis for a divine unit than the axis on which our globe rotates? If you divide the axis by 400,000, you obtain the sacred cubit. Taylor found other divine units of measurement in the Pyramid—in the capacity of a granite coffer in the King's Chamber, for example—all of which he thought had bases in nature and were therefore superior to those of other measuring systems.

In addition to all the supposed truths embodied in the Pyramid, Taylor also found a score of passages in both the *Old and New Testament* which, if wrenched from their contexts, can be interpreted as references to the stone monument. For example, we read in *Isaiah* 19: 19-20: "In that day shall there be an altar to the Lord in the midst of the land of Egypt . . . and it shall be for a sign and for a witness unto the Lord of hosts. . . ." And in *Job* 38:5-7: "Who hath laid the measures thereof, if thou knowest? Or who hath stretched the line upon it? Whereupon are the foundations thereof fastened? or who laid the cornerstone thereof, when the morning stars sang together, and all the sons of God shouted for joy?" Even St. Paul spoke of the Pyramid, Taylor believed, in such passages as ". . . Jesus Christ himself being the chief corner stone; in whom all the building, fitly framed together, groweth unto an holy temple of the Lord." *(Ephesians* 2:20-21). The Pyramid symbolized, he explained, the true Church with Christ as the topmost corner stone (this symbol, incidentally, has been popular in Christian mystical lore, and was adopted by the founding fathers as the reverse side of the United States Seal).

Taylor's speculaton would probably have soon been forgotten had it not been for the Astronomer-Royal of Scotland, a University of Edinburgh professor named Charles Piazzi Smyth. Fired with enthusiasm for Taylor's theory, Smyth soon convinced himself there were greater truths symbolized in the Pyramid than even Taylor suspected. His 664-page work, *Our Inheritance in the Great Pyramid,* is to Biblical Pyramidology what Donnelly's book is to Atlantology. The first edition, in 1864, was an immediate success. It went through four later editions (the last, greatly revised, in 1890), was translated into many languages, and has far exceeded all subsequent works on the topic in its influence. In 1865, Smyth went to Egypt at his own expense to make his own measurements of the Pyramid. The results of this research appeared in his three-volume

Life and Work at the Great Pyramid, 1867, and *On the Antiquity of Intellectual Man,* 1868.

Our Inheritance is a classic of its kind. Few books illustrate so beautifully the ease with which an intelligent man, passionately convinced of a theory, can manipulate his subject matter in such a way as to make it conform to previously held opinions. Unfortunately, space permits only the barest résumé of Smyth's sensational findings.

To begin with, Smyth discovered that the base of the Pyramid, divided by the width of a casing stone, equaled exactly 365—the number of days in the year. Casing stones originally composed the outside surface of the monument. The first of these stones was unearthed after Taylor's death, so its width had not been known to him. The stone measured slightly more than twenty-five inches, and Smyth concluded that this length was none other than the sacred cubit. If we adopt a new inch—Smyth calls it the "Pyramid inch"—which is exactly one twenty-fifth of the width of the casing stone, then we obtain the smallest divine unit of measurement used in the monument's construction. It is exactly one ten-millionth of the earth's polar radius. Somehow, it had been passed on through the generations, the Scottish astronomer believed, until it became the Anglo-Saxon inch, but in the process altered slightly, making the British inch a trifle short of the sacred unit. Many years later a number of other casing stones were dug up. They had entirely different widths. By that time, however, the Pyramid inch had become so firmly established in the literature of Pyramidology that devotees merely shrugged and admitted that the first casing stone just "happened" to be a cubit wide.

With incredible zeal, Smyth applied his Pyramid inch to every measurable portion of the Pyramid, inside and out, to see how many scientific and historical truths he could discover. These he found in great profusion. For example, when the height of the Pyramid is multiplied by ten to the ninth power, you obtain a distance which approximates the distance from the earth to the sun. Similar manipulations of Pyramid lengths give you the earth's mean density, the period of precession of its axis, the mean temperature of the earth's surface, and many other scientific facts only discovered in recent times. In addition to a system of sacred measuring units for length, weight, volume, and so on, Smyth even proposed a "Pyramid thermometer." It used freezing point as zero, and a fifty-degree mark based on the temperature inside the King's Chamber, which was on the fiftieth level of the monument's masonry.

Smyth's most spectacular contribution, however, was the elaboration of a theory proposed by one Robert Menzies—that there is a great outline of history symbolized by the Pyramid's internal passageways. When these passages are properly measured in Pyramid inches, counting an inch as equal to a year, and the symbolism correctly interpreted, you emerge with the principal dates in the earth's past and future. You discover, for instance, that the world was created about 4,004 years before Christ. The Flood, the time of the Exodus, and the date the Pyramid was built are also indicated. The beginning of a sloping passage called the Grand Gallery marks the birth of Christ. Other features indicate the Lord's Atonement (after 33 inch-years of life), his descent into Hell, and final Resurrection. Continuing upward along the gallery, one discovers that it terminates at a point between 1882 and 1911, depending on how the length of the Grand Gallery is measured. To Smyth this 29-year period is the great Tribulation which will precede the Second Coming of Christ.

It is not difficult to understand how Smyth achieved these astonishing scientific and historical correspondences. If you set about measuring a complicated structure like the Pyramid, you will quickly have on hand a great abundance of lengths to play with. If you have sufficient patience to juggle them about in various ways, you are certain to come out with many figures which coincide with important historical dates or figures in the sciences. Since you are bound by no rules, it would be odd indeed if this search for Pyramid "truths" failed to meet with considerable success.

Take the Pyramid's height, for example. Smyth multiplies it by ten to the ninth power to obtain the distance to the sun. The nine here is purely arbitrary. And if no simple multiple had yielded the distance to the sun, he could try other multiples to see if it gave the distance to the Moon, or the nearest star, or any other scientific figure.

This process of juggling is rendered infinitely easier by two significant facts. (1) Measurements of various Pyramid lengths are far from established. Competent archeologists in Smyth's day disagreed about almost all of them, including the most basic of all, the base length of the Pyramid. Later archeologists, after Smyth, made more accurate measurements and found still different figures. In many cases Smyth had a choice of several lengths to pick from. In other cases he used measurements made by himself. And sometimes he added together conflicting measurements and used the average. (2) The figures which represent scientific truths are equally vague. The distance to the sun, for example, was not known with great accuracy in Smyth's day, and besides, the distance varies considerably because the earth's path is not a circle but an ellipse. In such cases you have a wide choice of figures. You can use the earth's shortest distance to the sun, or the longest, or the mean. And in all three cases, you can choose between conflicting estimates made by different astronomers of the time. The same ambiguity applies to almost every scientific "truth" employed by Smyth.

The only Pyramid "truth" which cannot be explained easily in terms of such juggling is the value *pi*. The Egyptians may have purposely made use of this ratio, but it seems more likely that it was a by-product of another construction. Herodotus states that the Pyramid was built so the area of each face would equal the area of a square whose side is equal to the Pyramid's height. If such a construction is made, it fits the Pyramid perfectly, and the ratio of height to twice the base will automatically be a surprisingly accurate value for *pi*. (See *Popular Astronomy,* April, 1943, p. 185.)

Both Taylor and Smyth made a great deal of the fact that the number five is a key number in Pyramid construction. It has five corners and five sides. The Pyramid inch is one-fifth of one-fifth of a cubit. And so on. Joseph Seiss, one of Smyth's disciples, puts it as follows: "This intense *fiveness* could not have been accidental, and likewise corresponds with the arrangements of God, both in nature and revelation. Note the fiveness of termination to each limb of the human body, The five senses, the five books of Moses, the twice five precepts of the Decalogue."

Just for fun, if one looks up the facts about the Washington Monument in the *World Almanac,* he will find considerable fiveness. Its height is 555 feet and 5 inches. The base is 55 feet square, and the windows are set at 500 feet from the base. If the base is multiplied by 60 (or five times the number of months in a year) it gives 3,300, which is the exact weight of the capstone in pounds. Also, the word "Washington" has exactly ten letters (two times five). And if the weight of the capstone is multiplied by the base, the result is 181,500—a fairly close approximation of the speed of light in miles per second. If the base is measured with a "Monument foot," which is slightly smaller than the standard foot, its side comes to 56½ feet. This times 33,000 yields a figure even closer to the speed of light.

And is it not significant that the Monument is in the form of an *obelisk*—an ancient Egyptian structure? Or that a picture of the Great Pyramid appears on a dollar bill, on the side opposite *Washington's* portrait? Moreover, the decision to print the Pyramid (i.e., the reverse side of the United States seal) on dollar bills was announced by the Secretary of the Treasury on June 15, 1935—both date and year being multiples of five. And are there not exactly twenty-five letters (five times five) in the title, "The Secretary of the Treasury"?

It should take an average mathematician about fifty-five minutes to discover the above "truths," working only with the meager figures provided by the *Almanac*. Considering the fact that Smyth made his own measurements, obtaining hundreds of lengths with which to work, and that he spent twenty years mulling over these figures, it is not hard to see how he achieved such remarkable results.

Nevertheless, Smyth's books made a profound impression on millions of naive readers. Dozens of volumes appeared in all languages carrying on the great work and adding additional material. In France, the leading advocate of Pyramidology was Abbé F. Moigno, Canon of St. Denis, Paris. An International Institute for Preserving and Perfecting Weights and Measures was organized in Boston, in 1879, at a meeting in Old South Church. The purpose of the Society was to work for the revision of measuring units to conform to sacred Pyramid standards, and to combat the "atheistic metrical system" of France. President James A. Garfield was a supporter of the Society, though he declined to serve as its president.

A periodical called *The International Standard* was published during the 1880's by the Ohio Auxiliary of the Society, in Cleveland. The president of the Ohio group, a civil engineer who prided himself on having an arm exactly one cubit in length, had this to say in the first issue: "We believe our work to be of God; we are actuated by no selfish or mercenary motive. We depreciate personal antagonisms of every kind, but we proclaim a ceaseless antagonism to that great evil, the French Metric System. . . . The jests of the ignorant and the ridicule of the prejudiced, fall harmless upon us and deserve no notice. . . . It is the Battle of the Standards. May our banner be ever upheld in the cause of Truth, Freedom, and Universal Brotherhood, founded upon a just weight and a just measure, which alone are acceptable to the Lord."

A later issue printed the words and music of a song, the fourth verse of which ran:

> Then down with every "metric" scheme
> Taught by the foreign school,
> We'll worship still our Father's God!
> And keep our Father's "rule"!
> A perfect inch, a perfect pint,
> The Anglo's honest pound,
> Shall hold their place upon the earth,
> Till time's last trump shall sound!

The prophetic portions of Smyth's work appealed strongly to Protestant fundamentalists of all denominations—especially in England. One of the most popular early books, *Miracle in Stone*, 1877, by Joseph Seiss, ran through fourteen editions. A Col. J. Garnier produced a book in 1905 which proved by the Pyramid that Christ would return in 1920. Walter Wynn, in 1926, issued a similar work. Undaunted by the failure of its prophecies, he wrote another book in 1933 containing equally bad predictions. Bertrand Russell, in one of his essays, summed up this literature as follows:

"I like also the men who study the Great Pyramid, with a view to deciphering its mystical lore. Many great books have been written on this subject, some of which have been presented to me by their authors. It is a singular fact that the Great Pyramid always predicts the history of the world accurately up to the date of publication of the book in question, but after that date it becomes less reliable. Generally the author expects, very soon, wars in Egypt, followed by Armageddon and the coming of the Antichrist, but by this time so many people have been recognized as Antichrist that the reader is reluctantly driven to scepticism."

An American preacher enormously impressed by Smyth's researches was Charles Taze Russell, of Allegheny, Pa., founder of the sect now known as Jehovah's Witnesses. In 1891, Pastor Russell published the third volume of his famous series *Studies in the Scripture*. It is a book of Biblical prophecy, supplemented by evidence from the Great Pyramid. A letter from Smyth is reproduced in which the Scottish astronomer praises Russell highly for his new and original contributions.

According to Russell, the Bible and Pyramid reveal clearly that the Second Coming of Christ took place invisibly in 1874. This ushered in forty years of "Harvest" during which the true members of the Church are to be called together under Russell's leadership. Before the close of 1914, the Millennium will begin. The dead will rise and be given a "second chance" to accept Christ. Those who refuse are to be annihilated, leaving the world completely cleansed of evil. Members of the church alive at the beginning of the Millennium will simply live on forever. This is the meaning of the well known slogan of the Witnesses—"Millions now living will never die."

In England, two brothers, John and Morton Edgar, were so impressed by Russell's pyramid theories that they hurried to Egypt to make measurements of their own. There they found "beautiful confirmations" of the pastor's views "as day by day first one, and then the other, discovered fresh beauties in the symbolic and prophetic teaching of this marvellous structure." Their heroic research is recorded for posterity in two weighty tomes, from the first of which the above quotation is taken. The volumes appeared in 1910 and 1913, under the title, *The Great Pyramid Passages and Chambers*.

To the great disappointment of the Russellites, 1914 ushered in nothing more dramatic than the World War, and the sect lost thousands of members. New editions of Russell's Pyramid study were issued with the wording altered slightly at crucial spots to make the errors less obvious. Thus, a 1910 edition had read, ". . . The deliverance of the saints must take place some time before 1914. . . ." (p. 228) But in 1923, this sentence read, ". . . The deliverance of the saints must take place very soon after 1914. . . ." Morton Edgar (brother John died before the great disappointment) produced a series of booklets in the twenties which followed the then current Russellite line—namely, that in 1914 Christ (already on earth since 1874) had begun an *invisible* reign of righteousness.

Judge J. F. Rutherford, who succeeded Russell after the pastor died in 1916, eventually discarded Pyramidology entirely. Writing in the November 15 and December 1, 1928, issues of *The Watch Tower and Herald,* Rutherford releases a double-barreled blast against it, and advances many ingenious arguments that the so-called Altar in Egypt was really inspired by Satan for the purpose of misleading the faithful. Did Jesus ever mention the Pyramid? Of course not. To study it, the Judge writes, is a waste of time and indicates lack of faith in the all-sufficiency of the Bible. Whether Morton Edgar remained a faithful Witness after this date, renouncing his life-time work on the Pyramid, would be interesting to know.

The Judge did not remind his readers in these articles that he, too, had been guilty of a prophetic error. For many years he had taught that 1925 would mark the beginning of the great jubilee year. Alas, it also had passed without perceptible upheavals. The sect now discourages the sale and reading of Russell's writings, and although members still believe the Millennium is about to dawn, no definite dates are set.[1]

Another fundamentalist sect that has made even stronger use of the "Bible in Stone," is the Anglo-Israel movement. This cult regards the Anglo-Saxon and Celtic peoples as descendents of the ten lost tribes of Israel, and therefore heir to all the promises God made to Abraham. In the United States, the leading organization is the Anglo-Saxon Federation of America, with headquarters at Haverhill, Mass. Their handsome monthly magazine, *Destiny,* has been going now for more than twenty years.

Anglo-Israel's outstanding Pyramid work is a monumental tome, the size of a volume of the *Encyclopedia Britannica,* called *The Great Pyramid: Its Divine Message.* It was written by David Davidson, a structural engineer in Leeds, England, and first published in 1924. A revised eighth edition appeared in 1940. The book is based on Smyth, with important differences in the prophetic section. The "Final Tribulation" of the Anglo-Saxon peoples was to begin in 1928 and extend to 1936. From September 16, 1936 until August 20, 1953, the Anglo-Saxons—i.e., the "true" Israel—will be brought together and given divine protection against a coalition of world powers seeking to destroy them. This will be the Armageddon period, terminated by the return of Christ.

Numerous Anglo-Israel books and pamphlets have been based on Davidson's work, notably the books of Basil Stewart. In the United States, the Haverhill group is currently selling the tenth edition of *Great Pyramid Proof of God,* by George F. Riffert, of Easton, Pa. It likewise is a popular version of Davidson. The first edition, which appeared in 1932, placed great stress on the September 16, 1936 date. There was, in fact, considerable excitement among Anglo-Israelites in both England and the

United States when this day approached, but it slipped by without visible cataclysms.

The present edition of Riffert's book has an added chapter in which the author confesses, "A very real problem was, and still is, to ascertain the literal significance and character of the epoch whose crisis date was September 16, 1936." He suggests several events which took place on that day, the most important of which was "that the Duke of Windsor, then King of England, notified his prime minister, Mr. Baldwin, of his determination to marry Mrs. Simpson."

Riffert concludes: ". . . By 1953 the present Babylon-Beast-Gentile type of Civilization, the Capitalistic System of Money profits by exploitation and usury, the Armageddon Conflict, the Resurrection and Translation of God's spiritual Israel preparatory to their administrative service in the New Social or Economic Order, the overthrow of dictatorships, the regeneration and transformation of the Anglo-Saxon Nations into the world-wide Kingdom of God, and literal return of Jesus Christ as King of Kings to prepare and perpetuate the Millennial Age, will all have come to pass."

Adventist sects have a distressing habit of refusing to blow away when one of their major prophecies fails. It is too easy to discover "errors" in calculations and make appropriate revisions. Nevertheless, it will be interesting to observe the mental gyrations of the Haverhill leaders after August 20, 1953 slips by.[2]

It is perhaps worth mentioning that there is also a vast occult literature dealing with the Pyramid—especially in Rosicrucian and theosophical traditions. The Biblical prophecies of Smyth are rejected, but the authors find in the monument a great deal of mathematical, scientific, astrological, and occult symbolism which varies widely with individual writers. According to Madame Blavatsky, the interior of the Pyramid was used for the performance of sacred rituals connected with the Egyptian *Book of the Dead,* and most theosophists today assume there are vast mysteries of some sort connected with the stone monument that are known only to initiates. The best reference on this approach is a two-volume work by British theosophist William Kingland, *Great Pyramid in Fact and in Theory,* 1932-35. Another occult approach, connecting the Pyramid with the mystical Jewish writings known as the kabala, will be found in J. Ralston Skinner's *Key to the Hebrew-Egyptian Mystery,* 1875 (revised in 1931).[3]

As worthless as all this literature is, it is not entirely worthless if we can see in it an important object lesson. No book has ever demonstrated more clearly than Smyth's (the other Pyramid books, of course, to a lesser degree) how easy it is to work over an undigested mass of data and emerge with a pattern, which at first glance, is so intricately put together that it is difficult to believe it is nothing more than the product of a man's brain. In a sense, this is true of almost all the books of pseudo-scientists. In one way or another, they do not let the data speak for themselves. Consciously or unconsciously, their preconceived dogmas twist and mold the objective facts into forms which support the dogmas, but have no basis in the exterior world. Sir Flinders Petrie, a famous archeologist who made some highly exact Pyramid measurements, reports that he once caught a Pyramidologist secretly filing down a projecting stone to make it conform to one of his theories!

Perhaps this tendency to distort data operates in its subtlest forms in the great cyclical theories of history—the works of men like Hegel, Spengler, Marx, and perhaps, though one must say it in hushed tones, the works of Toynbee. The ability of the mind to fool itself by an unconscious "fudging" on the facts—an overemphasis here and underemphasis there—is far greater than most people realize. The literature of Pyramidology stands as a permanent and pathetic tribute to that ability.

Will the work of the prophetic historians mentioned above seem to readers of the year 2,000 as artificial in their constructions as the predictions of the Pyramidologists? Chesterton's hilarious fantasy of the future, *Napoleon of Notting Hill* (which opens, by the way, like Orwell's novel, in 1984) begins with these wise words:

> The human race, to which so many of my readers belong, has been playing at children's games from the beginning . . . and one of the games to which it is most attached is called, "Keep to-morrow dark," and which is also named (by the rustics in Shropshire, I have no doubt) "Cheat the Prophet." The players listen very carefully and respectfully to all that the clever men have to say about what is to happen in the next generation. The players then wait until all the clever men are dead, and bury them nicely. They then go and do something else. That is all. For a race of simple tastes, however, it is great fun.

CHAPTER 16

MEDICAL CULTS

IN NO OTHER field have pseudo-scientists flourished as prominently as in the field of medicine. It is not hard to understand why. In the first place, a medical quack—if he presents an impressive façade—can usually make a great deal of money. In the second place, if he is sincere, or partly sincere, the healing successes he is almost sure to achieve will greatly bolster his delusions. In some cases, of course, the doctor is an out-and-out charlatan. In other cases he is as sincere as was Piazzi Smyth about the Great Pyramid. In still other cases, there is that baffling mixture of sincerity and skullduggery which so often is found within a crackpot's brain.

There are two great secrets of the quack's success. One is the fact that many human ills, including some of the severest, will run their course and vanish without treatment of any sort. Suppose, for example, Mrs. Smith is unable to get rid of an annoying cold. She decides to try a new doctor she has heard about, whose methods are unorthodox, but who has been strongly recommended. The doctor proves to be a distinguished-looking man who talks with great authority about his work. Diplomas from several medical schools are on the wall, and he is apt to have a number of letters after his name. (Mrs. Smith doesn't know that these degrees were given by small schools no longer in existence, some of which the doctor himself may have founded.)

Mrs. Smith decides she has nothing to lose. In addition, she is lonely and enjoys talking to doctors about her troubles. So she takes off her shoes and stockings and lets the doctor shine infra-red light on her feet for ten minutes. It costs only five dollars, but of course she has to return for two or three additional treatments. After a week or so her cold has vanished. Incredible as it may seem, Mrs. Smith is now firmly persuaded that the infra-red light is responsible for the cure. She becomes one of the doctor's loyal boosters. Before the year is over, he has milked several hundred dollars from her bank account.

Charles Fort summed it all up succinctly. "Eclipses occur, and savages are frightened. The medicine men wave wands—the sun is cured—they did it." Half the successes of medical quacks are exactly of this sort.

The other half are due to the fact that many of life's ills are wholly or in part psychosomatic. If a patient with such complaints has faith in a doctor, regardless of how bizarre the doctor's methods may be, he often will be miraculously cured. And, of course, the larger the following the doctor has, the more the patient's faith is augmented. Moreover, if dozens of Mrs. Smith's friends are chattering about infra-red healing, the stronger will be her desire to become part of this trend—an initiate who can talk about *her* experiences with the new type of treatment. When everyone is seeing flying saucers, you naturally would like to see one yourself. If everyone is getting cured by infra-red, you want to be cured the same way. Regardless of what her more enlightened friends, or even the family doctor, may tell her, Mrs. Smith has one simple and irrefutable answer—it works.

And work it does. Every time the federal government drags a quack into court, he has no trouble at all finding scores of people willing to testify about miraculous cures. Just as every faith-healing revivalist, no matter how strange his doctrines, will have astonishing platform successes, so every modern witch doctor, no matter how preposterous his rituals, will always find patients he can heal.

In this chapter we shall glance at four outstanding medical cults, all of them founded by pseudo-scientists, which have won many millions of disciples in the United States. It will be followed by a survey of individuals whose views did not develop into "schools," a section on food fads, and finally a chapter on Dr. Bates and his methods of eye training.

The first medical cult of any importance in America—homeopathy—had its origin in the mind of a German doctor, Samuel Christian Hahnemann. He published his great opus, *The Organon,* in 1810. According to Hahnemann, there is a "Law of Similia" which states that "like cures like." In longer words, a drug will cure a disease if that same drug, taken by a healthy person, will produce symptoms similar to those of the disease.

Hahnemann and his followers set about "proving," as they called it, as many new remedies as possible. This involves giving the compound to a healthy person, in increasing amounts, until symptoms appear. The symptoms are then compared with those of known ailments, and if similar, the drug is deemed of value in treating that ailment. Although certain diseases have characteristic symptoms, and hence call for specific medication, actually each individual is considered unique and treated in terms of whatever complex of *symptoms* are found, regardless of the name of the disease.

Homeopathic remedies are administered in inconceivably small doses. It was Hahnemann's conviction that the more minute the dose, the more potent. Compounds are frequentiy diluted to one *decillionth* (i.e., a millionth of a millionth of a millionth, etc., up to ten of these millionths) of a single grain. One homeopath proved 1,349 symptoms from a dose of one decillionth of a grain of common table salt. Such dilution is like letting a drop of medicine fall into the Pacific, mixing thoroughly, then taking a spoonful. Hahnemann believed that as the drug became less "material" it gained "spiritual" curative powers, and in many cases recommended diluting until not a single molecule of the original substance remained! This produced remedies of extremely high potency. Moreover, the doctor believed, the full effect of such medicine may not be manifested until thirty days after being taken. In some cases, curative powers persist until the fiftieth day. Hahnemann also taught that seven-eighths of all chronic diseases were variations of psora, more commonly called the itch. This aspect of his views, however, was quickly discarded by his followers.[1]

Wrangling among homeopaths over the exact nature of the "homeopathic dose" soon split the movement into two factions—the purists who followed Hahnemann, and the "low potency" men who thought it of value to preserve at least *some* of the original compound, even though only a few molecules. Modern purists have discarded Hahnemann's "spiritual" effects for mysterious "radiations" which remain after the material substance has vanished, and which have a physical basis not yet understood. Just as the Law of Similia has an analogy in the vaccination principle, so the doctrine of the infinitesimal dose has a slight factual basis—but only in reference to a small number of drugs. The homeopathic error was to take both these limited truths, exaggerate them to the point of absurdity, and apply them universally to all medicines.

The *materia medica* of homeopathy is understandably much larger than that of "allopathy" (a homeopathic term, now obsolete, for orthodox medicine). About 3,000 distinctly different drugs have been "proved," and new ones are still being added. The Foundation for Homeopathic Research, headed by Dr. William Gutman, of New York City, recently "proved" that the metal cadmium, in a highly diluted dose, cured a certain type of severe migraine.

Some idea of the worth of homeopathic medicines may be gathered from the fact that one of them (no longer used) was called *lachryma filia*, and consisted of tears from a weeping young girl. Other curious remedies are made from such substances as powdered starfish *(asterias rubens)*, skunk secretion *(mephitis)*, crushed live bedbugs *(cimex lectularius)*, powdered anthracite coal, powdered oyster shells, and uric acid *(acidum uricum)* obtained from human urine or snake excrement. Most homeopathic medicines are obtained from plants, though in recent years there has been a trend toward proving metallic compounds. *Any* substance, organic or inorganic, is a potential homeopathic drug. A doctor announces that he has proved a new medicine, colleagues try it out, patients get well, and so a new remedy is added to the *materia medica*. Research by reliable pharmacologists has shown that all these weird drugs, in the diluted form in which they are given, are entirely harmless—producing neither symptoms nor cures (except, of course, psychosomatic ones).

As might be expected, however, millions of people took these infinitesimal doses of valueless drugs and were immensely benefited. Of course a few died, but then even allopaths can't save *all* their patients. The cult spread rapidly over Europe in the 1820's, reached England and America in the 1840's, and came to its pinnacle of success about 1880 in the United States. Emerson and William Cullen Bryant were believers, but Oliver Wendell Holmes, in his book *Homeopathy and Kindred Delusions,* 1842, delivered one of the earliest and most effective blasts against the movement. By 1900 there were twenty-two homeopathic colleges in the nation, an immense literature, and dozens of periodicals. A huge monument to Hahnemann was erected on Scott Circle, in Washington, D.C., where it stands to this day.

From 1900 on, the movement declined. One by one the American schools faded away. Some of the more prosperous—like New York Medical College, Manhattan, and the Hahnemann Medical College, Philadelphia—kept diluting the amount of homeopathy in the curriculum until they evolved into potent, first-rate medical schools. Today there are no homeopathic colleges in the States, though a few schools, like the two just cited, offer graduate courses in the subject. Several thousand doctors still consider themselves homeopaths, however, with Philadelphia leading in the number of practitioners. All these men have standard M.D.'s, and in matters of diagnosis, surgery, etc., make full use of orthodox medical science. It is only in the giving of drugs that they call upon the homeopathic tradition; though even here, especially in emergencies, they occasionally resort to allopathic remedies. The homeopathic drugs are obtained from special pharmacies which flourish in several large cities. *The Journal of the American Institute of Homeopathy,* published in Philadelphia, is the cult's leading periodical. The current issue at the time of writing (March, 1952) features an article on medicines made from various types of spider webs and poisons.

In the United States, homeopathy is still declining, although a number of prominent people, like Marlene Dietrich (who also believes in astrology), are enthusiastic patrons. In Europe, on the other hand, there has been a marked revival since the war, especially in Germany and France. In France it is in competition among faddists with a recent rebirth of interest in "acupuncture," the ancient Chinese art of curing ills by puncturing the body at various spots with gold and silver needles. Homeopathy continues to be respectable in England, where it is traditional for the Royal Family to maintain a homeopath as family physician, and where the Royal Homeopathic Hospital of London is one of the finest in the world. The cult has always been popular among the nobility of England and the Continent. India and South America are other areas where the movement is flourishing.

The Law of Similia has been distorted in many weird ways by quacks. Thirty years ago Dr. Loyal D. Rogers, of Chicago, widely advertised his methods under the slogan "without use of drugs or bugs." Rogers' medicine was obtained by "attenuating, hemolizing, incubating and potentizing" a few drops of the patient's blood. His book *Auto-hemic Therapy* was published in 1916. In New York City, Dr. Charles H. Duncan went a step further. His *Autotherapy,* 1918, explains how he cured boils by giving the patient an extract from the boil, tuberculosis by an extract from the sputum, dysentery by injecting a fluid obtained by filtering the excretions, and so on.

Naturopathy, like homeopathy, is a world-wide medical cult which had its origin in Europe. Unlike homeopathy, however, it has no single founder. It simply grew. In essence, it is a complete reliance on "nature" for healing. Medicine and surgery are used as little as possible or not at all. As might be expected, hundreds of strange methods of therapy clustered about the movement, so it is not easy to say exactly what the tenets of naturopathy are.

The earliest naturopaths were European doctors of the eighteenth and nineteenth centuries. Vincenz Priesnitz and Father Sebastian Kneipp were pioneers of hydrotherapy (water cures). Adolph Just's *Return to Nature* recommended sleeping on bare ground, walking barefooted on wet lawns and sand, and using clay compresses. Louis Kuhne's *The New Science of Healing* opposed all drugs, recommending instead the use of steam baths, sunlight, a vegetarian diet, and whole wheat bread. Heinrich Lahmann was against putting table salt on foods and drinking water at mealtimes. Antoine Bechamp defended the view that disease produces bacteria rather than the other way around.

An early pioneer of naturopathy in the United States was John H. Kellogg, a Seventh Day Adventist who founded the Battle Creek

Sanitarium.[2] He was responsible for the great importance nature therapy plays in present Adventist beliefs. Another American, Henry Lindlahr, made the "discovery" that disease, instead of being the result of invasion of the body by harmful microbes, was really the body's natural way of healing something. Finally, there was Benedict Lust, a disciple of Father Kneipp, who perhaps should be regarded as the most important early figure in American naturopathy. He established a school in New York, resort spots in Butler, New Jersey and Tangerine, Florida, wrote many books, edited several magazines (one of which, *Nature's Path,* in the hands of Lust's descendants is still going lustily), and managed to get himself arrested about sixteen times in battles against the "drug trusts." His advertisements often appeared in Bernarr Macfadden's health magazines.

Macfadden himself was a great promoter of naturopathy. His monumental five-volume *Encyclopedia of Physical Culture,* 1912 (subtitled *A work of reference, providing complete instructions for the cure of all diseases through physcultopathy),* is one of the greatest of all pseudo-medical works. Volume 4 contains 572 pages devoted to an alphabetical listing of all major diseases—including Bright's disease, polio, cancer, etc.—together with Macfadden's methods of home treatment. The treatments involve, in most cases, special diets, exercises, and water therapy. Cancer, for instance, is treated with a fast, followed by exercises and a "vitality building regimen." There is no suggestion that the patient should consult a physician. In fact there is a "Word of Warning" at the beginning of the section which states, "It positively must be remembered that the methods recommended in this work cannot be combined with the internal use of drugs or medicine. An attempt to use drugs while pursuing the treatments here advocated may lead to very serious results, and is to be depended upon under no circumstances."

In fairness to Macfadden it should be said that in later years he has become less extreme in his medical opinions. But not much so.[3] He is firmly convinced, for example, that cancer can be cured by a diet of nothing but grapes, and a few years ago offered $10,000 to anyone who could prove it wasn't so. (The theory that a grape diet can cure all kinds of ailments has long been popular in grape growing areas of Europe, and has a literature as extensive as the literature extolling the "virtues" of goat's milk.)

Hundreds of schools calling themselves naturopathic sprang up here and there in the early years of the century. They were as frowsy as could be imagined. Most of them consisted of a few rooms in a walk-up apartment, with classes at night, and gave handsomely engraved diplomas at the close of brief instruction periods. Sometimes several diplomas were given, bearing the names of different schools, all using the same premises and teaching the student simultaneously. When framed, they made an impressive-looking wall for the graduate's office. There was little unity of beliefs behind these schools. In addition to their strange diets, massages, and water cures, dozens of curious little crackpot movements found their way into the curriculum. We shall glance briefly at two of them—iridiagnosis and zone therapy.

Iridiagnosis is the diagnosis of ills from the appearance of the iris of the eye.[4] This great science was discovered by Ignatz Peczely, a Budapest doctor, who published a book about it in 1880. The art found an immediate response among homeopaths in Germany and Sweden, and was introduced into the United States by Henry E. Lahn, who wrote the first book in English on the subject in 1904. Naturopath Henry Lindlahr, a pupil of Lahn's, produced a definitive study, *Iridiagnosis and other Diagnostic Methods,* in 1917, although a few more recent works have appeared.

According to Lindlahr, Dr. Peczely discovered the new science at the age of ten when he caught an owl and accidentally broke the bird's foot. "Gazing straight into the owl's large, bright eyes," writes Lindlahr, "he noticed at the moment when the bone snapped, the appearance of a black spot in the lower central region of the iris, which area he later found to correspond to the location of the broken leg." Young Ignatz kept the owl as a pet. As the leg healed, the black spot developed a white border, indicating the formation of scar tissue in the bone.

According to iridiagnosticians, the iris is divided into about forty zones which run clockwise in one eye, counterclockwise in the other. The zones connect by nerve filaments to various parts of the body, much in the manner that chiropractic theory connects the body to parts of the spine. Spots on the iris are called "lesions." They indicate malfunctioning of the corresponding body part. J. Haskell Kritzer, in his *Textbook of Iridiagnosis,* fifth edition, 1921, carefully explains how to recognize artificial eyes, thus avoiding the embarrassment of basing a lengthy diagnosis on them.

Anyone who thinks no medical movement could be more insane than iridiagnosis, is much mistaken. Zone therapy is even worse. This point of view assumes that the body is divided vertically into exactly ten zones, five on each side of the body, and each zone terminating in an individual finger and toe. Exactly how the parts of each zone are connected is one of the mysteries of this cult, since the ten divisions completely ignore the nerve and blood vessel systems. Zone therapists suspected that some hitherto undiscovered submicroscopic network was involved.

Without going into detail, the zone therapists believed that almost every type of body pain could be checked, and in many cases the cause of the pain removed, by putting pressure on the proper finger or toe, or some other part of the affected zone. This pressure was applied by various means—chiefly rubber bands (worn on a finger or toe until it turned blue), spring clothes pins, or the teeth of a metal comb pressed into the flesh.

The inventor of zone therapy was Dr. William H. Fitzgerald, a graduate of the University of Vermont, and for many years the senior nose and throat surgeon of St. Francis Hospital, Hartford, Connecticut. His associate, Dr. Edwin F. Bowers, first introduced the new science in a series of popular articles in *Everybody's Magazine,* where they were enthusiastically endorsed by the editor, the late Bruce Barton. Later, in 1917, Fitzgerald and Bowers collaborated on a book titled *Zone Therapy.* Many subsequent books appeared by other authors, of which the most notable was *Zone Therapy,* by Benedict Lust, the father of American naturopathy.

Lust's book describes how to treat most of the common ills, including cancer, polio, and appendicitis. Goiter requires pressure on the first and second fingers, but if "the goiter is very extensive, reaching over into the fourth zone, it may be necessary to include the ring finger. . . ." Eye pains and disorders also call for pressure on first and second fingers, but deafness requires a squeezing of the ring finger or the third toe. "One of the most effective means of treating partial deafness," Lust writes, "is to clamp a spring clothes pin on the tip of the third finger, on the side involved in the ear trouble."

Nausea is relieved by pressing a metal comb against the backs of both hands, and childbirth is rendered painless if the mother clasps a comb in each hand so she can press the tips of all fingers against the teeth. "Also," Lust adds, "rubber bands around the great toe and the second toe afford a gratifying help." For dentists, zone therapy is invaluable. No need to use anesthetics—merely attach tight rubber bands to whatever fingers relate to the zone including the tooth, and the tooth becomes insensitive to pain.

For falling hair, Lust recommends a method which he correctly describes as "simplicity itself." It consists of "rubbing the fingernails of both hands briskly one against the other in a lateral motion, for three or four minutes at a time, at intervals throughout the day. This stimulates nutrition in all the zones, and brings about a better circulation in the entire body, which naturally is reflected in the circulation of the scalp itself."

The tongue, throat, and roof of the mouth also have the tenfold division, hence many therapeutic measures involve pressure in these regions. Headaches, for instance, are cured by pressing the thumb against the roof of the mouth. Menstruation difficulties are relieved by pressures on certain parts of the tongue. Whooping cough is cured in three to five minutes by pressing a certain spot on the back of the throat. "In an experiment with several hundred cases of whooping cough," wrote Fitzgerald and Bowers, "we have not yet seen a failure from the proper application of zone therapy."

It seems inconceivable that this cult would attract a following, nevertheless hundreds of naturopaths took it seriously and reported astonishing results. Books dealing with the art are filled with case histories of patients who found immediate relief from intense pain, and eventually a complete recovery from long-standing and serious ailments.

Today, the shrewder schools of naturopathy have abandoned iridiagnosis, zone therapy, and other wilder aspects of the movement, but all of them agree that the chief cause of disease is not bacteria from outside but a violation of natural laws of living. And all of them believe that drugs are harmful. Dr. Robert A. Wood, a Chicago naturopath and former president of the American Naturopathic Association of Illinois, expresses the creed as follows:

> Naturopaths do not use drugs of any kind, nor do they use inorganic substances that may injure the system. Instead, they rely on vitamins, minerals, chlorophyll, vegetable and fruit juices, raw cow's milk, and a balanced diet. With all diseases, the allopath suppresses but does not cure. His sole aim is to kill pain, to give temporary relief. He uses antipyretics to suppress all fevers— quinine, aspirin, salicylate, and other coal-tar products which suppress the fever, but leave the toxins which cause the fever inside the system. The naturopath aids the fever, however, or reduces it physiologically with natural methods—distilled water with lemon juice; lots of fresh, raw fruit juices; wrapping the body in hot, wet bath towels; and last, but not least, the greatest fever reducer of all—colonic irrigation."

Naturopaths are fond of using enemas to rid the body of poisons. Apparently they feel it is quite "natural" to insert a tube into the rectum and pour large quantities of water into the lower intestines. On the other hand, it is "unnatural" to take a drug which in many cases is merely a compound found in nature, but purified so its effects are stronger.

More than 85 per cent of all cases of appendicitis, according to Dr. Wood, can be relieved by fasting for a short period, then taking a cold-water enema every day for four days, followed by a special diet. The use of drugs to cure syphilis, he says, not only does not cure, but also causes locomotor ataxia. In an article in the *American Mercury,* May, 1950 (from which the above views and quotations are taken), he boasts of having cured a man of sixty-five who contracted syphilis at the age of sixteen and had not been treated since. "I used no mercury, nor any other allopathic 'guesswork' remedies," he states.

On another occasion, Dr. Wood went to work on a boy of five who suffered from tuberculosis of the pelvic bone. The disease had eaten two holes through the bone. "With the exception of the tuberculosis holes," Dr. Wood writes, "a slight spinal curvature, and the fact that one leg had become shorter than the other, he was in fairly good condition. His diet was given priority, with natural foods supplemented by natural calcium. Sitz baths, sun baths, cold sprays, hot packs, infra-red, vibrations, exercise, manipulations, colonic irrigations, and other natural remedies were used. The last X-rays showed the condition completely healed."

The naturopath's opposition to the bacterial theory of disease is, of course, shared by many religious groups. Christian Science, New Thought, and Unity head the list, not to forget Jewish Science, founded in 1922 by Rabbi Morris Lichtenstein, of New York City. He is the author of numerous works on the subject, including *Jewish Science and Health,* a fair imitation of Mrs. Eddy's famous text. Among prominent individuals who opposed the germ theory, George Bernard Shaw was perhaps the most notable. In one of his last books, *Everybody's Political What's What,* 1944, you will find a witty defense of naturopathy. Drugs merely suppress symptoms, he writes. The disease usually breaks out again unless the person is healthy enough to let nature cure him in spite of the drugs. Like Antoine Bechamp, a French chemist and contemporary of Pasteur, Shaw rejects the theory that diseases are caused by air-borne germs. Germs—according to Bechamp, Shaw, and most of the men mentioned in this chapter—are the *products* of disease. They develop within ailing body cells. Once developed, however, they are infectious. Shaw was convinced that most epidemics were traceable to laundries where microbes from a sick person's handkerchiefs and clothing would infect the clothes of others. As might be expected, Shaw was also a lifelong opponent of vaccination, vivisection, the eating of meat, Caeserean operations, and the removal of the tonsils and appendix.[5]

Eugene Debs, the famous Socialist labor leader, died in the Lindlahr Naturopathic Sanitorium, Elmhurst, Illinois. Morris Fishbein, in his *Fads and Quackery in Healing,* 1932 (from which much of this chapter is drawn), tells the tragic story. Debs had recently been released from prison, and finding himself ailing, had gone to the sanitorium for a rest. One day he visited Carl Sandburg, who lived nearby, and on his way back lapsed into unconsciousness. After two days of treatment at the sanitorium, Debs' brother asked Dr. Fishbein to check on the labor leader's condition. Fishbein found Debs in a coma, the pupil of one eye dilated and the other contracted—a fact not noticed by the staff, and which indicated a brain disturbance. His body was badly dehydrated. Being unconscious, he had not asked for a drink in two days, and so no one had given him one. He was suffering from malnutrition, having been on a fasting cure then being recommended by Bernarr Macfadden and Upton Sinclair. When Debs' heart began to falter, the "doctors" administered a nature remedy—totally worthless—made from cactus. After this failed, they tried electric treatment, badly burning Debs' skin. In final desperation, they made a crude attempt to inject digitalis, a drug which properly administered might have had beneficial effects. But Debs was beyond help and died the following day. His treatment was typical, Fishbein reports, of naturopathic methods.

It is difficult to say how many naturopaths are operating in the United States today—probably not more than a few thousand. Several health magazines follow the naturopathic line, and dozens of mail order pharmacies continue to supply "natural" remedies in the form of mineral salts, vitamin compounds, yeast foods, herb products, and so on. Many of these firms operate in a semi-undercover fashion, blaming their "persecution" on the American Medical Association and the "drug cartels."

It is equally difficult to estimate how much harm naturopaths do. Their opposition to vaccination, fortunately, has not influenced enough people to prevent the astounding health gains it has made in recent years. Another decade, and smallpox, diphtheria, and whooping cough may be completely wiped out as native diseases. Tens of thousands used to die annually of these preventable scourges. The return to raw milk would bring with it new epidemics of scarlet fever, typhoid, tuberculosis, and other disease against which pasteurization has made unbelievable strides. Rejection of sulfa drugs and penicillin (which attack germs, not symptoms) has probably caused the deaths of untold numbers of naturopathic patients for whom colonic irrigations caused only harm.

Perhaps the best insight into the medical knowledge of a naturopath can be gained from the following statement by Dr. Wood in a letter to the *American Mercury,* August, 1950: "If atmospheric bacteria bring about disease as claimed by the medical profession, then why is it millions of Indians . . . bathe daily in the filthy Ganges river, a river teeming with billions of germs? . . . To my knowledge, there has never been a serious epidemic outbreak of any disease." To which Dr. Joseph Wassersug politely replied by pointing out that the death rate from infectious diseases in India is higher than almost anywhere else in the world, and that deadly epidemics of cholera, in some cases great enough to become world-wide, have been directly traced to bathing festivals in the Ganges. There is no indication, however, that this newly gained knowledge of India has induced naturopaths to stop flushing colons and pocketbooks.

A number of cults have been built around an exaggerated notion of the effects of bad posture and poor muscular co-ordination on health, and although they are not strictly part of the naturopathy movement, this is an appropriate spot to mention them. The late philosopher John Dewey, who lived to the age of ninety-two, always attributed his longevity to his practice of the highly dubious theories of a self-educated Australian named Frederick Matthias Alexander. Alexander was a professional reciter until his voice mysteriously failed him. After orthodox doctoring failed to help, he began to experiment on himself with the result that he regained his voice, and found he had developed an elaborate tecnique of sensory and muscular training. In 1904 he established a school in London, where he won many distinguished converts, including Aldous Huxley and the late Sir Stafford Cripps. The school was moved to Stow, Massachusetts, at the outbreak of the last world war, then back to England a few years later. At present, Alexander directs a school near Bexley, in Kent. He has written four books explaining his methods, three of which contain enthusiastic introductions by Dewey.[6] A German system of posture and exercise, developed by Bess M. Mensendieck, also acquired a considerable vogue in this country during the thirties (see her work, *The Mensendieck System of Functional Exercises,* 1937, as well as the similar theories of Mabel E. Todd, *The Thinking Body,* 1937).

The belief that all sorts of human ills spring from an inability to relax properly also has a large literature and following. Edmund Jacobson's *Progressive Relaxation,* 1929, is the most important reference to this point of view. It has been revived recently by a reissue of a more popular work of Jacobson's, *You Must Relax,* 1934, and by Dr. David H. Fink's *Release From Nervous Tension,* 1943. Thirty years ago a Chicago doctor named E. H. Pratt decided that an inability to relax certain openings of the body caused many common ailments. His practice of what he called "orificial therapy" apparently brought relief to thousands.

Osteopathy, America's third great medical cult, was the brain-child of a medical illiterate named Andrew Taylor Still. Very little is known about Still's life in spite of the fact that he published in his old age, at his own expense, a lengthy autobiography. It is a rambling, contradictory, egomaniacal work—one of the unintentionally most amusing autobiographies ever penned by a self-declared genius. It is difficult to understand how any osteopath with a sense of humor could read this book and remain in his profession. Almost all the biographical data has since been found false, although most osteopaths continue to accept the book as a valid account of the author's career.

There is no evidence Still had any early medical education other than helping his father, a Methodist missionary, take care of ailing Shawnee Indians. He claimed to have been a cavalry officer for the northern army during the Civil War. A picture in his book shows him riding his faithful mule, brandishing a sword, and leading a battle charge on one occasion when a musket ball passed through his vest lapels. The picture is captioned *Osteopathy in danger*. Six illustrations are devoted to a dream allegory in which the author is severely butted by the *Ram of Reason*. His osteopathic school had been losing money and the ram was trying to butt some business sense into him. Another picture, in full color (when pseudo-scientists publish a book they spare no expense), shows a peacock with tail feathers spread. It is titled *Professor Peacock*. The text explains how we can learn from the bird's feathers how God governs the body.

After a ten year gap in his later life, about which Still says nothing, he writes that "In June, 1874, I flung to the breeze the banner of osteopathy." The Swedish massage movement was then popular, and it is likely that this gave him his basic idea, although Still attributes it to divine inspiration. The word osteopathy, which he coined, means literally "sick bones." (He also invented the word "diplomate" for a graduate of his school, but osteopaths have not cared for the word.) It was Still's theory that diseases are caused by a malfunctioning of the nerves or blood supply, in turn chiefly due to the dislocation of small bones in the spine. These dislocations are called "subluxations of the vertebrae." Their pressure on nerves and blood vessels prevents the body from manufacturing its own curative agents. The osteopath's job is to find these subluxations and "adjust" them, though what makes them stay adjusted remains an osteopathic mystery. Actually, subluxations are as elusive as the canals on Mars. Only believers are able to locate them. Other doctors consider them entirely mythical. In extremely rare cases the small bones of the spine may get out of place, but if they do, they cause almost none of the symptoms or ailments osteopaths say they cause.

In his autobiography Still records some astonishing cures brought about by spinal rubbing. In one case he grew hair three inches long, in a week, on a head completely bald. On another occasion, in Hannibal, Missouri, he "set 17 dislocated hips in one day." The reader is not told how it happened that Hannibal, which had a population then of 7,000, had so many dislocated hips.

It was not until 1894 when Still, then sixty-six, founded his first school, at Kirksville, Missouri. At eighty-two he wrote the school's first textbook. In it he reveals the methods of spinal manipulation by which he had been able to "cure" yellow fever, malaria, diphtheria, rickets, piles, diabetes, dandruff, constipation, and obesity. He never admitted the existence of germs. "I pay no attention to laboratory stories of microorganisms," he wrote. "We have but little time to spare in analyzing urine, blood lymph, or any other fluid substance of the body because we think life is too precious to dilly-dally in laboratory work. . . ."

It was Still's considered opinion that he was "possibly the best anatomist now living." One of the most fascinating pictures in his autobiography is one of Still seated under a tree in his back yard, studying an anatomical chart and dictating to his wife. The bones of a human arm are draped over one leg. Other parts of the skeleton are leaning against the tree and hanging from the trunk.

Much of the book is devoted to religious speculation, Still having been a convert to the Millerite movement (forerunner of Seventh Day Adventism), although he later went back to his original faith of Methodism.

Osteopathy made little headway abroad, but in the States it quickly blanketed the continent. It seems that every graduate of Still's school wanted to open up his own college. Today there are about 11,000 practitioners. Outside of Los Angeles, where the concentration is greatest, they flourish best in small cities. As Dr. Fishbein points out, in a large city they meet too much resistance from a well-established medical profession, and in the small town they lack the knowledge to take care of emergencies. In small cities, however, they can operate quietly and catch patients with minor ills and psychosomatic ailments who perhaps benefit from the massage. Actually, a brisk walk or a warm bath would be even better for the circulation, but the back rub feels pleasant—especially for patients with repressed sexual longings (or homosexual if the practitioner is of the same sex).

Since Still's day there has been considerable evolution of osteopathic doctrines as its doctors try, in Fishbein's phrase, to enter orthodox medicine by the back door. They now massage other parts of the body than the spine, give water and electrical therapy, administer anesthetics, prescribe drugs, and even perform elementary surgery, like yanking tonsils or delivering babies. In recent years they have turned their attention toward mental ills. One osteopathic study has shown that schizophrenia is due to subluxations in the upper bones of the neck! The viewpoint of the more enlightened present-day osteopath is that medicine has a function, but does only half the job. Manipulation is still needed to restore the body's "structural integrity." In the six so-called "accredited" schools of osteopathy there are now four-year courses in basic medicine, and all but eight states permit the graduates to give drugs and perform surgery.

Chiropractic—the fourth and greatest of American medical follies—arose about twenty years after osteopathy, from which it borrowed heavily. Like the original osteopaths, its practitioners concentrate on adjusting spinal subluxations which they believe cause most human ills, and they share with the naturopaths a rejection of drugs and the germ theory of disease. Unlike osteopathy, it has shown small tendency to change. Today its some 20,000 practitioners (almost all of them in the United States) are about as scientific in their methods as the Chinese and French acupuncturists.

The founder—Daniel D. Palmer—was originally a grocer and fish peddler in Davenport, Iowa. In 1895 he discovered he could cure people by "animal magnetism," so he closed his store and for ten years practiced magnetic healing. Then one day he made an even more momentous "discovery." Here is an account of the event, as told by his son B. J. Palmer in a Wisconsin court:

> Harvey Lillard came in thoroughly deaf. Father looked him over, and there was a great subluxation of the back. Harvey said he became deaf within two minutes after that popping occurred in his spine, and had been deaf for seventeen years. Father thought of this thing, which was that if something went wrong in the back and caused deafness, then reduction of this subluxation should cure it. The bump was adjusted, and within ten minutes Harvey regained his hearing.

That was "B. J.'s" account. Those who have studied the history of chiropractic, however, suspect it is legendary, and that "D. D." probably picked up the notion of subluxations from the osteopaths. His textbook, *The Science, Art, and Philosophy of Chiropractic*, 1910, contains little original speculation. At any rate, there is no question that it was his son, B. J., who developed chiropractic into the thriving business it is today.

B. J., like many chiropractors, ended his formal education with the ninth grade. His school at Davenport, founded in 1895, originally gave only a two-week course. At present its period of training is eighteen months. Hundreds of other schools are scattered over the United States, differing widely in theories and methods. For example, B. J. taught that diphtheria came about by the subluxation of the sixth dorsal vertebra. But a textbook used by a college in Chicago says that diphtheria must be treated by manipulating the third, fifth, and seventh cervical vertebra, the first, second, third, fourth, fifth, seventh, tenth, and twelfth dorsals, and the fifth, ninth, tenth, and eleventh cranial nerves. The same textbook says that in treating scarlet fever "particular attention must be given to the second to fifth cervicals and the tenth to twelfth dorsals." According to B. J., scarlet fever is due to subluxations between the sixth and twelfth dorsals.

If you are a devotee of chiropractic, here is a simple test you can make. Call on a practitioner, name a few symptoms, memorize carefully the exact spots where he finds subluxations, then leave on some pretext before he gives you a treatment. Go to a second chiropractor, name a different set of symptoms, and see if he finds the same spinal spots in need of adjustment. If he doesn't, try a third chiropractor, and continue until you obtain duplicate diagnoses. This may be an expensive experiment, but it should prove illuminating.

The current Manhattan *Red Book* lists over 200 chiropractors (they are not licensed in New York) to about 80 osteopaths. One advertisement shows the familiar spinal chart with parts labeled to indicate the areas of the body they control. These charts have about the same relation to modern anatomy as a phrenological chart of head bumps bears to modern brain study. Palmer graduates usually identify themselves as such in the ads, and call attention to their neurocalometer service.

The neurocalometer is one of B. J.'s most lucrative inventions. It is an electrical device which allegedly makes a spinal diagnosis. The National Chiropractic Association, the cult's leading organization, does not think much, however, of either Palmer or the neurocalometer. The N.C.A. publishes a magazine with a much larger circulation than B. J.'s, and they have not included the Palmer school on their list of "accredited" colleges.

Another chiropractic invention fills an entire room at the Palmer clinic. I am not sure exactly what it does, but its name is magnificent. It is called an electroencephaloneuromentimpograph.

CHAPTER 17

MEDICAL QUACKS

INDIVIDUAL MEDICAL quacks in the United States, who have not founded a "school," but nevertheless achieved a wide following, are legion. It would be impossible to survey completely even the major ones—such a task would require a lifetime of research and many volumes. A few men, however, stand out as particularly unusual or amusing, and it is with these individuals that this and the following two chapters will be concerned.

America's first great quack was Dr. Elisha Perkins (1740-1799). The doctor had a theory that metals draw diseases out of the body, and in 1796 patented a device consisting of two rods, each three inches long. One rod was supposed to be an alloy of copper, zinc, and gold; the other—iron, silver, and platinum. By drawing "Perkins' Patented Metallic Tractor" downward over the ailing part, the disease was yanked out.

Perkins sold his tractors for five guineas each to such notables as George Washington, whose entire family used it, and Chief Justice Oliver Ellsworth. His son, Benjamin D. Perkins (Yale, class of '94), made a fortune selling the tractors in England. In Copenhagen, twelve doctors published a learned volume defending "Perkinism." Benjamin himself wrote a book about it in 1796, containing hundreds of stirring testimonials by well-educated people. They included doctors, ministers, university professors, and members of Congress. Most historians of the subject think the old man actually believed in his tractors, but that the son—who retired in New York City as a wealthy man—was simply a crook promoter.

It is worth noting that orthodox medical opinion, by and large, ignored

Perkinism, regarding it as not worthy of serious refutation. One doctor, however, did trouble to make some tests with phony tractors. They looked like the genuine article, but actually were nonmetallic. His results, of course, were excellent. Oliver Wendell Holmes, in an amusing discussion of Perkinism, relates that one woman was quickly cured of pains in her arm and shoulder by using a fake tractor made of wood. "Bless me!" the woman exclaimed. "Why, who could have thought it, that them little things could pull the pain from one!"

Our next quack can best be introduced by the following quotation from a chapter on medical charlatans, in a book published fifty years ago. "The physician is only allowed to think he knows it all, but the quack, ungoverned by conscience, is permitted to know he knows it all; and with a fertile mental field for humbuggery, truth can never successfully compete with untruth." Who is the author of these wise sentences? Ironically, they were written by Dr. Albert Abrams, of San Francisco—a man who eventually became one of the most fantastic quacks in history!

Dr. Abrams was a distinguished-looking man. He had a dark Vandyke beard and sported a pince-nez attached to a long black ribbon. His early medical career was quite orthodox. After obtaining a degree at Heidelberg in 1882, he returned to California where he practiced general medicine, held important medical posts, and wrote a dozen reputable textbooks. In 1909 and 1910 he published two works which suggested he was venturing into uncharted waters. They dealt with methods of diagnosis by means of rapid percussion (tapping) on the spine. It was not long until Abrams discovered it was even better to tap the abdomen. His theory was that every ailment had its own "vibratory rate," hence the sounds produced by the tapping were clues to the person's condition.

Dr. Abrams' first invention was a diagnosing machine called a "dynamizer." It was a box containing an insane jungle of wires. One wire ran to an electrical source, and another was attached to the forehead of a *healthy* person. A drop of blood was obtained from the patient, on a piece of filter paper, and placed inside the box. Abrams would then percuss (tap) the abdomen of a healthy person, who was stripped to the waist and always—for a reason never made too clear—facing *west*. By listening to the sounds, the doctor was able to diagnose the ills of the patient who provided the blood sample!

This makes more sense than one might at first imagine. The spine, one must understand, has nerve fibers which "vibrate" at different rates. The dynamizer picked up "vibrations" from the blood, transmitted them to the healthy person's spine, which in turn sorted out the different wave lengths and sent them to various parts of the abdomen where they were detected by the doctor's expert rapping.

In addition to determining the nature of the blood donor's illness, Dr.

Abrams could ascertain the exact part of the body where the ailment was localized, and its severity. Later he discovered he could also determine the patient's age, sex, and whether he belonged to one of six religious groups—Catholic, Protestant, Jewish, Seventh Day Adventist, Methodist, and Theosophist. Eventually, he found he could diagnose from a sample of handwriting as well as blood.

The handwriting angle opened up fascinating possibilities. Abrams began experimenting with signatures of people no longer living. His disciples blandly accepted the discovery that Samuel Johnson, Poe, Wilde, and Pepys all suffered from syphilis, but found it hard to believe this same diagnosis when the doctor obtained it from the signature of Henry Wadsworth Longfellow.

In 1920 Dr. Abrams announced the completion of a new invention called an "oscilloclast." It made use of vibratory rates for healing. "Specific drugs," Abrams declared, "must have the same vibratory rate as the diseases against which they are effective. That is why they cure." But why use drugs? All you need do is direct the proper radio waves toward the patient and kill the bacteria even more effectively. Abrams also invented a "reflexophone" for diagnosing via telephone, and several other ingenious electrical devices. He ran a school to teach his disciples how to use these machines, published a periodical, and lectured widely throughout the country.[1]

When the doctor died, in 1923, he left a two-million-dollar estate. The money had been made chiefly by leasing out oscilloclasts (they were never sold) for $250, with an additional $200 for a course on how to operate them. Hundreds of lesser quacks rented the machines, bringing Abrams an estimated $1500 every month. The doctors were forced to swear they would never look inside the tightly sealed box. Shortly before the doctor's death, however, a committee of scientists opened one of the magic boxes and issued a report on what they found. It contained an ohm-meter, rheostat, condenser, and other electrical gadgets all wired together without rhyme or reason.

One would think no one in his right mind would fall for such nonsense; nevertheless many intelligent people did. The most distinguished convert was Upton Sinclair, who wrote many magazine pieces praising the doctor. In his *Book of Life*, 1921, Sinclair gave an enthusiastic description of Abrams' diagnosing machine, and added, "So is opened to our eyes a wonderful vision of a new race, purified and made fit for life. . . . Take my advice, whoever you may be that are suffering, and find out about this new work and help make it known to the world."

When the *Journal of the American Medical Association* attacked Dr. Abrams, Sinclair replied angrily: "He has made the most revolutionary discovery of this or any other age. I venture to stake whatever reputation

I ever hope to have that he has discovered the great secret of the diagnosis and cure of all major diseases. He has proved it by diagnosing with taps of his own sensitive fingertips over fifteen thousand people, and my investigation convinces me he has cured over ninety-five percent.''

Using the name of Miss Bell, the doubting A.M.A. sent to an Abrams practitioner in Albuquerque a blood specimen of a healthy male guinea pig. They received a report saying that Miss Bell had "cancer to the amount of six ohms," an infection of the left frontal sinus, and a streptococcic infection of her left fallopian tube. A Michigan doctor sent Abrams himself a blood sample from a Plymouth Rock rooster, obtaining a diagnosis of malaria, cancer, diabetes, and two venereal diseases.

The 1926 revised edition of the *Book of Life,* published after Abrams' death, is Sinclair's final word on the subject. The total failure of the doctor to make good, when subjected to tests by the medical profession, does not worry Sinclair in the least. When Abrams was perfecting his devices, Sinclair points out, he had the air-waves all to himself. But by the time the challenges came, the air was filled with the "complex vibrations of I know not how many radio stations." Naturally this interference played havoc with his machines, and so, as Sinclair puts it, "the old man died, literally, of his bewilderment and chagrin."

To those who accuse Abrams of deliberate deception, Sinclair writes: "The idea that Albert Abrams was a conscious fraud I consider out of the question. I have known many scientists, but never one . . . more passionately convinced of the truth of his teachings. Abrams worked all day and most of the night with never a rest; he literally killed himself in this way. His books are a mine of strange and suggestive ideas, and now that he is gone, hardly a week passes that I do not come upon a record of some new discovery . . . that causes me to say: 'There is Abrams again!' . . . Men said that Albert Abrams was 'insane'; but I predict that when the future comes to trace the leaps that his mind took, it will see that he had a reason for every one.''

One could not ask for a more clinically perfect statement of the persistence of irrational belief on the part of a convert to a totally worthless set of theories hatched in the brain of a brilliant paranoid.[2] It is surprising that Sinclair did not attribute Abrams' successes to faith healing. In 1914 Sinclair did a series of articles for *Hearst's Magazine* on his experiments in this art, and later wrote: ". . . If you will lay your hands upon a sick person, forming a vivid mental picture of the bodily changes you desire, and concentrating the power of your will upon them, you may be surprised by the results, especially if you possess anything in the way of psychic gifts.'' But radio had just been developed, and it was as easy to imagine that radio waves might have some sort of therapeutic effect as it was easy for Perkins' followers to suppose that his tractors had some connection with the newly discovered electrical currents.

Since Abrams' day, hundreds of similar electrical devices have reaped fortunes for their inventors. In Los Angeles, for example, Dr. Ruth B. Drown is currently operating an Abrams-type machine which diagnoses ailments from the "vibrations" of blood samples. She keeps a huge file of blotting papers on which are preserved samples of the blood of all her patients. By placing a sample in another machine, she can tune the device to the patient, then broadcast healing rays to him while he remains at home! An issue of her periodical, *Journal of the Drown Radio Therapy*, has a picture of her "Broadcasting Room," showing dozens of dials around the walls, by means of which dozens of patients can be treated simultaneously regardless of where they are at the moment. When Tyrone Power and his wife had an auto accident in Italy a few years ago, they were treated in Italy by short-wave therapy from Drown in California. Eventually, of course, Dr. Drown makes a more material contact with the patient. She sends a bill.

Exactly how the Drown devices operate is none too clear, but you can read about them in the doctor's books—*The Science and Philosophy of the Drown Radio Therapy*, 1938, and *The Theory and Technique of the Drown Radio Therapy and Radio Vision Instruments*, 1939. These works also explain two other Drown machines—one for taking "radio photographs" of body organs, and another which uses radio waves to stop bleeding. A more recent book by the doctor, published in 1946, is titled *Wisdom from Atlantis*.

Mrs. Drown—a handsome, mannish-looking woman—acquired her knowledge of electronics by working in the electrical assembly department of Southern California Edison Company. Her first invention was made in 1929. She has been practicing radio therapy ever since and selling her machines to chiropractors, osteopaths, and naturopaths all over the nation. Mrs. Drown herself is a licensed osteopath, and a member of the American Naturopathic Association.

Several years ago a number of prominent Chicagoans were so impressed by Mrs. Drown's work that they persuaded the Dean of the Biological Sciences Division, University of Chicago, to conduct a careful investigation. Dr. Drown personally operated her machines during the test, and with spectacular lack of success. (See the *Journal of the American Medical Association*, Feb. 18, 1950, for details.) She was given blood specimens from ten patients. Her diagnoses of the first three were so erroneous that she did not even attempt the remaining seven. Here is her report on a patient who had tuberculosis: ". . . a type IV cancer of the left breast with spread to ovaries, uterus, pancreas, gallbladder, spleen and kidney." In addition, the patient was diagnosed as blind in her right eye, blood pressure of 107/71, ovaries not producing ova, and the following organs not functioning properly—pancreas, adrenal, pituitary, uterus, right ovary, parathyroid, spleen, heart, liver, gall bladder, kidneys, lungs, stomach, spinal nerves, intestines, and ears.

The committee investigating Dr. Drown summed up her diagnosing methods as follows: "It is our belief that her alleged successes rest solely on the noncritical attitude of her followers. Her technic is to find so much trouble in so many organs that usually she can say 'I told you so' when she registers an occasional lucky positive guess. In these particular tests, even this luck deserted her."

Concerning Mrs. Drown's "radio photography," the committee said: "We find that the film images which have intrigued Mrs. Drown and her disciples are simple fog patterns produced by exposure of the film to white light before it has been fixed adequately. These images are significantly identical regardless of whether or not the film is placed in Mrs. Drown's machine before being submitted to the highly unorthodox processing which has been devised by her. In the numerous old films shown us by Mrs. Drown we can see no resemblance to the anatomical structures, appliances, bacteria, etc., that Mrs. Drown professes to see. In short, it is our opinion that the so-called Drown radio photographs are mere artifacts and totally without clinical value."

In a final test, Mrs. Drown attempted to stop, by means of radio waves, an anesthetized laboratory animal from bleeding. The animal bled until, as the committee reports, "her friends found the sight beyond their capacities." Nevertheless, Mrs. Drown continues to have thousands of followers. Her machines are particularly popular with spine thumpers in Chicago, where several osteopaths are currently using them.

Another popular form of quack radiation therapy is based on the "discovery" 35 years ago, by an obscure doctor named Abbott E. Kay, of a mysterious substance which he called "vrilium." The name comes from "vril," the cosmic energy used by the super-race in Bulwer-Lytton's utopian fantasy, *The Coming Race*, 1871. Madame Blavatsky often wrote about vril, which she said had been mastered by the Atlanteans, and was the motivating power of John Keely's perpetual motion machine.

During the twenties Robert T. Nelson, a businessman, sold a small brass cylinder, about two inches long, which was supposed to contain vrilium. You pinned it to your lapel or wore it around the neck. It radiated for a distance of twenty feet, Nelson claimed, keeping off bacteria and killing germs inside the body. After Nelson died, his son, Robert Nelson, Jr., developed the manufacture of these "magic spikes," as they were called, into big business—selling them for as much as $300 each. A number of political bigwigs in Chicago, including former Mayor Kelly, wore the cylinders. "I don't pretend to know how it works,"

Kelly told newsmen, "but it relieves pain. It has helped me and my wife." When the government took action against Vrilium Products Co. in 1950, it revealed that the cylinders contained nothing but a cheap rat poison. "I believe we have an *unrecognized* form of radioactivity," Nelson said when the government demonstrated that the cylinders had no effect on Geiger counters.[3]

Currently operating another remarkable device is Dr. Fred Urbuteit, a Florida naturopath. He has a Sinuothermic Machine which shoots a mild electrical current into the body to perform miraculous cures of incurable diseases. The machine sells in various models at prices from $1500 to $3000. Although arthritis is an ailment the machine is supposed to help, Dr. Urbuteit is confined to a wheel chair by this affiiction—a fact which has no effect on the intense loyalty of patients at the doctor's Sinuothermic Institute in Tampa. "Professor" William Estep, author of a 742-page medical treatise called *Eternal Wisdom and Health* (published by himself in 1932), is the inventor of still another therapy machine and has been in and out of southern jails for years.[4] The device is called an Estemeter, but I have been unable to find out how it operates or what it does.

Perhaps the greatest quack of them all is eighty-year-old Colonel Dinshah Pestanji Framji Ghadiali, of Malaga, New Jersey, who has been treating people for thirty years with colored lights. The colonel is a white-goateed, bespectacled little old man who was born in Bombay in 1873. He came to the United States in 1911, and became a citizen in 1917. During World War I he served without pay as commander of the New York Police Reserve Air Service—a civilian organization of pilots organized to protect New York harbor. This is where the "colonel" in his title comes from. The government cracked down on him a few years ago, with fines totaling $20,000, and a suspended three-year prison sentence.

In 1920 Ghadiali "discovered" Spectro-Chrome Therapy. In theory it is exceedingly simple. Every ailment can be cured by proper diet plus colored lights of the right "tonation." All you need is a Spectro-Chrome machine, built by Ghadiali. The machine contains a strong light source over which you slide the appropriate panes of colored glass. If you are diabetic, for example, you eat lots of starches and brown sugar, and bathe the body alternately with yellow and magenta light. Early stages of gonorrhea require green or blue-green rays. Scarlet increases sexual desires. Purple dampens them. In addition to special diets for each ailment, one must also abstain from tobacco, alcohol, meat, tea, coffee— and sleep with the head pointing north. All this and much more is explained in fascinating detail in Ghadiali's three-volume *Spectro-Chrome Metry Encyclopedia,* in shorter books, and in his monthly magazine, *Spectro-Chrome.*

In 1925, when Ghadiali was on a lecture tour, he was arrested in

Seattle and sentenced under the Mann Act to five years in the Atlanta Penitentiary. He later published a two-volume work, *Railroading a Citizen,* in which he blamed this unjust "persecution" on the medical trusts, the KKK, Catholics, Negroes, Henry Ford, the Department of Justice, and Great Britain. The book reprints the more sensational parts of the trial in which his teen-age secretary accuses him of rape, forcing her into "unnatural practices," and later performing an abortion. Ghadiali's purpose in reprinting this testimony is to allow himself a chance to interject comments accusing the girl of lying. Unfortunately, the impression left on the reader is that the girl was telling a straightforward story.

Since 1924 Ghadiali's Spectro-Chrome Institute has been located on a fifty-acre estate at Malaga, New Jersey. Signs on the fences read OUR AIM: SPECTRO-CHROME IN EVERY HOME. More than 10,000 people bought memberships in the Institute for $90, which included a lease on a Spectro-Chrome, plus a Favoroscope which showed the best time of day to use the machine. In addition, the new member paid $250 for two weeks of study at Malaga. Photographs in Ghadiali's books, showing him at work in his Malaga laboratory, are indistinguishable from stills of a grade D movie about a mad scientist.

During his recent trial, Ghadiali had no trouble at all finding 112 witnesses to testify they had been miraculously healed by colored lights. One witness, an epileptic, had a courtroom fit just after shouting, "I tell you I had fits all my life till Doctor Ghadiali cured me!" The government introduced relatives of many patients who died under color therapy. A son told how Ghadiali had advised his father, a diabetic, to stop taking insulin and give himself doses of colored light. The father lived three weeks.[5]

Ghadiali is not alone in advocating color therapy. It has a long and confused history, chiefly in occult traditions. In America in 1861, a General Augustus J. Pleasanton became convinced that sunlight shining through blue glass had curative properties. His work, *The Influence of the Blue Ray of the Sunlight and the Blue Color of the Sky,* was printed in 1871 on blue paper. It was followed by a book titled *Blue and Red Light,* by Dr. Seth Pancoast of Philadelphia, printed in blue letters on white paper, with a red border on each page. Dr. Pancoast thought blue good for some ills, red for others. During the 1870's these speculations led to a minor "blue glass craze" among New Englanders.[6]

In more recent times, the "I am" cult, founded by the Ballards, has placed a large emphasis on the spiritual and bodily effects of color. Dr. George Starr White, a Los Angeles homeopath and occultist, has long been recommending "Rithmo-Duo-Color Therapy" in his many books, as well as "biodynamochromatic diagnosis" in which the patient's abdomen is bathed with colored light, then thumped in the manner of Dr.

Abrams. (See Macfadden's *Physical Culture* magazine, February, 1918, for an article on this system. It is by Edwin F. Bowers, who co-authored the work on zone therapy cited in the previous chapter.) The Rainbow Lamps of Dr. Charles Littlefield were discussed earlier in the book. In England, some operators of pendulums are combining radiesthesia with color therapy. The pendulum is used in various ways to determine what color the body needs. "Color healing is a science of the future," writes Bruce Copen, in *The Pendulum*, January, 1952, "and coupled with radiesthesia it becomes a wonderful healer because the effects of any one color can be constantly checked for reaction."[7]

Some idea of the difficulty in obtaining convictions against quacks is revealed by the government's failure, after two sensatioal trials in 1943 and 1946, to convict Dr. William F. Koch (he pronounces it "Coke") of Detroit. Dr. Koch has the reputation of being the best educated and most successful cancer quack in the nation's history. Born in Detroit in 1885, he was graduated in 1909 at the University of Michigan, where he later (1917) obtained a Ph.D. degree in chemistry. In 1918 he received a medical degree from the Detroit College of Medicine, Wayne University. From 1910 to 1913 he taught histology and embryology at the University of Michigan, and from 1914 to 1919 was professor of physiology at Detroit Medical College. His two books, *Cancer and Its Allied Diseases,* 1929 (revised in 1933), and *The Chemistry of Natural Immunity,* 1938, are among the best counterfeits of sound medical writing in the entire annals of pseudo-science.

In 1919, Dr. Koch first announced his "discovery" of a cure-all which he calls "glyoxylide." It is, he claims, a catalyst synthesized by obscure methods, and when injected into a person suffering from *any known disease,* including cancer, tuberculosis, and leprosy, there is complete recovery in over 80 per cent of the cases. The catalyst does not attack the disease directly, Koch explains. It merely raises the body to such a frenzy of health that the body manufactures its own remedies.

Usually one injection is all that is given. Dr. Wendell G. Hendricks, a California osteopath who claims to have treated 3,000 patients with the Koch method, likens the injection to the starter button on a car. "The Koch catalysts may be considered to be the starter," he writes in a pamphlet, "and after getting the recovery process started, all that is necessary to keep it going is to provide the body with fuel. . . ." In some cases a second injection is given, and in difficult cases, like cancer, several may be administered. For decades, doctors using the Koch cure have demanded $300 and up for a single injection. During the late forties, Koch was charging $25 for a two-cubic-centimeter ampoule of glyoxylide, bringing him an estimated income of $100,000 annually.

There is not the slightest doubt about the complete worthlessness of the

"Koch treatment." Government chemists testified in 1943 that Koch's glyoxylide was indistinguishable from distilled water. Yet scores of "doctors" throughout the country, most of them osteopaths and chiropractors, are still giving Koch injections. In 1949, Senator William Langer of North Dakota actually placed in the *Congressional Record* a report of the alleged success of Koch injections in curing ailing cows! The report was promptly reprinted by Koch and widely distributed.

In recent years Koch has adopted a Protestant fundamentalist front for his activities. The Koch Cancer Foundation has been abandoned for the Christian Medical Research League, Detroit, which currently supplies the catalysts. Gerald B. Winrod, a fundamentalst rabble-rouser in Wichita, Kansas, who was one of the Nazi seditionists on trial during the last war, has been promoting Koch vigorously in his hate-sheet, *The Defender*. In 1950, Winrod published a book called *The New Science in the Treatment of Disease,* which hails Koch as a great medical genius and compares his "persecution" by the American Medical Association to the persecution of Semmelweis. A Detroit organization called The Lutheran Research Society (no connection with the Lutheran church) has also been issuing literature promoting Koch, including a magazine called *The Eleventh Hour*. The magazine is also against communism and the Jews.

The government dismissed its case against Koch in 1948 after failing to secure conviction in its two trials. The Federal Trade Commission did, however, manage to secure a temporary injunction against him in 1942 which still prevents him from advertising his drug as a cure. At present he is operating in Rio de Janeiro, and indications are that South America has proved as lucrative a pasture as the northern continent.

This chapter would not be complete without some mention of the occult areas of medical quackery. The field is much too large even to summarize, but a few individuals stand out as being of special interest.

A recent work by British surgeon Kenneth Walker, *Venture with Ideas,* 1951, has revived interest in George Ivanovitch Gurdjieff—a bald, walrus-mustached, Russian-born Greek whom *Time* once characterized as a "remarkable blend of P. T. Barnum, Rasputin, Freud, Groucho Marx, and everybody's grandfather." His only published work is an allegory about Beelzebub, modestly titled *All and Everything,* and almost as unreadable as Madame Blavatsky's writings. Gurdjieff's Institute for the Harmonious Development of Man, near Paris, attracted hundreds of intelligent disciples during the twenties, including the British writer Katherine Mansfield, who died while there. Another enthusiast was Margaret Anderson, the American editor of an *avant garde* magazine. Her recently published autobiography, *The Fiery Fountains,* gives a valuable picture of the Gurdjieff movement.

Gurdjieff's medical views are hard to pin down. They seem to be a blend of Yogi and other occult systems, with some original material mixed in. His therapy included tree chopping and complicated dance exercises (at one time he directed an Oriental ballet troup in Moscow) accompanied by tunes written by Gurdjieff. Mansfield's death may have been hastened by her own "cure," which involved living in a cowloft where she could breathe air exhaled by cows.

Gurdjieff's most active disciple was Peter D. Ouspensky, who founded a Gurdjieff Institute in London and wrote several large books elaborating the Master's "system." Like Madame Blavatsky, Gurdjieff claimed to have obtained his "system" from "initiates" in obscure Eastern monasteries. As a consequence, Ouspensky's speculations are so mixed with esoteric revelation, and so far removed from science, that we have not discussed them in this book. He died in 1947, two years before the death of Gurdjieff.

As far as I know, the only occult resort of recent times which surpassed Gurdjieff's in madness was the infamous monastery established near Cefalu, in Sicily, by the fabulous British occultist, Aleister Crowley. It was supposed to teach Yoga and self-discipline but actually gave instruction largely in drinking, drug addiction, and sex. When an English poet died there it precipitated a Sicilian explosion, and Crowley was forced to emigrate to Tunis. One of Somerset Maugham's novels, *The Magician,* is based on Crowley, with whom Maugham and Arnold Bennett once shared an apartment in Paris. Crowley was as satanic a mixture as has ever been thrown together of poet, painter, occultist, mountain climber, chess player, mountebank, psychotic, drug addict, and satyr—and the less said of him here the better. If you want to know more about him, you can read John Symond's *The Great Beast,* or try to read some of Crowley's mystical poetry or his tomes on black magic.

The diagnosis of ailments and the giving of medical advice have long been the stock in trade of certain psychics. The information may come from God, or departed spirits, or simply from clairvoyant insight. Andrew Jackson Davis, known as the "Poughkeepsie Seer," practiced this type of medicine for thirty-five years around New England during the latter half of the past century. He even wrote a five-volume work called *The Great Harmonia* about his visions. But the greatest psychic diagnostician of all time was, without a doubt, Edgar Cayce, of Hopkinsville, Kentucky. When he died in 1945, he left full stenographic accounts of 30,000 medical "readings" given over a period of forty-three years.

The best reference on Cayce (pronounced "Casey") is *There is a River,* 1943, by the American Catholic writer, Thomas Sugrue.[8] The book is similar in many ways to Kenneth Roberts' recent tribute to the dowsing abliities of his friend Henry Gross. Sugrue had been a friend of Cayce's, was completely sold on his psychic powers, and, like Roberts, tells the story of his friend's abilities in a form resembling fiction. The book does not give an objective account of Cayce's work, but it does give a vivid picture of the man and his history.

As a child, Cayce was dreamy and introverted. He played with imaginary playmates, had visions of his dead grandfather, and on one occasion talked to an angel with wings. He was deeply religious (the family belonged to the Christian Church). Once a year he read the Bible from cover to cover. Although he never went beyond the ninth grade, he did a vast amount of miscellaneous reading, and at one time worked in a bookstore. Sugrue emphasizes the fact that Cayce was a simple, untutored man who could not possibly have possessed the information he gave during his trances, but a far more reasonable supposition is that he absorbed large quantities of knowledge from reading and contacts with friends—knowledge he may have consciously forgotten.

There is no question about the genuineness of Cayce's trances. His technique was to lie on his back with his head facing south (later he changed it to north) and place himself in a state of self-induced hypnosis. The patient was usually present, though not necessarily, as Cayce gave thousands of readings for people who wrote to him for help. The reading commenced with the statement, "Yes, we have the body." He would then proceed to give a rambling diagnosis of the cause of the disorder, in terms borrowed largely from osteopathy and homeopathy.

Most of Cayce's early trances were given with the aid of an osteopath who asked him questions while he was asleep, and helped later in explaining the reading to the patient. There is abundant evidence that Cayce's early association with osteopaths and homeopaths had a major influence on the character of his readings. Over and over again he would find spinal lesions of one sort or another as the cause of an ailment and prescribe spinal manipulations for its cure. Here is a portion of a Cayce reading on his wife, who was suffering at the time with tuberculosis:

> The condition in the body is quite different from what we have had before . . . from the head, pains along through the body from the second, fifth and sixth dorsals, and from the first and second lumbar . . . tie-ups here, and floating lesions, or lateral lesions, in the muscular and nerve fibers which supply the lower end of the lung and the diaphragm . . . in conjunction with the sympathatic nerve of the solar plexus, coming in conjunction with the solar plexus at the end of the stomach. . . .

This is talk which makes sense to an osteopath, and to almost no one else. Sugrue records the case of a priest who wrote to Cayce for advice on a condition resembling epilepsy. Cayce recommended osteopathic treatment, "with particular reference to a subluxation as will be found indicated in the lower portion of the 9th dorsal center, or 9th, 10th, and 11th. Co-ordinate such correction with the lumbar axis and the upper dorsal and cervical centers. There should not be required more than six adjustments to correct the condition."

In addition to spinal massage, Cayce advocated a bewildering variety of remedies borrowed from homeopathy and naturopathy, with occasional inventions of his own subconscious tossed in. There were special diets, tonics, herbs, electrical treatments, and such medicines as "oil of smoke" (for a leg sore), "peach-tree poultice" (for a baby with convulsions), "bedbug juice" (for dropsy), "castor oil packs" (for the priest mentioned above), almonds (to prevent cancer), peanut oil massage (to forestall arthritis), ash from the wood of a bamboo tree (for tuberculosis and other diseases), and fumes of apple brandy from a charred keg (for his tuberculous wife to inhale). It is something of an innocent and thumping understatement when Sugrue confesses, "There was apparently no elaborate medical system, or theory, to be got from the cures."

In later years Cayce and his associates actually manufactured and sold some of the remedies he invented while in trance. These included Ipsab (for pyorrhoea), Tim (for hemorrhoids), a hay fever inhalant, and various devices for radioactive and electrical treatments. One of his readings advised attaching the copper anode of a battery to the third dorsal plexus center, and the nickel anode first to the left ankle, then to the right ankle. None of these remedies has, of course, the slightest value from the standpoint of medical science.

Eventually Cayce developed an interest in occult literature, and by answering metaphysical questions while in the trance state, a complex occult philosophy slowly emerged. From the summary given by Sugrue, it seems to be a confusing hodge-podge of Christianity, astrology, Pyramidology, theosophy, and other occult traditions.[9] The conscious mind is located in the pituitary gland. Arcturus is the next stop for souls leaving the solar system. And so on *ad nauseam*—little bits of information gleaned from here and there in the occult literature, spiced with occasional novelties from Cayce's unconscious.

Cayce became a confirmed occultist, having no difficulty harmonizing the new outlook with his Christianity. Christ was simply one of the "initiates" who did not preach all he knew. In addition to medical readings, Cayce began giving "life readings" in which he described the subject's past incarnations. Osteopathic causes for ailments gave way to causes rooted in the subject's "Karma" (the accumulation of good and bad effects from previous lives). A magazine called *The New Tomorrow* was published quarterly by a society formed to study these revelations. Later, the Association for Research and Enlightenment, Inc., at Virginia Beach, Virginia, took over the Cayce records and is now issuing material on it. Typical of this literature is a booklet by Cayce titled *Auras,* published in 1945 with a preface by Sugrue. Cayce describes his lifelong ability to see a colored aura surrounding every person's head and shoulders, tells how he diagnoses character and health from the colors, and predicts that healing by colors will eventually become part of medical science. (The literature of occultism is rich with books and articles describing techniques for making visible, measuring, photographing, and even weighing the so-called "human aura."[10] Unfortunately, only occultists succeed with the techniques.) Many study groups devoted to Cayce are currently meeting in large cities, and if interest continues to grow, he may become as important a figure in modern occultism as Madame Blavatsky.

There seems no doubt about Cayce's sincerity. He was a kindly, gentle man—with a round boyish face, gray-blue eyes behind rimless glasses, and a receding chin. He seemed constantly surprised and baffled by his unique gift, fearful it might be a source of evil but convinced until his death that it came from God.

Although thousands of people believed themselves cured by Cayce's trance-given remedies, in many cases even the initial diagnosis was widely off the mark. Rationalizations, however, always come easily. If the patient had "doubts" about the procedure, Sugrue naively writes, the diagnosis would not be a good one. Since almost anyone would have *some* doubts, which he would be quick to express if the reading was obviously bad (but would not mention otherwise), it is hard to see how evidence could be found that would shake the faith of a Cayce believer.

Sugrue does record, however, that Dr. Joseph Rhine of Duke University was unimpressed by Cayce when a reading the psychic made for Rhine's daughter failed to fit the facts. Presumably, if anyone would be favorably disposed to find Cayce practicing clairvoyance it would be Dr. Rhine, but no doubt Sugrue feels the professor had "doubts" which disturbed the diagnosis.

CHAPTER 18

FOOD FADDISTS

ENORMOUS PROGRESS HAS been made since the turn of the century in the scientific understanding of diet and its effects on health. Like any other science, the science of nutrition advances by slow, painstaking steps. Research studies are undertaken. Results are reported in meetings and in the journals. There is much debating, repetition of experiments, checking and double-checking. Unfortunately, this process takes place on a technical level beyond the understanding of the general public. Some of it, of course, reaches the layman through news reports, magazine articles, and books by writers who try to do a conscientious job of popularization. Alas, these voices are drowned out by the louder voices of the charlatans and faddists.[1] It is so easy to take a truth, or half-truth, then magnify its importance at the expense of other truths. The result may be exciting, and a convenient gimmick for a cult, but with so little reference to the facts that it becomes more a health menace than a panacea.

The cult of fasting is a good example. Some body ills are accompanied by nausea and loss of appetite, which naturally enforce a temporary fast. From this fact it is an easy but completely false step to the belief that there is some sort of magic therapeutic value in the fast itself, even for a person in good health. Actually, a prolonged fast by a healthy person can cause nothing but harm. There is such a general weakening of the entire body, and lowering of resistance to disease, that only an extremely vigorous and healthy person can stand a lengthy fast. Yet in spite of all medical evidence, the cult persists.

Hereward Carrington (author of many books on spiritualism and kindred topics) wrote a 648-page volume titled *Vitality, Fasting, and Nutrition,* in 1908, which is probably the best introduction in English to this once highly touted fad. Another advocate of fasting, who has written with a dogmatism even surpassing that of Carrington, is Upton Sinclair. His book *The Fasting Cure,* 1911, tells how prolonged starvation will combat tuberculosis, syphilis, asthma, cancer, liver trouble, Bright's disease, colds, and even locomotor-ataxia—the result of a destroyed nerve!

"I have known of two or three cases of people dying while they were fasting," Sinclair writes in his *Book of Life,* "but I feel quite certain that the fast did not cause their death, they would have died anyhow." And he adds, "I would not like to guess just what percentage of dying people in our hospitals might be saved if the doctors would withdraw all food from them. . . ." No statement could be more typical of a scientific Philistine. Sinclair "feels quite certain" about it, but it never occurs to hin to seek competent advice from men who know. Obviously, in his opinion, professional physicians are not men who know.

For a while, Sinclair was a Fletcherite. It was at the Battle Creek Sanitarium that he met Horace Fletcher, the handsome, athletic, prematurely white-haired author of *Fletcherism: What it is, or, How I Became Young at Sixty.* Like Fletcher's earlier works, the book had quite a vogue when it was published in 1913.

The motto of the Fletcherites was "Nature will castigate those who don't masticate." The idea was to eat only when you are hungry, choose the foods that appeal to you most, and chew each mouthful thirty to seventy times. It was bad for the digestion, Fletcher argued, to swallow food before it had been reduced to such a liquid state that it "swallowed itself." Even soup and milk had to be "Fletcherized" by rolling it about the mouth until it was thoroughly mixed with saliva. John D. Rockefeller was an ardent Fletcherite, and the philosopher William James once gave the system a three-month trial. "I had to give it up," James later testified. "It nearly killed me."

There are still some Fletcherites around. Recently the Sunday cartoon page, *Grin and Bear It,* had a drawing of a matronly wife returning from a lecture and saying to her husband, "And that diet lecturer said that if we all chewed our food properly, we wouldn't have all these wars. . . ."

Dozens of ingenious food fads have been built around the conviction that certain foods should never be eaten together. For example, fruit and milk is considered bad because the acid curdles the milk, in spite of the fact that as soon as milk reaches the stomach it promptly meets with acid secretions. Milk and fish is another combination against which faddists often warn. According to the "Hayites," meat and potatoes form a similar evil partnership.

The Hayites are worth a few paragraphs. The late Dr. William Howard Hay, who started it all, held a medical degree from the University of the

City of New York (class of 1891), and from 1932 until his death, directed his own resort at Mount Pocono, Pa. His views are set forth at length in his famous text, *Health via Food,* published in 1933.

According to Dr. Hay, almost all bodily ills are the result of "acidosis." This in turn is caused by (1) too much protein, (2) too much adulterated food, like white bread, (3) combinations in the diet of protein and carbohydrates, (4) retention in the bowels of food beyond twenty-four hours after eating. He also recommends frequent fasting, apparently unaware that fasting *really* causes acidosis.

Dr. Hay's reason for objecting to meat and potatoes is interesting. Starches need alkaline for digesting, proteins need acid, and "no human stomach can be expected to be acid and alkaline at the same time." Actually, most foods contain mixtures of proteins and starches. But such facts are much too complicated for the self-appointed diet experts. It's more fun to invent new recipes which avoid the poisonous mixtures— such as Hay's Happy Highball, Pale Moon Cocktail, Easter Bunny Salad, and Parcel Post Asparagus, which are described in the doctor's cookbooks.

Many food fads are built around the view that certain foods should be avoided entirely. A recent cult of this sort stems from the strange dietary views of Dr. Melvin Page, a dentist in St. Petersburg, Florida, and head of the Biochemical Research Foundation. When Page was operating in Muskegon, Michigan, in 1940, the government stopped him from selling a nostrum called Ce-Kelp which cured everything from cataracts to cancer. At the moment Dr. Page is, among other things, against milk. "As far as I know," he writes in his learned opus, *Degeneration— Regeneration,* published by himself in 1949, "he [man] and a certain species of ant are the only ones who use an animal secretion after the age of weaning."

Page thinks milk is fine for babies before being weaned, but after that it is a dangerous food and a frequent cause of colds, sinusitis, colitis, and cancer. The doctor points out that more people die of cancer per capita in Wisconsin than any other state, and of course Wisconsin takes the lead in milk production.[2] If we don't stop drinking this animal secretion, and reform our diet in other curious ways proposed by Dr. Page, he fears the Anglo Saxon race will continue to degenerate faster than certain "primitive" races which he does not specify.

The principal "don't eat this" cult has long been, of course, vegetarianism.[3] It is particularly popular with Hindu and occult groups, the Trappist monks, and small Protestant sects like the Seventh Day Adventists. Tolstoy, Gandhi, and George Bernard Shaw were vegetarians. Upton Sinclair, who at one time or another has embraced almost every food fad of the century, once wrote a book defending it.

Not all vegetarians agree. The "lacto-ovo vegetarians" are willing to take milk, eggs, and cheese, but stricter sects regard these foods as forms of meat. There is even a group calling themselves "fruitarians" who confine their diet to fruit. All these groups are loosely organized into the American Vegetarian Union, affiliated with the older International Vegetarian Union of Europe. A monthly magazine, *The American Vegetarian,* is currently published in New York City.

We need not be concerned here with the ethical arguments for avoiding meat, but the medical arguments are worth citing. Vegetarians are fond of pointing out that meat produces in the body harmful deposits of uric acid and "necrones," which in turn play a role in causing disease. Oddly enough, no medical doctor has ever found out what a necrone is. It sounds terrible but is completely without scientific meaning. As for uric acid, it is true that its increase in the blood is associated with some ailments, such as gout, but only because the body itself produces more of the acid. The notion that uric acid in the diet is a *cause* of such ailments is a myth long ago punctured by nutritional science.

No doctor denies that it is possible to obtain a balanced diet without meat, but the fact is that it is difficult to do this, and totally unnecessary. Among the amino-acids which are essential to health, about ten must be supplied by the food we eat. It is extremely hard to obtain all ten from a plant diet, and if even one is missing, there are nutritional deficiencies. On the other hand, amino-acids are the products of protein digestion, and the addition of meat to the diet is a simple method of getting all the needed ones. It is true that in some ailments, such as gout, meat in the diet must be limited. On the other hand, in other ailments diets rich in protein are necessary to restore health. There is no evidence whatever that meat plays a significant role in *causing* bodily disorders, least of all cancer, which some vegetarians trace to meat by means of wildly distorted statistics. Some food fads, it is interesting to note, advocate high protein diets even for the healthy! See, for example, Daniel C. Munro's *You Can Live Longer Than You Think,* 1948, in which he argues that Methuselah lived 969 years because he ate mostly meat.

Even more extreme than the vegetarians are the "raw food" fanatics who rail against the eating of cooked "dead" foods. "No animal eats cooked food," writes Jerome I. Rodale, of Emmaus, Pennsylvania, a statement with which one must heartily agree. "Man is the only creature that does," he continues. "It is a known fact that cats thrive much better on raw rather than cooked meat."

A manufacturer of electrical wiring devices, Rodale is also the leader in this country of a movement known as "organic farming." Not only does cooking devitalize food, according to Rodale, but food also loses in health value if it is grown in soil that has been devitalized by chemical fertilizers. Farming must be "God's way." The soil is like a living organism, Rodale claims, and only animal or vegetable fertilizers preserve its vitality. Soil and nutrition experts tell us that if plants grow at all, their composition tends to remain essentially the same, with respect to mineral and vitamin content, as plants grown in "rich" soil. A depleted soil will produce fewer or smaller plants, and if completely depleted, none at all. According to Rodale, however, the use of "artificial" fertilizers and sprays has caused almost all the nation's health disorders, including cancer. He has written many books about it (published by himself), of which *The Organic Front,* 1948, is the best summary of his opinions. In addition, he edits three monthly magazines— *Organic Gardening, The Organic Farmer,* and *Prevention,* the latter devoted to preventing disease by organic farming. In these magazines one may find many advertisements for Sunflower seeds. Rodale regards them as the great "forgotten food," of enormous health value when added to the diet."

Closely related to the organic farming movement is the German anthroposophical cult founded by Rudolf Steiner, whom we met earlier in connection with his writings on Atlantis and Lemuria. The anthroposophists go one step further than Rodale, and regard the earth as an actual living organism. It "breathes" twice a day, and its soil is "living" in much more than a metaphorical sense.

"Bio-Dynamic Farming" was established by Steiner at his School of Spiritual Science, in Dornach (near Basle), Switzerland. (Dornach is now the anthroposophical world center—a city with its own curious architecture and populated almost entirely by anthroposophists.) His two chief researchers were Lili Kolisko, whose works have not been translated, and Ehrenfried Pfeiffer. Many of Pfeiffer's books are available in English. His *Bio-Dynamic Farming and Gardening* was issued by the anthroposophists in 1938, and *The Earth's Face and Human Destiny* was printed in 1940 by Rodale.

In essence, the anthroposophists' approach to the soil is like their approach to the human body—a variation of homeopathy. (See Steiner's *An Outline of Anthroposophical Medical Research,* English translation, 1939, for an explanation of how mistletoe, when properly prepared, will cure cancer by absorbing "etheric forces" and strengthening the "astral body.") They believe the soil can be made more "dynamic" by adding to it certain mysterious preparations which, like the medicines of homeopathic "purists," are so diluted that nothing material of the compound remains. In 1923, Lili Kolisko experimented with progressively rarefied salt solutions on germinating wheat. She found that the effect of the solution faded when the dilution passed the tenth and twelfth decimal, but after that, appeared again! In writing about this in *Organic Gardening,* December, 1950, anthroposophist Dr. Herman Poppelbaum states: "By a simple calculation it can be figured out that in such high dilutions nothing 'material' of the ponderable solute is left. The effect therefore may be called imponderable, that is, not based on the physical presence of the material salt in the solvent. The substance then exercises an effect which is merely dynamic.''

Dr. Pfeiffer was born in Munich in 1899, and was graduated at the University of Basle. He holds an honorary degree from the Hahnemann Medical College, Philadelphia, for discovering a method of diagnosing human ills from crystal patterns which form when a drop of blood, mixed with chloride of copper, is crystallized. In the thirties he directed the anthroposophists' Biochemical Research Laboratory, at Dornach, and also the cult's 800-acre experimental farm at Loverendale, Holland. When the Nazis took over Holland in 1940, he escaped with his family and came to the United States.

After creating a model farm at Phoenixville, Pennsylvania, he bought his own farm near Chester, New York. It was here he made what he regards as his most momentous discovery—special blends of bacteria strains (the exact formulas are highly secret) which he claims will convert ordinary garbage into rich organic fertilizer. Only a tablespoon of the bacteria need be added to each ton of garbage. In a week the garbage is transformed into an odorless supercompost.

With funds provided by the owner of a wastepaper business in Buffalo, a company has recently been formed in Oakland, California, which buys garbage from the city, processes it with Pfeiffer's wonder germs (his Biochemical Laboratory at Spring Valley, New York, gets a royalty for supplying the bacteria), then sells it as organic fertilizer. Laboratory tests (by Pfeiffer) show that vegetables grown with this compost weigh 25 percent more than vegetables grown with ordinary fertilizers, and have one to three times as much vitamin A. Grain grown in the treated soil has higher protein content. Even sand can be made into rich farm land, Pfeiffer says, if water is available. You can find the details of this revolutionary project in an article, ''The City with the Golden Garbage,'' *Collier's*, May 31, 1952. The article neglects to inform the reader of the anthroposophical views which underlie Pfeiffer's work.

At the present time, one of the most popular eating fads in America is built around the personality of Gayelord Hauser—a handsome, virile-looking man with dark wavy hair, whose face appears prominently in the advertisements of his book. His latest work, *Look Younger, Live Longer,* 1950, was condensed by *Reader's Digest,* serialized in the Hearst papers, and is still far outselling its closest competitor, *Eat and Grow Younger,* by Lelord Kordel, in spite of the fact that Kordel looks even younger and handsomer.

Hauser was born in Tübingen, Germany, in 1895, as Helmut Eugene

Benjamin Gellert Hauser. He came to the United States at the age of sixteen, where he contracted tuberculosis of the hip. A Chicago hospital decided his case was incurable, and shipped him back to Europe to die. "There, high up among the snow-capped peaks," Hauser writes, "a miracle happened." An old man who was visiting the family said to him, "If you keep on eating dead foods, you certainly will die. Only living foods can make a living body." Young Hauser took his advice and began eating fresh fruits and vegetables. His hip began to heal.

Intensely curious about what was happening to him, Hauser developed a strong interest in naturopathy. Benedict Lust, whose role as a pioneer naturopath has previously been discussed, advised him on dietary matters. From naturopathy Hauser went to naprapathy—a Chicago-born offshoot of chiropractics—and eventually enjoyed a full recovery. He returned to the United States in the early twenties, had his name officially changed, and began practicing naprapathy at a small office on Michigan Boulevard, in Chicago. Outside of his attendance at the Chicago College of Naprapathy, he has had no formal schooling in either medicine or the science of nutrition.[4]

Eventually, Hauser gave up naprapathy for writing and lecturing. His success was so immediate that he invaded Hollywood in 1927, where his dietary views quickly became a movie colony craze. His most enthusiastic convert was Greta Garbo. She became his constant companion in a friendship which lasted many years. England and Europe later provided pastures as green as those of Hollywood. Lady Elsie Mendl, who until her death at ninety-four regularly stood on her head as a therapeutic measure, was one of his strong supporters. So was the Duchess of Windsor, who wrote an introduction for the French edition of his current best-seller. Queen Alexandra of Yugoslavia, Baron Philippe de Rothschild, Cobina Wright, Sr., and Paulette Goddard are other notables included among those whom Hauser likes to call "my people."

What is the Hauser system?[5] Basically it is a naturopathic approach with special emphasis on what Hauser calls his five wonder foods—skim milk, brewer's yeast, wheat germ, yogurt, and blackstrap molasses. "Any one of these five foods, used daily," he writes, "can probably add five youthful years to your life." Although sold chiefly through health food stores in large cities, regular grocers now stock them, and there is no question that the astonishing growth of the Hauser cult has enormously increased the sales of these products. There is considerable question, however, about the virtues of the wonder foods. The best medical opinion is that they offer nothing one cannot obtain less expensively from ordinary foods. Yogurt, for example, is a specially fermented milk of no more health merit than buttermilk, but it costs a great deal more.

Blackstrap molasses is the dark, sticky dregs that remain after the process of sugar refining is completed. According to Hauser, it has enormous medicinal properties. In his book he states that it will help cure insomnia, nervousness, menopause troubles, baldness, and low blood pressure. In addition, Hauser claims, it will help restore gray hair to its former color, aid the digestion, prevent many changes due to old age, help the functioning of glands, and strengthen the heart. Government nutrition experts have called these claims "false and misleading."[6]

In addition to income from his lectures, radio and TV appearances, books, and a magazine called *Gayelord Hauser's Diet Digest,* Hauser also is a partner in Modern Products, Inc., a Milwaukee firm of more than twenty years standing. It is from Modern Products that one can obtain the special foods and medicines which Hauser promotes in his books and lectures. In the past there have been several brushes with the government over advertising claims for many of these products.

Due to the leniency of courts, the government has great difficulty in preventing the sale of magic foods and nostrums. It took the Food and Drug Administration four years, for example, to restrict the sale of Nutrilite by 15,000 door to door salesmen. Nutrilite was a cheap mixture of alfalfa, parsley, and watercress, that was supposed to cure 57 different diseases at a cost of $200 a year to the client. The government did not win its case. It succeeded only in forcing the manufacturers of Nutrilite to soften their claims. The product is still widely advertised and sold.

It would take many volumes to discuss all the tonics, vitamin products, mineral salts, and other miracle foods which in recent years have made fortunes for their promoters. One manufacturer even put vitamins in soap, where they are about as useful as the hormones that appeared in a nationally advertised brand of face cream or the magnetic properties recently acquired by a certain make of razor blade. Chlorophyll seems to be the latest wonder compound turning up everywhere to do everything.

If one wants insight into the rise and promotion of a worthless tonic, he should read H. G. Wells' amusing novel, *Tono-Bungay.* The title is from the name of a patent medicine that not only had none of the virtues advertised but also was slightly harmful to the kidneys. And if you think Wells' narrative is dated, compare it with the recent fabulous rise of Hadacol, a cure-all vitamin and mineral tonic that smells and tastes terrible, but is high in alcoholic content.

*   *   *

Groucho Marx, on his radio program a few years ago, interviewed the inventor of Hadacol, Louisiana State Senator Dudley J. LeBlanc. When Groucho asked him what Hadacol was good for, LeBlanc gave an answer of startling honesty. "It was good," the senator said, "for five and a half million for me last year."

CHAPTER 19

THROW AWAY YOUR GLASSES!

ONE of the nation's most influential medical eccentrics, whose work is still accepted by many thousands of intelligent but ill-informed people, was Dr. William Horatio Bates, an eye, ear, nose, and throat specialist of New York City. He was the first important figure in the modern cult of revolt against spectacles and the reliance on eye exercises for the treatment of visual defects.

Dr. Bates was born in Newark, New Jersey, in 1860. His early medical record is impressive—graduate of Cornell in 1881; medical degree from the College of Physicians and Surgeons, 1885; clinical assistant at Manhattan Eye and Ear Hospital; attending physician at Bellevue Hospital and later at the New York Eye Infirmary. From 1886 to 1891 he taught ophthalmology at the New York Postgraduate Medical School and Hospital.

In 1902, Dr. Bates suddenly vanished. Six weeks later his wife learned he was working as an assistant at Charing Cross Hospital, in London, where he had been taken in as a patient. She went to London immediately and found him in a state of exhaustion, with no memory of what had happened. Two days later he vanished again. Mrs. Bates searched for him on the Continent. She came back to America and searched. She was still looking for him when she died.

Exactly how Dr. Bates was found again is still wrapped in obscurity. According to his obituary in the *New York Times*, July 11, 1931, a fellow oculist found him by accident in 1910. He had been practicing in Grand Forks, North Dakota, for six years. Apparently Bates was persuaded to return to Manhattan, where he shared offices with his discoverer and served as attending physician in Harlem Hospital until 1922. He remarried, and was survived by his wife (it was his third) when he died in 1931.

In 1920, Dr. Bates issued at his own expense a book titled *Cure of Imperfect Eyesight by Treatment Without Glasses*. It was published with the imprint of the Central Fixation Publishing Company, New York City, and opens with a tribute to the author by the Rev. Daniel A. Poling. (Bates attended Rev. Poling's church, and also treated his eyes. The Reverend today credits Bates for the fact that at sixty-seven he still gets along without glasses.) A revised, condensed version of the book appeared twenty years later.

Dozens of books by other writers have been little more than restatements of Dr. Bates' views. *Sight Without Glasses*, 1944, by Harold M. Peppard,[1] has been a popular American work, and in England, *The Improvement of Sight*, 1934, by Cecil S. Price. As late as 1948 Random House issued *See without Glasses*, by Ralph J. MacFayden—another book based on Bates' theories.

The original Bates work is a fantastic compendium of wildly exaggerated case records, unwarranted inferences, and anatomical ignorance. Much of the material had appeared earlier in articles which Bates contributed to several medical journals, and in a correspondence course on eye-training written in collaboration with Bernarr Macfadden. (The course was heavily advertised in Macfadden's *Physical Culture* magazine.) More than fifty photographs are reproduced in the book, many of them exceedingly curious. One, for example, is titled *Myopes who never went to school or read in the subway*. It pictures the faces of four nearsighted animals—an elephant, buffalo, monkey, and pet dog.

At the heart of Dr. Bates' views is his theory of accommodation. "Accommodation" is a term for the focusing process which takes place within each eye when you shift attention to objects at varying distances. One of the best established facts in eye anatomy is the fact that this adjustment involves an alteration in the shape of the lens. A tiny muscle called the ciliary muscle causes the lens to become more convex as the eye is focused on closer objects. This change of the lens has been photographed in detail, and measured with a high degree of accuracy.[2] Dr. Bates, however, denied all this categorically. The lens, he stated, is "not a factor in accommodation." Instead, the focusing is accomplished by an alteration in the entire length of the eyeball, in turn brought about by two muscles on the outside of the eye!

To support this odd theory, Dr. Bates records (with many photographs) some experiments he performed on the eye of a fish. After the lens of the fish eye had been removed, the eye was still able to accommodate. Dr. Bates was unconcerned with the fact that a fish eye has very little resemblance to a human eye. Some mammalian experiments, chiefly on rabbits and cats, are also described. Unfortunately, what the description reveals most clearly is an almost total lack of laboratory competence.

When an eye doctor puts drops in your eyes, the power of accommodation is temporarily killed. This has been proved to be due to paralysis of the muscles controlling the lens and pupil, but Dr. Bates thought otherwise. The loss of accommodation, he asserted, does not occur until the drug affects the muscles outside of the eye. Moreover, he claimed to have observed patients whose lenses had been removed by a cataract operation and who still accommodated with ease! Curiously, this has never been observed by any other eye man. Actually, Bates' theory of accommodation (so necessary to explain the value of his exercises) is so patently absurd that even most of his present-day followers have discarded it.

The cause of all refractive errors (nearsightedness, farsightedness, and astigmatism), according to Bates, is simply "strain," in turn due to an "abnormal condition of mind." "The origin of any error of refraction," he wrote, "of a squint, or of any other functional disturbance of the eye, is simply a thought—a wrong thought—and its disappearance is as quick as the thought that relaxes. In a fraction of a second the highest degree of refractive error may be corrected, a squint may disappear, or the blindness of amblyopia may be relieved. If the relaxation is only momentary, the correction is momentary. When it becomes permanent, the correction is permanent."

Glasses cannot cure this strain, Bates believed. In fact, they make the cure impossible because the eye adjusts to them, then the strain makes the eye get steadily worse so that stronger and stronger glasses are necessary. In Bates' opinion, glasses are simply "eye crutches." They should be tossed away.

The Bates system is designed to relieve strain. This involves "central fixation"—learning to see what is in the center of vision, without staring. Here is what Dr. Bates had to say about its importance:

"Not only do all errors of refraction and all functional disturbances of the eye disappear when it sees by central fixation, but many organic conditions are relieved. I am unable to set any limits to its possibilities. I would not have ventured to predict that glaucoma, incipient cataract, and syphilitic iritis (inflammation of the iris of the eye) could be eliminated by central fixation, but it is a fact that these conditions have disappeared when central fixation was attained. Relief was often obtained in a few minutes, and, in rare cases, this relief was permanent. . . . Infections, as well as diseases caused by protein poisoning and the poisons of typhoid fever, influenza, syphilis and gonorrhea, have also been benefited by it. Even with a foreign body in the eye there is no redness and no pain so long as central fixation is retained."

To achieve central fixation, or learning to see without strain, Dr. Bates proposed the following exercises:

(1) "Palming." The patient puts the palms of both hands over his eyes (without pressing or rubbing), and tries to think of "perfect black." When the patient is able to see a pure blackness, there is an immediate improvement of sight. Dr. Bates tells of one patient, a man of seventy with astigmatism and incipient cataract, who was completely cured after having palmed continuously for 20 hours!

"The smaller the area of black which a person is able to remember," wrote Bates, "the greater is the degree of relaxation indicated." To achieve this, he recommended thinking of a large black area, such as the top letter on an eye test chart, then proceeding to smaller and smaller letters until a period is reached. "Instead of a period, some people find it easier to remember a colon," he wrote, " . . . or a collection of periods, with one blacker than all the others, or the dot over a small I or J. Others, again, prefer a comma to a period."

If black bores or depresses you, then a color with more pleasant associations is best. "One woman's sight," Bates declared, "was corrected by the memory of a yellow buttercup, and another was able to remember the opal of her ring when she could not remember a period."

2) The "shift" and the "swing." By shifting, Dr. Bates meant moving the eye back and forth so that one gets an illusion of an object "swinging" from side to side. The shorter the shift, the greater the benefit. You can even close the eyes and make a mental image swing. Alternating visual and mental swinging is particularly beneficial. After one masters the art of shifting and swinging, he finally attains what Bates called the "universal swing." Here are his own words about it:

"When swinging, either mental or visual, is successful, a person may become conscious of a feeling of relaxation which is manifested as a sensation of universal swinging. This sensation communicates itself to any object of which a person is conscious. The motion may be imagined in any part of the body to which the attention is directed. It may be communicated to the chair in which a person is sitting, or to any object in the room, or elsewhere, which is remembered. The building, the city, the whole world may appear to be swinging."

In addition to palming, shifting, and swinging, Bates also recommended reading under unusually adverse conditions—such as when lying on the back, riding a bus or train, in dim light, or in bright sunlight. The eyes are also strengthened by looking directly at the sun for short moments so that the beneficial rays may bathe the retina (a practice, by the way, which may easily cause permanent retinal damage).

Every eye doctor except Bates had assumed that presbyopia was a normal inability to accommodate that accompanies the aging process. Dr.

Bates soon set them straight. "The truth about presbyopia," he wrote, "is that it is not 'a normal result of growing old'. . . . It is caused not by a hardening of the lens but by a strain to see at the nearpoint. It has no necessary connection with age. . . . It is true that the lens does harden with advancing years, just as the bones harden . . . but since the lens is not a factor in accommodation, this fact is immaterial. Also, while in some cases the lens may become flatter or lose some of its refractive power with advancing years, it has been observed to remain perfectly clear and unchanged in shape up to the age of ninety. Since the ciliary muscle is not a factor in accommodation either, its weakness or atrophy can contribute nothing to the decline of accommodative power."

Squinting, specks in the eye, and even the twinkling of stars are all due to eye strain, according to Bates. To physicists, there is little mystery about star twinkling. It is due to constantly changing currents of air with different densities, and hence different refractive powers. But to Bates, twinkling is in the mind. If you look at the stars without strain, they cease to twinkle. " . . . when the illusion of twinkling has been produced, one can usually stop it by 'swinging' the star. On the other hand, one can make the planets or even the moon twinkle, if one strains sufficiently to see them."

In England, Dr. Bates' most distinguished convert was Aldous Huxley, the victim of an early eye infection which left his corneas permanently scarred. After trying the Bates system he felt that his vision had greatly benefited, and in 1942 wrote a book called *The Art of Seeing*—a book destined to rank beside Bishop Berkeley's famous treatise on the medicinal properties of "tar water." Huxley summarizes Bates' theories, and adds a few additional forms of therapy such as juggling, and playing dice and dominoes. Sitting far in the rear of a movie theater is valuable exercise, he declares, for extremely *nearsighted* eyes. Also good for the same ailment is closing the eyes and imagining that you are holding a rubber ring between the thumb and finger. You squeeze the ring into an ellipse, then let it spring out again, and continue until it becomes tiring.

The most interesting practice recommended by Huxley is what he calls "nose writing." You imagine that your nose is extended forward about eight inches, like Edward Lear's "Dong with the luminous nose." Then you close your eyes and pretend the nose is a pencil. By moving the head you write an imaginary signature in the air. "A little nose writing, followed by a few minutes of palming . . . will result in a perceptible temporary improvement of defective vision," he writes. If you keep it up, it gradually becomes permanent.

Huxley's explanation of why palming works, sounds like something Mesmer, the discoverer of "animal magnetism" might have said. ". . . all parts of the body carry their own characteristic potentials; and it is possible that the placing of the hands over the eyes does something to the electrical condition of the fatigued organs—something that reinvigorates the tissues and indirectly soothes the mind.''

Dr. Bates died in 1931, but his eccentric methods have been kept alive by numerous disciples throughout the country. In England and Germany during the twenties, dozens of schools sprang up to teach the Bates method. Under Hitler it blossomed into a widespread cult. There is no question that thousands of patients imagined they were benefited. How can this be explained in view of the fact that there is not the slightest factual foundation for Bates' theory of accommodation or for the value of his exercises?

The answer is many-factored. In the first place, there are many optometrists who will sell a pair of glasses to a patient who does not really need them. After wearing the glasses for a time, their eyes adjust to them, so that when the glasses are first removed, the vision is noticeably poorer. After a week or so without the spectacles, the vision will become more normal. If the person is palming and shifting during this period, he will attribute the improved vision to the exercises. In some eye exercise clinics operated by quacks, the procedure with a new patient is to remove his glasses and immediately test his vision with an eye chart. Naturally, his vision will be at its worst. Then after a half-hour of exercises without glasses, he is tested again. Naturally his vision has improved. What he fails to realize is that the same improvement would have occurred without the exercises, as his eyes slowly adjusted to seeing without the spectacles.

A second point to consider is that a number of eye defects may change for the better as a person ages. Astigmatism sometimes alters beneficially. An incipient cataract may be heralded by a temporary phase of much improved vision. Many diseases of the eye go through cyclical changes. If the patient is following the Bates system when a natural change of this sort occurs, he will tend to credit the system for it.

Finally—and this is probably the most important factor of all—the process of ''seeing'' is deeply involved with one's mental attitude. If there is any value in Bates' work it is his strong insistence on the mental side of vision. He claimed, for example, he could tell when a person was lying by observing his refractive error through a retinoscope. This is highly doubtful, but there is no question that mental factors may greatly aggravate or alleviate a person's discomfort in seeing, even though an examination of his eyes will show no organic changes. One person may be driven almost crazy by eyestrain and headaches, and another person, with exactly the same type and amount of refractive error, will not be bothered at all. The eye can take far more punishment than people realize. Given the right frame of mind—which may be induced by faith in any kind of treatment—a person with even a large visual error may be able to toss away his glasses and get along comfortably. Bates himself wrote about how some persons found their vision much improved after they had been given glasses which were almost plain glass. What he did not realize was the possibility that his own system might operate along similar lines.

In many cases, the nature of the refractive error is an aid in enabling a person to feel comfortable without his glasses. An extremely nearsighted patient, for example, may have little difficulty reading with unaided eyes, and if he is not bothered by hazy distant vision he can manage fairly well without spectacles. Naturally, after going for a while without them he will learn to "see" blurred objects in the distance better than he could immediately after discarding his glasses. He could see a great deal better *with* glasses, but if he gets psychic satisfaction out of not leaning on "eye crutches," and fancies his vision is improving, he will be able to give stirring testimonials about the value of Bates' methods. Another example is that of the elderly man who has a natural loss of accommodating power due to the hardening of his lenses. If one eye happens to be nearsighted and the other farsighted he can throw away his spectacles and get along without much trouble for the rest of his life. One eye does the distant seeing, and the other eye does the close-up work. With a proper pair of bifocals he could use both eyes all the time, but if he is sufficiently sold on Bates he will take great pride in the fact that even at his age glasses are unnecessary.

A few eye disorders, it should be pointed out, do respond to exercises—but these are disorders which involve the exterior muscles, such as partially crossed eyes, or walled eyes. Exercises for ailments of this sort are prescribed by orthodox eye doctors, and have nothing in common with the exercises of Dr. Bates. Most eye defects, however, are refractive errors due to the shape of the eye, lens, or cornea, and no amount of shifting or swinging will produce an *organic* change. During the past war the Air Force conducted a series of experiments with eye exercises. Many pilots thought their sight much improved even though examinations showed that the image on the retina remained unchanged. For many years a Manhattan eye doctor had a standing and unclaimed offer of $1,000 to any patient with a refractive error who practiced the Bates system, and whose eyes showed organic improvement when tested by a competent doctor.

For several years Bates edited a magazine called *Better Eyesight*. A collection of articles from it was made by Emily C. Lierman, the doctor's assistant, and published in 1926 with the title, *Stories from the Clinic*. Its case records are the best evidence in print of the "faith cure" aspect of Bates' methods, and the careless diagnosing procedures of the doctor and his helpers.

In one case an aged woman had glaucoma (a hardening of the eyeball) in one eye and "absolute glaucoma" in the other, rendering it totally blind. She was shown how to palm. "In just a few minutes the pain ceased and the eyeball became soft." Through the blind eye she was able to read the top letter of the test chart. This is unquestionably the fastest "cure" of glaucoma on record. How can such a miracle be explained?

The answer is suggested in the next sentences. "She was very happy and wanted to talk, which I encouraged her to do. She said she was living in a small furnished room and . . . had no one to look after her." Such a patient—lonely, old, neurotic—has little to do except meditate on her infirmities. Often a new type of treatment, by a friendly doctor, will offer new hope and with it a great change in mental outlook. At the beginning, such patients will exaggerate the poorness of their vision, and as the therapy progresses, if they have faith in its curative power, will exaggerate in the opposite direction. Later, this particular lady suffered a relapse and was visited in her room by the author, who writes: "Her thin face was lined with pain. . . . I began to talk to her about the days when she did not suffer. . . . She began to palm . . . and became able to imagine a daisy waving in the breeze. I asked her to imagine that her body was swinging with the flower. She did this, and within a few minutes her pain left her and she smiled."

One chapter of Miss Lierman's book describes eight cataract "cures." In these cases there can be no doubt that no organic changes occurred in the clouded lenses of the patients, yet all of them improved in their ability to read the test card. In some cases, perhaps, they had memorized the card, but in most cases, the improvement was probably real enough in the patient's mind to result in an actual improvement of "seeing." If such improvement can take place, you may ask, then is not the Bates system of value? The answer is "yes" if you are willing to grant an equal value to the "cures" which follow the drinking of Hadacol, or the use of Colonel Ghadiali's colored light machine.

Bernarr Macfadden was one of the first self-styled health authorities to get on the Bates bandwagon. We have already mentioned his promotion of an early correspondence course by Bates. Later this was formed into a book, ghosted for Macfadden by Hereward Carrington, and published in 1924 under the title, *Strengthening the Eyes*. Evidently Macfadden thought the shift and swing too complicated, because he substitutes for it the much simpler exercises of moving the eyes up and down, from side to side, and rolling them about like Eddie Cantor. Another Macfadden exercise is to hold up a pencil and alternately focus on the tip of the pencil and a distant object. Frequent bathing of the eyes, by opening them while the face is submerged in a bowl of water, is also recommended. Dr. Bates had strongly condemned any type of eye massage, but

Macfadden worked out a number of finger massages for the eyeball, the type of treatment varying with the nature of the refractive error.

The all-time low, however, in books on the Bates system was achieved in 1932 by none other than Gayelord Hauser. It was called *Keener Vision without Glasses,* and ran through many editions, including a cut version called *Better Eyes without Glasses,* 1938. "Thanks to the research of this great man [Bates]," Hauser writes, "a permanent correction of defective eyesight has been developed." In addition to the Bates methods, he adds some new exercises of his own which he calls "gymnastiques," and also the "Seven-day elimination diet." Important ingredients of this diet are Hauser Potassium Broth, Nu-Vege-Sal, Swiss Kriss, Peppermint and Strawberry Teas, and Santay Meatless Bouillon, all of which were obtainable from Hauser's firm in Milwaukee.

In the back of the book are listed the major eye ailments along with Hauser's methods for correcting them. Here is his treatment for astigmatism. "To overcome strain and tension, first of all, follow the Seven-day Elimination diet . . . and do the eye gymnastiques, especially Nos. 1 and 2 and rolling the eyes in all directions. . . . Do at least twenty to thirty minutes of palming each day, use the cold water eye bath in the morning, and apply the herbal eye-pads noon and night. . . ."

According to Hauser, nearsightedness is due to nervousness, bad diet, and too many acid foods. Crossed-eyes, as well as presbyopia, are attributed to bad diet. For cross-eyes he recommends swinging on a rope swing. "Twisting the rope and letting it unwind makes the world whirl around. It is a very relaxing exercise as it forces constant shifting of the eyeball."

The dread eye disease of glaucoma is due, says Hauser, to "autointoxication, faulty foods, and worry." It is corrected by relaxation and a new (Hauser) diet. "Palming and swinging are valuable exercises in glaucoma because they are so very relaxing. Have the neck worked on by an osteopath. . . ."

Cataract, Hauser admits, "is a most troublesome condition. It occurs when the lens becomes congested with waste. The entire body is toxic, because of faulty diet." The fact that there is no known cure for cataract does not daunt the intrepid diet expert. He recommends (1) the Hauser Seven-day Elimination Diet, (2) flushing the bowels daily with a warm water enema, to which the juice of one lemon is added, (3) osteopathic treatments for the neck, (4) palming thirty to sixty minutes a day ("Palming seems to have a magnetic effect on circulation. It actually seems to flush and reflush the eyes with more lymph, and carries away waste"), (5) sun treatments, (6) soaking little pads of cotton in lime juice and leaving them on the eyes "as long as possible."

Not one of the above six procedures will have the slightest effect on a cataract. Nor will any of Hauser's treatments have a significant effect on any of the eye ailments he discusses. To mislead the reader into tossing away his glasses, stop seeing his eye doctor, and relying on such magic is surely an act of arrogance hard to match in the literature of pseudo-ophthalmology.

At the time of writing, several organizations in New York City are teaching the Bates system—notably the Margaret D. Corbett School of Eye Education, a branch of the parent school in Los Angeles. Mrs. Corbett is the author of *How to Improve Your Eyes*, 1938, and was the teacher responsible for Huxley's improved vision.[3] A Mrs. Robert W. Selden advertises that she is "incorporating the Bates system" in her work. The American Association for Eye Training, under Clara Allison Hackett, has schools in Los Angeles, San Diego, and Seattle, as well as in Manhattan. Miss Hackett was teaching history in a Tacoma, Washington, high school when her eyes went bad. The Bates system "cured" her and she has been teaching Bates ever since.[4]

A. E. van Vogt, one of the most popular of American writers of science fiction, fell for Dr. Bates about the same time he fell for General Semantics and Gayelord Hauser. He threw away his pince-nez and wrote a science-fiction novel called *The Chronicler* in which the views of Bates play an important role. A friend of mine who played chess with him during this period reports that van Vogt frequently picked up an opponent's piece instead of his own. The prince-nez is back at the moment, and van Vogt is currently preoccupied with the promotion of west coast dianetics.

Although Aldous Huxley's "seeing" may have improved considerably since he discovered Bates, there have been no changes in the opacities of his corneas. This was dramatically revealed recently when he spoke at a Hollywood banquet. The following quotation is from Bennett Cerf's column in *The Saturday Review*, April 12, 1952:

"When he arose to make his address he wore no glasses, and evidently experienced no difficulty in reading the paper he had planted on the lectern. Had the exercises really given him normal vision? I, along with twelve hundred other guests, watched with astonishment while he rattled glibly on. . . . Then suddenly he faltered—and the disturbing truth became obvious. He wasn't reading his address at all. He had learned it by heart. To refresh his memory he brought the paper closer and closer to his eyes. When it was only an inch or so away he still couldn't read it, and had to fish for a magnifying glass in his pocket to make the typing visible to him. It was an agonizing moment. . . ."

The real tragedies occur, however, when a Bates enthusiast suffers from glaucoma, atrophy of the optic nerve, or some other ailment which may demand immediate medical attention before it leads to blindness.

Such tragedies cluster about the work of every medical pseudo-scientist. And they serve to point up, for any intelligent reader, a very simple and obvious moral.

The moral is that when you encounter a new medical theory, universally condemned by the "orthodox," you will do well to take their word for it. It is always possible, of course, that the self-styled genius *may* be what he claims to be—another Pasteur, years ahead of stubborn colleagues. But the odds are heavily against it. For every quack who later proves to be a genius, there are ten thousand quacks who prove later only to be quacks. Many of them, as we have seen, are brainy men who write and speak with great authority and persuasion. As a medical layman, however, your health is much too precious to trust to your own faulty judgment. You may keep your mind open, but to rely on the consensus of informed medical men is the soundest and sanest course of action.

CHAPTER 20

ECCENTRIC SEXUAL THEORIES

THE FIELD OF sexual pseudo-science is understandably vast and grotesque. Hardly an aspect of sex has escaped the theorizing of eccentric biologists.

Consider, for instance, the problem of how the sexes came into being. Theosophists and anthroposophists believe that early "root races" were hermaphroditic, the male and female united in each individual (a kind of modern occult version of the theory defended by Aristophanes in Plato's *Symposium*). A strongly contrary theory is suggested by the title of an obscure book published in 1927 by William H. Smyth, the British engineer who invented technocracy. It is called *Did Man and Woman Descend from Different Animals?*

Arabella Kenealy, the well-known English anti-feminist, tackled the problem in 1934 in her book, *The Human Gyroscope*. The subtitle reads, "A consideration of the gyroscopic rotation of earth as a mechanism of the evolution of terrestrial living forms, explaining the phenomenon of sex: its origin and development and its significance in the evolutionary process."

Miss Kenealy's marathon opening sentence states her thesis as follows: "In presenting the consideration that, as plastic clay on the rotary disc of a little potter's wheel of industry is shapen and moulded in varieties of symmetrical three-dimensional form, increasingly uprising in the vertical, so upon the rotating surface of the great terrestrial potter's wheel of Creative Evolution, the plastic matter of terrestrial organisms has been shapen and moulded in the countless diversities of increasingly complex, structurally differentiated three-dimensional forms of living species, progressively uprising in the vertical in the terms of increasingly complex elevated posture—I have ventured to base my argument upon the Gravitation of great Newton, instead of on the later Einstein theory.''

The cosmos, according to Kenealy, exhibits a dual male-female aspect from the lowly atom to the lofty galaxy. Maleness and femaleness are, in fact, the warp and weft in the fabric of creation. She also thinks that northern races are masculine and southern races feminine (owing somehow to the earth's rotary movements), and that the right side of everyone's body is more masculine than the left. Her book contains excellent photographs of a giraffe, fish, dog, camel, spiral nebula, crab, and gyroscope, and an interesting section on her experiences with telepathy. ''During my marriage engagement, years ago,'' she writes, ''to a man to whom I was deeply attached, we were in continual telepathic communication with one another. It was as natural as breathing.''

Curious books and articles about the relative superiority of the sexes form another interesting body of literature. It ranges all the way from early works that prove the innate *inferiority* of women (smaller brain capacity and the like) to an article by anthropologist Ashley Montagu, in the March 1, 1952, *Saturday Review,* which proves innate female *superiority.*[1] The question of whether the fair sex is improving or deteriorating as a result of efforts to achieve equality with men has likewise produced many unusual volumes, beginning with Arabella Kenealy's famous attack on feminism, *Feminism and Sex Extinction,* 1920, and continuing down to more recent studies.

Another branch of sexual pseudo-science worth looking into concerns methods of determining the sex of children before they are conceived. Folk superstitions on this matter have linked the sex of unborn infants to almost everything—weather, phases of the moon, diet, relative ages of parents, position during coitus, and so on—many of these theories having been the subject of learned treatises. Aristotle thought the direction of winds had something to do with it. Aquinas stated his belief that if there had been no Fall, parents would have been able to produce a child of whatever sex they desired, but he did not elaborate on the method by which this might be accomplished. In more recent times, the theory that eggs on one side of a woman's body produce males, and those on the other side, females, prompted considerable European laboratory work, later found unreliable.

One German sexologist maintained that the right testicle and right ovary were male, and the left testicle and ovary female, but did not specify what resulted when a sperm from the right testicle united with a left ovum.[2] Bernarr Macfadden, who has eight children of whom six are girls, favors another German theory that children tend to be of the sex opposite that of the parent with the greatest passion and virility. Other authorities, who fathered mostly sons, have vigorously defended the view that children are of the *same* sex as the more vigorous parent. There is also the theory that a telepathic influence tends to produce offspring of a sex contrary to that most firmly desired by the father.

Methods for overcoming impotence have likewise been the object of considerable quasi-scientific investigation. An authority can be found for almost every folk belief about the sexually stimulating qualities of certain foods—in most cases, foods of an uncommon variety which are somehow associated with sex. Eggs and caviar (fish eggs), for example; or foods which suggest or resemble sex organs (asparagus, celery, onions, clams, oysters, and so forth). Hundreds of quack medicines and devices have been devoted to stimulating potency. In London, in the mid-eighteenth century, John Graham, O.W.L. (the initials stood for "Oh Wonderful Love!") made a fortune charging couples for the privilege of sleeping on his "celestial bed." The bed had curious coils attached to it, soft music was played, incense burned, and colored lights bathed the sleeper.

The subject of rejuvenation has a long, insane history. In spite of the fact that there is not the faintest evidence that goat glands, or the glands of any other animal, can be transplanted successfully into a human male, "Doctor" John R. Brinkley of Kansas became a millionaire during the twenties performing such operations on thousands of innocent oldsters. His clinic at Milford, Kansas, charged a minimum fee of $750, but the glands of a very *young* goat cost as high as $1,500. Even the late E. Haldeman-Julius, the publisher of Little Blue Books, was fooled by Brinkley. For several years the publisher ran Brinkley's advertisements and favorable pieces on the doctor in a periodical devoted chiefly to debunking American life. Later, however, Haldeman-Julius realized his mistake and publicly apologized.

After the license of Brinkley's radio station was taken from him, the doctor bought a station in Mexico, across the Rio Grande from Del Rio. There he pretended to perform prostate operations, and hawked a medicine which contained nothing but a blue dye and a little hydrochloric acid. The doctor had four cars, several yachts, and a private plane in which he flew back and forth between Del Rio and Little Rock, Arkansas. He had founded a hospital in Little Rock, in 1937, shortly before his death. Alf Landon narrowly defeated him on one of the three occasions when the doctor ran for governor of Kansas. Brinkley polled thousands of write-in votes in adjacent Oklahoma where he was not even on the ballot. This is even more frightening in light of the fact that the doctor contributed funds to William Dudley Pelley's Silver Shirts, a native fascist organization. In Milford, a touching inscription on the Brinkley Memorial Church reads: *Erected to God and His Son Jesus in appreciation of the many blessings conferred upon me, by J. R. Brinkley.*

*   *   *

Pseudo-scientific writing on the subject of homosexuality is voluminous and varied, especially the German crank literature which is larger than that of any other country. In many ways it parallels the literature of racism. Writers who for one psychological reason or another were violently prejudiced against homosexuals have produced books in which inversion is regarded as a form of evil degeneracy. Other sex authorities, themselves homosexual, have argued that inversion is a superior way of life—that most of the world's great men and women were inverts, and that cultural heights were achieved only by societies in which inversion was prominent.

Eccentric theories of homosexuality range all the way from those of the occultists who think a male soul becomes incarnated in a female body, or vice versa, to more scientific authorities who find "homosexual centers" in the brain. The American writer Charles G. Leland produced a curious book in 1904, *The Alternate Sex,* which argued that the subconscious mind was always of the opposite gender. Leland had frequent dreams in which he imagined he was a woman, and from this, made his generalization. One of the strangest of all homosexual theories was advanced by Sir Richard Burton. He thought there was a geographical strip circling the globe, called the "Sotadic zone," in which inversion was concentrated.

Odd sexual theories, without factual foundation, turn up where you least expect them.[3] In *Ideal Marriage,* by the Dutch gynecologist T. H. Van de Velde, you will learn that celery is an aphrodisiac, and that there is a semen odor in a woman's breath after coitus. Frank Harris, who considered himself one of the world's foremost lovers and sex authorities, reveals, in the first volume of his notorious *My Life and Loves,* an astonishing amount of misinformation. Among other things, he believed that a woman's "safe" period was exactly midway between menstruations, that there was no danger of impregnation from second and third repetitions of the sex act, that a water douche would kill sperm, and that nocturnal emissions were debilitating. As a young man, he once "cured" himself of such emissions by a piece of string, tightly tied. I have not yet read Harris' second volume, which promises to reveal love secrets of Europe and the Orient.

Of many strange theories relating to the sexual act, few have been stranger than the view of Congregationalist John Humphrey Noyes, founder of the Oneida Community. Noyes graduated from Dartmouth in 1830, and studied for a time in the Yale Divinity School. There he became convinced that Christ had returned in 70 A.D., with the Fall of Jerusalem, and since that time expected moral "perfection" of all his followers. Had the Lord not said, "Be ye perfect"? Noyes returned to his birthplace in Vermont, and in 1843, founded at the village of Putney a society called the Putney Corporation of Perfectionists. The group's views aroused so much local opposition, however, that in 1848 they moved to Oneida Creek, in Madison County, New York. In a few years the colony grew to several hundred members.

"Bible Communism" was Noyes' term for the colony's organization, in which all property was communally owned. The group's sex life was governed by two principles—Male Continence and Complex Marriage. It was the founder's belief that every member of the colony should love every other member equally, and that marital fidelity was a sin of selfishness. Sex relations were permitted, therefore, between any man and woman who mutually desired them. On the other hand, births were strictly regulated in accord with eugenic laws designed to improve the community's stock. This was before the word "eugenics" had been coined. Noyes called it "stirpiculture," and apparently his colony was the world's first practical experiment in applied eugenics. In this respect, Noyes perhaps deserves the high praise bestowed upon him by Havelock Ellis and George Bernard Shaw. Noyes' pamphlet, *Male Continence,* was certainly one of the earliest recognitions of the fact that the pleasures of sex and the bearing of children were events which might be separated in the interests of the community.

Noyes is mentioned here, however, because of his unique views concerning the best manner of accomplishing this separation. His method, later termed *coitus reservatus,* involved the complete withholding of male orgasm. After the woman had achieved climax, the man permitted a gradual subsidence of desire. It seems to have worked out fairly well, without apparent injury to the men. It was, in fact, Noyes' curious belief that this practice conserved male energy and led to increased health and virility.

Boys were initiated into the art of *reservatus* at puberty by older women of the colony. Girls were similarly instructed, though at a slightly later age, by the older men. Children were permitted to carefully chosen couples and were raised by the community (instead of by the parents) in the manner first proposed by Plato. This was to promote equality and to prevent what Noyes called the "idolatrous love of mother and child." A system of "mutual criticism," in which each member was publicly criticized by others, took the place of trials and punishments.

The community thrived for some thirty years, issuing a huge amount of propaganda in the form of books, pamphlets, and periodicals, and deriving its income chiefly from the manufacture of high-quality steel hunting traps. Later, the plating of silver flourished as a business enterprise, growing eventually into the now famous Community Plate industry. The colony finally split into factions, and Noyes was forced to flee to Canada. The experiment came to a formal end in 1879.

The view that *coitus reservatus* energizes and prolongs male life has since been held by a number of religious cults and defended by several writers—notably Mrs. Alice Bunker Stockham, an American physician. She called it "Karezza" in a book with that title which she published in 1896. According to Mrs. Stockham, an "exquisite exaltation" is experienced by a couple when the sex act is limited to a "quiet motion," and the climax avoided by both parties (unless, of course, they wish children). "In the course of an hour," she writes, "the physical tension subsides, the spiritual exaltation increases, and not uncommonly visions of a transcendent life are seen and consciousness of new powers experienced."

Mrs. Stockham thought there should be an interval of two to three weeks between performances, and added, ". . . Many find that even three or four months afford a greater impetus to power and growth as well as more personal satisfaction; during the interval the thousand and one lover-like attentions give reciprocal delight, and are an anticipating prophecy of the ultimate union." The book was very popular with women readers, and ran through many editions.

Almost all modern sex authorities are opposed to Karezza, though some for reasons almost as curious. Dr. Marie Stopes, the famous British promoter of birth-control, held that when the fluid from the male prostate was absorbed by the woman, it had a tonic, health-giving effect.[4] Joseph Yahuda, in a recently published English work, *New Biology and Medicine,* 1951, carries this view one step further, and asserts that the male likewise acquires and absorbs health-promoting secretions—a kind of mutual "grafting" process, he terms it.[5] Dr. Alfred Kinsey, in his famed report, takes the view that it is "normal" for a healthy male to reach climax in an extremely short period of time. His reason for thinking this is that chimpanzees ejaculate in ten to twenty seconds. "It would be difficult to find another situation," the doctor writes, "in which an individual who was quick and intense in his responses was labeled anything but superior, and that in most instances is exactly what the rapidly ejaculating male probably is, however inconvenient and unfortunate his qualities may be from the standpoint of the wife in the relationship." Dr. Kinsey makes no attempt to explain how or why evolution would produce such a tragic disparity.

Another modern sex authority, now living in the United States, whose views are even more violently opposed to Karezza, is the German psychiatrist, Wilhelm Reich. In Reich's thinking, the orgasm (for both sexes) assumes a more central and greater importance than in the thinking of any other authority. No neurotic, according to Reich, is capable of experiencing a full and normal orgasm. ". . . There is only one thing wrong with neurotic patients," he writes in his best known work, *The Function of the Orgasm,* "the *lack of full and repeated sexual satisfaction.*" (Italics his.) From this point of view, the orgasm becomes the best index of a patient's mental health, and consequently the achievement of a true "Reichian orgasm," as some of his followers call it, is one of the main objectives of Reichian therapy.

In view of the fact that Reich has in recent years acquired a devoted band of disciples, chiefly in *avant garde* literary and art circles[6] in New York and California, his theories are worth a more extended treatment. The next chapter will discuss some of his remarkable discoveries.

CHAPTER 21

ORGONOMY

WILHELM REICH, the discoverer of orgone energy (or "life energy"), was born in Austria in 1897. He received the M.D. in 1922 from the University of Vienna Medical School, became a protégé of Freud, and for the next eight years rose rapidly in psychoanalytic circles. He held several important teaching and administrative posts in Vienna psychoanalytical organizations, and contributed to their periodicals. You will find many references to him scattered among the footnotes and bibliographies of early Freudian writings.

Politically, Reich was active in the Austrian Socialist Party until he broke with them in 1930, and moved to Berlin where he joined the Communists. Arthur Koestler, in his contribution to *The God that Failed,* 1949, reveals that he and Reich served in the same Party cell. "Among other members of our cell," writes Koestler, "I remember Dr. Wilhelm Reich, founder and director of the *Sex-Pol* (Institute for Sexual Politics). He was a Freudian Marxist; inspired by Malinowski, he had just published a book called *The Function of the Orgasm,* in which he expounded the theory that the sexual frustration of the Proletariat caused a thwarting of its political consciousness; only through a full, uninhibited release of the sexual urge could the working-class realize its revolutionary potentialities and historic mission; the whole thing was less cock-eyed than it sounds."

Reich failed, however, to convince the comrades of the revolutionary importance of his views. Moscow branded his writings "un-Marxist rubbish," and it was not long until he had severed his connections with the Communist movement. Differences with Freud and his followers led eventually, in 1934, to Reich's formal expulsion from the International Psychoanalytical Association.

Having written in 1933 a book attacking German fascism as the sadistic expression of sex-repressed neurotics, Reich was not looked upon kindly by the Nazis when they came to power. He fled to Denmark, then to Sweden, and finally settled in Oslo, Norway, where he continued his research for several years. Here, however, a furious press campaign against his work was instigated, and Reich came to the United States in 1939 to regain the quiet necessary for undisturbed work.

For two years, Reich was an associate professor at the New School for Social Research, in Manhattan. He established the Orgone Institute, a laboratory in Forest Hills, Long Island, and a press in Greenwich Village which began issuing English translations of his books. The books were favorably reviewed in liberal, Socialist, and anarchist periodicals, and cited frequently in such works as Fenichel's *The Psychoanalytic Theory of Neurosis,* 1945, and *Modern Woman, the Lost Sex,* by Lundberg and Farnham, 1947.

At the moment, Reich is a ruddy-faced, distinguished looking man, living in semi-retirement on his estate near Rangeley, Maine. There he directs the multifarious activities of the Orgone Institute and the Wilhelm Reich Foundation. In addition to the publishing of Reich's books, the Foundation also issues the *Orgone Energy Bulletin* (a quarterly which superseded the *International Journal of Sex-Economy and Orgone Research*), *The Annals of the Orgone Institute,* and other literature.

Reich's early books (*The Function of the Orgasm,* 1927; *The Sexual Revolution,* 1930; *The Mass Psychology of Fascism,* 1933; and *Character Analysis,* 1933) were fairly close to the Freudian tradition. Although they contain much debatable material—presented in a repetitious, heavy-handed, totally humorless style—they also contain many fresh and impressive ideas which have become a permanent part of the analytic literature. *Character Analysis,* probably his most significant book, is still used (in the unrevised edition) by many analysts who deplore Reich's later thinking.

Particularly valuable were Reich's early insights into the neurotic aspects of social and political forces, and his stress on sexual health as a prerequisite for genuine morality and political progress. According to Reich, happiness and goodness are the products of sexual well-being, and unless a culture is sexually healthy, all attempts to build a good society are bound to fail. The "change of heart" or "rebirth" that Christian Socialists and Tolstoyan anarchists find essential to political reform is replaced by the Reichian concept of "orgastic potency."

Orgastically potent individuals, in turn, are the product of proper rearing by their parents and society, or they are former neurotics who have successfully undergone orgone therapy. Since there are so few such individuals around (outside of primitive cultures), it follows that most political action is useless. Regardless of how institutions are changed, the same sick individuals take control of them, and the same sick impulses quickly corrupt good intentions. This is why, according to Reich, the Russian Revolution failed so miserably. Not until we have a society of healthy (orgastically potent) citizens will we be able to achieve a decent political order. And when the order is achieved, it will be largely self-regulating, with no need for "compulsive" laws and morality. "Work democracy" is Reich's term for such a society. It is not hard to understand why these views have combined so easily with anarchist sentiments in England and the United States.

It would be out of place to describe here at any greater length Reich's early contributions to psychiatric theory. Many of them are complex and technical, and in order to be understood, would require a mastery of the elaborate and cumbersome Reichian terminology. What has been said, however, should give a faint indication of the importance of the topics which Reich tackled courageously during the German phase of his career.

From this point onward, you may take your choice of one of three possible interpretations of Reich's development. (1) He became the world's greatest biophysicist. (2) He deteriorated from a competent psychiatrist into a self-deluded crank. (3) He merely switched to fields in which his former incompetence became more visible. Critics who favor the last view point out that psychoanalysis is still in such a confused, pioneer state that writings by incompetent theorists are easily camouflaged by technical jargon and a sprinkling of sound ideas borrowed from others. When Reich turned to biology, physics, and astronomy—where there is a solid core of verifiable knowledge—his eccentric thinking became easier to detect.

Whatever the correct explanation may be, there is no doubt about the great turning point in Reich's career. It came in the late thirties when he discovered, in Norway, the existence of "orgone energy." Freud had earlier expressed the hope that some day his theory of the libido, or sexual energy, might be given a biological basis. Reich is convinced that his discovery of orgone energy fulfilled this hope—a discovery which he ranks in importance with the Copernican Revolution. Since coming to America, he has considered himself less a psychiatrist than a biophysicist, probing deeper into the mysteries of orgone energy, and applying this strange new knowledge to the treatment of bodily and mental ailments.

Exactly what is orgone energy? According to Reich it is a nonelectromagnetic force which permeates all of nature. It is the *élan vital* or life force, of Bergson, made practically accessible and usable. It is blue in color. To quote from one of Reich's booklets, "Blue is the specific color of orgone energy within and without the organism. Classical physics tries to explain the blueness of the sky by the scattering of the blue and of the spectral color series in the gaseous atmosphere. However, it is a fact that blue is the color seen in all functions which are related to the cosmic or atmospheric or organismic orgone energy.'' Protoplasm, says Reich, is blue with orgone energy, and loses its blueness when the cell dies. Orgone also causes the blue of oceans and deep lakes, and the blue coloration of certain frogs when they are sexually excited. ''The color of luminating, decaying wood is blue; so are the luminating tail ends of glowworms, St. Elmo's fire, and the aurora borealis. The lumination in evacuated tubes charged with orgone energy is blue.'' (The latter has been photographed on color film and forms the cover photo of the booklet from which the above quotations are taken.)

The so-called ''heat waves'' you often see shimmering above roads and mountain tops, are not heat at all, Reich declares, but orgone energy. These waves do not ascend. They move from west to east, at a speed faster than the earth's rotation. They cause the twinkling of stars. All phenomena which orthodox physicists attribute to ''static electricity'' are produced by orgone energy—e.g., electric disturbances during sunspot activity, lightning, radio interference, and all other forms of static discharges. ''Cloud formations and thunderstorms,'' he writes, ''—phenomena which to date have remained unexplained—depend on changes in the concentration of atmospheric orgone.'' That is why thunder clouds and hurricanes are deeply blue. ''One of the hurricanes which was personally experienced by the writer [Reich] in 1944 was of a deep blue-black color.'' In an article in the *Orgone Energy Bulletin,* July 1951, Reich reports on some experiments made by himself which prove that dowsing rods operate by orgone energy![1]

In the human body, orgone is the basis of sexual energy. It is the *id* of Freud in a bio-energetic, concrete form. During coitus it becomes concentrated in the sexual parts. During orgasm, it flows back again through the entire body. By breathing, the body charges its red blood cells with orgone energy. Under the microscope, Reich has detected the ''blue glimmer'' of red corpuscles as they absorb orgone. In 1947, he measured orgone energy with a Geiger counter. A recent film produced by his associates demonstrates how motors may some day be run by orgone energy.

The unit of living matter, Reich tells us, is not the cell but something much smaller which he calls the ''bion,'' or ''energy vesicle.'' It consists of a membrane surrounding a liquid, and pulsates continually with orgone energy. This pulsation is the dance of life—the basic convulsive rhythm of love which finds its highest expression in the pulsation of the ''orgasm formula.'' Bions propagate like bacteria. In fact, Reich's critics suspect, what he calls bions really *are* bacteria.

According to Reich, bions are constantly being formed in nature by the disintegration of both organic and inorganic substances. The bions first group themselves into clumps, then they organize into protozoa! In Reich's book, *The Cancer Biopathy,* 1948, are a series of photomicrographs showing various types of single-cell animals, such as amoebae and paramecia, in the process of formation from aggregates of bions.

Needless to say, no "orthodox" biologist has been able to duplicate these revolutionary experiments. The opinion of bacteriologists who have troubled to look at Reich's photographs is that his protozoa found their way into his cultures from the air, or were already present on the disintegrating material in the form of dormant cysts. Reich is aware of these objections, of course, and vigorously denies that protozoa could have gotten into his cultures in any way other than the way he describes.

In 1940, Reich invented a therapeutic box. Technically called an Orgone Energy Accumulator, it consists of a structure resembling a short phone booth, made of sheet iron on the inside, and organic material (wood or celotex) on the outside. Later, three to twenty-layer accumulators were made of alternate layers of steel wool and rock wool. The theory is that orgone energy is attracted by the organic substance on the outside, and is passed on to the metal which then radiates it inward. Since the metal reflects orgone, the box soon acquires an abnormally high concentration of the energy. In Reich's laboratory in Maine, he has a large "orgone room" lined with sheet iron. When all the lights are turned out, he claims, the room takes on a blue-gray luminescence.

According to Dr. Theodore P. Wolfe, Reich's former translator, "The Orgone Energy Accumulator is the most important single discovery in the history of medicine, bar none." In 1951, Reich issued a booklet (there is no author's name on the title page) called *The Orgone Energy Accumulator,* which is the best available reference on the accumulator's construction and medical use. Most of the following material is taken from this work.

Orgone accumulators can be bought, but the Foundation holds rights to their medical use, and rents them to patients on a monthly basis, the charge varying with ability to pay. By sitting inside, lightly clothed, you charge your body with orgone energy. At first you feel a prickling, warm sensation, accompanied by reddening of the face and a rise in body temperature. There is a feeling that the body is "glowing." After you have absorbed as much orgone as your system demands, you begin to feel a slight dizziness and nausea. When this happens, you step out of the accumulator, breathe some fresh air, and the overcharge symptoms quickly vanish. "Under no circumstance," Reich's booklet reads, "should one sit in the accumulator for hours, or, as some people do, go to sleep in it. This can cause serious damage (severe vomiting, etc.). It is better, if necessary, to use the accumulator several times a day at shorter intervals than to prolong one sitting unnecessarily. At this stage of research, no accumulator over 3-layers should be used without medical supervision."

For people who are bedridden, Reich has developed an "orgone energy accumulator blanket." This is a curved structure which can be placed on the bed, over a reclining figure, while a set of flat layers goes beneath the mattress. There also are tiny orgone boxes, called "shooters," for application to local areas. A flexible iron cable, from which the inner wires have been removed, carries the energy from the box to the part of the body being irradiated. If the body area is larger than the end of the cable, a funnel is attached. "Only *metal* (iron) funnels can be used," the booklet warns, "funnels made of plastic are ineffective."

It is Reich's belief that the natural healing process of a wound is greatly accelerated by applying the shooter. "Even severe pain will be stopped soon after the accident if orgone energy is applied locally through the shooter," the booklet states. "In severe cases of burns, experience has revealed the amazing fact that no blisters appear, and that the initial redness slowly disappears. The wounds heal in a matter of a few hours; severe ones need a day or two. Only chronic, advanced degenerating processes require weeks and months of daily irradiation. But here, too, severe lesions, as for instance *ulcus varicosus,* will yield to orgone energy irradiation."

In addition to speeding up healing, the energy also sterilizes a wound. "Microscopic observation shows that, for example, bacteria in the vagina will be immobilized after only one minute of irradiation through an inserted glass pipe filled with steel wool. . . . *Do not mix orgone irradiation with other, chemical applications. Orgone energy is a very strong energy.* We do not know as yet what such a mixture can do." (Italics his.)

The following ailments are listed in Reich's booklet as ills to which orgone treatment can be applied with great benefit: fatigue, anemia, cancer in early stages (with the exception of tumors of the brain and liver), acute and chronic colds, hay fever, arthritis, chronic ulcers, some types of migraine, sinusitis, and any kind of lesion, abrasion, or wound. *"Neuroses cannot be cured with physical orgone energy,"* the booklet states. "Only the biopathic somatic background and certain somatic consequences of severe neuroses can be alleviated or diminished." In Reich's opinion, disease-producing bacteria are often formed by body bions in a degenerate state because of a patient's neuroses. This "auto infection" can be cleared up by sitting in the accumulator, though many chronic ailments require several years of treatment. The body's slow progress toward a higher energy level is observed by the Wilhelm Reich Blood Tests.

Cancer cells, according to Reich, are protozoa which develop from the bions coming from disintegrating tissues. "Many cancer cells," he has observed, "have a tail and move in the manner of a fish." If the formation of these protozoa were not stopped by early death, he writes, "the cancer mouse or the cancer patient would change completely into protozoa." These quotations are from *The Cancer Biopathy* where you will find it all explained in detail.

Orgone therapy includes a type of treatment called "character analysis," exact details of which have not been printed for fear they would be misunderstood and abused. In most cases, the patient lies in a bathing suit on the couch. This is to give the orgonomist an opportunity to observe the patient's muscular reactions. It is Reich's belief that every neurosis is linked to "muscular armor"—a rigidity such as a furrowed brow, tense neck muscles, hunched shoulders, tight anus, and so on. "There is no neurotic . . . who does not show a tension in the abdomen," Reich has written.

The orgonomist tries to make the patient understand the cause of his muscular tensions, and there are certain technical procedures to help him get rid of them. If, for example, he has a tenseness around the jaws, because of an unconscious desire to bite someone, he may be given a towel and told to bite it. Parallel with this "orgone therapy" is the "character analysis." The latter involves free association and other standard devices which seek to penetrate the patient's "character armor."

An important part of orgone therapy is breathing, it being another Reichian belief that "There is no neurotic . . . capable of exhaling . . . deeply and evenly." This is owing to abdominal tenseness. The patient must overcome his inhibition against breathing out properly, often with the therapist assisting by applying pressure on the abdomen. As the breathing therapy advances, a curious phenomenon appears. The patient has an involuntary impulse to move his pelvis. A "dead pelvis," according to Reich, is a rigidity due to "pleasure anxiety," in turn rooted in childhood punishments for wetting the bed, playing with genitals, and so on. It prevents the neurotic from moving his pelvis naturally during the sex act, and also causes lumbago and hemorrhoids. The forward movement which appears spontaneously as the therapy proceeds is an instinctive motion. It is the motion made by the hips during normal coitus—the "orgasm reflex." A *voluntary* effort to move the pelvis during the sex act is considered neurotic.

The final goal of the therapy is the development of the patient's ability to have a full and complete orgasm—this being possible only to the "genital" or non-neurotic personality. The normal sex act follows the Reichian four-beat "orgasm formula"—mechanical tension, bio-electrical charge, bio-electrical discharge, and mechanical relaxation. During the orgasm, orgone energy raises the bio-electrical potential of the skin, especially on erogenous zones. One of the oscillograph photograms reproduced in *The Function of the Orgasm* is captioned, "Mucous membrane of the anus in a woman in a state of sexual excitation."

In recent years, Reich's sense of personal greatness and bitterness against colleagues who dismiss his orgone research as evidence of the tragic disintegration of a once brilliant thinker have grown alarmingly. "Emotional plague" is his term for the social manifestation of sexual sickness, and as might be expected, he treats all opposition to his work as signs of the plague. In 1947, this aspect of the plague reached a climax when the Pure Food and Drug Administration began to investigate his orgone accumulators, and Mildred Brady wrote two magazine articles about him ("The New Cult of Sex and Anarchy," *Harpers*, April, 1947, and "The Strange Case of Wilhelm Reich," *New Republic*, May 26, 1947). The following year Dr. Wolfe penned a pamphlet rebuttal titled *Emotional Plague Versus Orgone Biophysics*, from which the following statement by Reich is taken:

> It is an old story. It is older than the ancient Greeks whom we consider the bearers of a flourishing culture. . . . It was no different two thousand years later. Giordano Bruno, who fought for scientific knowledge and against astrological superstition, was condemned to death by the Inquisition. It is the same psychic pestilence which delivered Galileo to the Inquisition, let Copernicus die in misery, made Leeuwenhoek a recluse, drove Nietzsche into insanity, Pasteur and Freud into exile. It is the indecent, vile attitude of contemporaries of all times. This has to be said clearly once and for all. One cannot give in to such manifestations of the pestilence.

Even stronger language appears in Reich's *Listen Little Man!*, an angry volume issued in 1948 and delightfully illustrated by William Steig (Steig is an associate member of the Wilhelm Reich Foundation). The book purports to attack the neurotic, sexually-sick "little man" who fails to see his own sickness, and is responsible for the rise of all varieties of fascism. Actually, the book is a violent outburst against the world for its failure to recognize Reich's greatness.

". . . When the discoverer has just found out why people die of cancer," Reich writes, ". . . and . . . you, Little Man, happen to be a Professor of Cancer Pathology, with a steady salary, you say that the discoverer is a faker; or that he does not understand anything about air germs . . . or you insist that you have a right to examine him, in order to find out whether he is qualified to work on "your" cancer problem, the problem you cannot solve; or you prefer to see many, many cancer patients die rather than admit that *he* has found what *you* so badly need if you are to save your patients. To you, your professional dignity, or your bank account, or your connection with the radium industry means more than truth and learning. And that's why you are small and miserable, Little Man.''

Like all other ''decent writing of today,'' Reich declares, his book is addressed ''to the culture of 1000 or 5000 years hence as was the first wheel of thousands of years ago to the Diesel locomotive of today.'' It is true, he writes, that a new ''era of atomic energy'' has arrived. '' . . . But not in the way you think. Not in your inferno, but in my quiet, industrious laboratory in a far corner of America.''

Reich compares himself to a lonely eagle trying to hatch chicken eggs in the vain hope that he may hatch eagles. ''But no, at the end they are nothing but cackling hens. When the eagle found out this, he had a hard time suppressing his impulse to eat up all the chicks and cackling hens. What kept him from doing so was a small hope. The hope, namely, that among the many cackling chicks there might be, one day, a little eagle capable like himself, to look from his lofty perch into the far distance, in order to detect new worlds, new thoughts and new forms of living. It was only this small hope that kept the sad, lonely eagle from eating up all the cackling chicks and hens. . . . But he thought about it and began to pity them. Sometime, he hoped, there would be, there would have to be, among the many cackling, gobbling and short-sighted chickens, a little eagle capable of becoming like himself. The lonely eagle, to this day, has not given up this hope. . . .''

Finally Reich concludes: ''Whatever you have done to me or will do to me in the future, whether you glorify me as a genius or put me in a mental institution, whether you adore me as your savior or hang me as a spy, sooner or later necessity will force you to comprehend that *I have discovered the laws of the living*. . . . I have disclosed to you the infinitely vast field of the living in you, of your cosmic nature. That is my great reward.''

Reich's latest book, *Cosmic Superimposition,* 1951, 30 years after his first steps in natural science, carries orgonomy into the realm of astrophysics. ''We are moving into the open spaces,'' he writes, ''to find, if possible, what the newborn infant brings with him onto the stage inside.'' As he expresses it, he is turning from the microscope, which he used in exploring the microcosmos, to the telescope to explore the macrocosmos—in search of some common element which will unite human love with all of nature.

Obviously the four-beat orgasm formula is not the answer. This convulsion is confined only to the living, and Reich carefully warns against the danger of finding analogies for it in the inorganic world. Earthquakes, for example, are convulsions—but not, he points out, of the orgasm type.

The true answer lies in something simpler. It is the "sexual embrace" which precedes the orgasm, and which Reich terms "superimposition." "Whence stems the overpowering drive toward superimposition of male and female orgonotic systems?" he asks. The reply is that it is a drive which runs through all of nature, from the lowest pre-atomic level to the reaches of the stars. It is the basic pattern both of love and natural law.

Orgone energy units, Reich explains, move in a spiral path. When two or more such units approach each other, they superimpose in a way analogous to a sexual embrace. The result is the creation of "offspring"—in this case, a particle of matter!

On the astrophysical level, a similar phenomenon accounts for the birth of galaxies. Structureless streams of orgone energy (the Magellanic Cloud is cited by Reich as an example) are attracted to each other. They superimpose in a great cosmic embrace. The result—a galaxy! The book contains many excellent photographs of nebulae, showing the spiral arms in whose luminous embrace the galactic suns emerge.

But this is not all. Space is not empty. It is in reality a vast ocean of orgone energy, and its movements are responsible for the motions of the heavenly bodies. "The function of gravitation is real," he writes, "It is, however, not the result of mass attraction but of the converging movements of two original orgone energy streams. . . ." Again: "The sun and the planets move in the same plane and revolve in the same direction due to the movements and direction of the cosmic orgone energy stream in the galaxy. Thus, the sun does not 'attract' anything at all. It is merely the biggest brother of the whole group."

And so Reich, searching the skies from his observatory in Maine, has found the secret of gravitation, the origin of matter, and the cause of the shape of spiral nebulae. "As the process of functional reasoning unfolded more and more," he confesses, "the observer [i.e., Reich] . . . experienced most vividly his own amazement at his own power of reasoning which was in such perfect harmony with the natural events thus disclosed."

At the moment Reich is hard at work on an even more important problem—an antidote to nuclear destruction of life. Back in 1945, shortly after the first atomic bomb had fallen, he had written in his journal that "orgone energy is in fact nothing but 'atomic' energy in its original and natural form." Unlike atomic energy, however, it creates matter and strengthens life. It operates in a slow, constructive fashion. Atomic energy destroys in an explosive fashion. The two energies are, in fact, the underlying principles of love and hate, good and evil, God and Satan. "The horror," Reich wrote, "at the 'discovery' of the atom bomb has its counterpart in the quiet but glowing enthusiasm of anyone who works with orgone energy or experiences its therapeutic effects."

Since atomic energy and orgone energy have such contradictory properties, it was only natural for Reich to suppose that orgone might be useful as an antidote for nuclear radiation. "If, against any expectation, I should ever discover any murderous potentiality of the orgone energy," he wrote in 1945, "I would keep the process secret. We shall have to learn to counteract the murderous form of the atomic energy with the life-furthering function of the orgone energy and thus render it harmless."

In January, 1951, Reich established his ORANUR project (the letters stand for "Orgonomic Anti-Nuclear Radiation") to work out the details of this stupendous undertaking. Reports on his progress are currently appearing in the *Orgone Energy Bulletin,* and other publications. The early experimental work reads like a comic opera. Reich purchased some radium, brought it into his orgone room, then before anyone knew what was happening, the OR (orgone energy), in combating the NR (nuclear energy), ran amok. It unexpectedly turned into DOR ("deadly orgone energy") and the entire laboratory crew came down with "ORANUR sickness."

The latest report on all this is a brief notice titled "Emergency at Orgonon." It is dated May 12, 1952, and reads as follows:

> Since March 21, 1952, an acute emergency exists at Orgonon [the name of Reich's headquarters in Rangeley]. The emergency is due to severe Oranur activity. This activity set in a few hours after the tornado developed in the Middle West on March 21. The details of the emergency which developed from high-pitched Oranur activity will be reported in the second report on the Oranur experiment which is due to be published sometime during October, 1952, or sometime in 1953.
>
> The routine work at Orgonon has collapsed. Several workers had to abandon their jobs. Most buildings at Orgonon became uninhabitable. No work could be done in these buildings up to date. It is uncertain when circumstances will return to normal, if at all, and it is also uncertain what exactly has caused the emergency.
>
> The work had to be contracted to the necessary minimum and priority has been given to activities which promise elucidation and mastery of the situation.

Of course, one way to find out exactly what happened would be to call in a nuclear physicist, but it is doubtful if Reich will deem this necessary. Presumably, the situation will eventually be mastered, the mystery will be fully explained, and the work, let us hope, will proceed with greater caution.

CHAPTER 22

DIANETICS

DIANETICS (from a Greek word meaning "thought") is a new science of the mind discovered by Lafayette Ronald Hubbard, a popular writer of science fiction. According to the opening sentences of his first book on the subject, "The Creation of dianetics is a milestone for Man comparable to his discovery of fire and superior to his inventions of the wheel and arch. . . . The hidden source of all psychosomatic ills and human aberration has been discovered and skills have been developed for their invariable cure."

That word "invariable" is not a typographical mistake. "Dianetics is an exact science," Hubbard writes, "and its application is on the order of, but simpler than, engineering. Its axioms should not be confused with theories since they demonstrably exist as natural laws hitherto undiscovered." Dianetic therapy operates with mathematical precision. It never fails. These are claims worth looking into, but before surveying the basic tenets of dianetics, let us first glance at the fabulous rise of the movement.

The founder, L. Ron Hubbard, is a large, good-looking man with flaming red hair and a tremendous energy drive that keeps him in a constant state of high gear. Friends vary widely in estimates of what makes Ronald run. To some he is an earnest, honest, sincere guy. To others he is the greatest con man of the century. Still others regard him as basically sincere, with just a touch of the charlatan, and now a tragic victim of his own psychoses.

Hubbard was born in 1911 at Tilden, Nebraska. Exact details about the rest of his life are hard to come by. He seems to have been in the Marines when a young man. For a few years in the early thirties, he attended the

George Washington University Engineering School, in the nation's capital, but did not graduate. He never held an engineering job, but evidently this schooling gave him the engineer's outlook that underlies so much of dianetics. For the past twenty years, he has been an enormously prolific writer of pulp fiction, with occasional stints at radio and movie scripting. He holds a glider pilot's license. He is an expert small-boats mariner. For a while, he sang and played the banjo on a radio program in California (he has a deep, rich voice). He considers himself an explorer, having made numerous jaunts around the globe, including a sojourn in Asia where he studied mysticism. During the war, he was a naval officer on destroyer escort duty, and was severely wounded in action.

According to Hubbard, it was in 1938 that he first discovered the basic axioms of dianetics and began his twelve years of research. Many of his friends insist, however, that these twelve research years are entirely mythical, and that it was not until 1948 that dianetics was hatched. At any rate, one of his earliest patients was John Campbell, Jr., editor of *Astounding Science Fiction.* Campbell was suffering, among other things, from chronic sinusitis. His treatment by Hubbard so impressed him, that in May 1950, he published in his pulp magazine the first public report on dianetics. It was an article by Hubbard, written in a few hours, and in a style resembling the broadcast of a football game. The article apparently aroused science-fiction fans to such a pitch of anticipation that when Hubbard's book, *Dianetics: The Modern Science of Mental Healing,* was published a few weeks later by Hermitage House, they grabbed the first copies they could lay their hands on.

Dianetics is a book of impressive thickness, written in a repetitious, immature style. Hubbard claims he wrote it in three weeks. This is believable because most of his writing is done at lightning speed. (For a while, he used a special electric IBM typewriter with extra keys for common words like "and," "the," and "but." The paper was on a roll to avoid the interruption of changing sheets.) Nothing in the book remotely resembles a scientific report. The case histories are written largely out of Hubbard's memory and imagination. Like the later works of Wilhelm Reich, his book is simply a Revelation from the Master, to be tested and confirmed by lesser men. It is dedicated, curiously, to Will Durant. An appendix on "The Scientific Method," is signed *John W. Campbell, Jr., nuclear physicist.* (Campbell attended the Massachusetts Institute of Technology for three years, then transferred to Duke University where he was graduated. For a short time he worked in the laboratory of Mack Trucks, Inc., New Brunswick, N. J.)

The book was a tremendous success. Early purchasers were science-fiction fans, but it was not long until the volume launched a nation-wide cult of incredible proportions. Dianetics became a fad of the movie colony. It struck the colleges. Students held "dianetic parties" at which they tried the new therapy on each other. At Williams College, in Williamstown, Massachusetts, a distinguished professor of political science, Frederick L. Schuman, was drawn into the movement. He visited Hubbard, lectured on dianetics in Boston, wrote indignant letters to periodicals that reviewed Hubbard's book unfavorably (see the *New Republic,* September 11,[1] and *New York Times Book Review,* August 6, 1950), and even contributed an enthusiastic article on the subject to *Better Homes and Gardens,* April, 1951.

A Dianetic Research Foundation was established at Elizabeth, New Jersey, with centers in the nation's major cities. Hundreds of practitioners trained by the foundation put up their shingles in Hollywood, on Park Avenue, and in the Gold Coast of Chicago. Hubbard flew back and forth across the continent giving lecture demonstrations. A *Dianetic Auditor's Bulletin* made its appearance. Later the foundation expanded and moved to Wichita, Kansas, where it took the name of the Hubbard Dianetic Foundation, Inc. For $500 the institution offered thirty-six hours of therapy, and for the same fee gave four to six weeks of instruction. Those who passed the tests became certified dianetic "auditors."

What, precisely, is dianetics?

Briefly, it is the view that all mental aberrations (neuroses, psychoses, and psychosomatic ills) are caused by "engrams." To make this clear, however, we must first make a journey through the jungle of Hubbard's elaborate terminology.

The conscious mind is called by Hubbard the "analytical mind." It operates like a gigantic computing machine. The working is flawless. It may, however, direct the body in an aberrated manner if it is fed false data by the unconscious mind.

The unconscious mind is termed the "reactive mind." Actually, it is always conscious—even when a person is sleeping, or "unconscious" from some other cause. The reactive mind is incapable of "thinking" or "remembering." It is a moron. But when the analytical mind becomes unconscious or semi-conscious, in a manner associated with bodily pain or painful emotion, the reactive mind starts to make "recordings." These recordings are called "engrams." They are like phonograph records except that they record, in addition to sounds, *all* the perceptions received by the reactive mind while the analytical mind is "turned off."

Hubbard illustrates this with the following example: "A woman is knocked down by a blow. She is rendered 'unconscious.' She is kicked and told she is a faker, that she is no good, that she is always changing her mind. A chair is overturned in the process. A faucet is running in the kitchen. A car is passing in the street outside. The engram contains a running record of all these perceptions: sight, sound, tactile, taste, smell, organic sensation, kinetic sense, joint position, thirst record, etc. The engram would consist of the whole statement made to her when she was 'unconscious': the voice tones and emotion in the voice, the sound and feel of the original and later blows, the tactile of the floor, the feel and sound of the chair overturning, the organic sensation of the blow, perhaps the taste of blood in her mouth or any other taste present there, the smell of the person attacking her and the smells in the room, the sound of the passing car's motor and tires, etc.''

Engrams, then, are perceptual recordings made when the analytical mind is turned off in a manner associated with pain or painful emotion. Unconsciousness because of injury, anesthetics, illness, drugs—even an alcoholic stupor—are sufficiently "painful" to produce engrams. Since the reactive mind is an idiot, incapable of evaluating, everything it experiences goes into the engrams. These engrams are filed away in the "reactive bank." Hubbard has classified and labeled them in various ways—such as bouncers, denyers, groupers, holders, and misdirectors—but we need not go into these distinctions. Nor will we have space to discuss his "demon circuits"—commanding demons, critical demons, listen-to-me demons, tell-you-what-to-say demons, and so on. A glossary of the major Hubbardian terms will be found at the back of *Dianetics.*

All neuroses, psychoses, and psychosomatic ailments (including the common cold and possibly diabetes and cancer) are caused by engrams. In most cases, the trouble-making engrams are recorded *before one is born.* This introduces us to Hubbard's most revolutionary concept—the prenatal engram.

In *Dianetics,* you learn that the embryo is capable of recording engrams immediately after conception. How these records are made, since the embryo does not develop sense organs until late in its history, remains a profound mystery. They take place on a cellular level, involving some unknown type of change within the protoplasm. According to Hubbard, life in the womb is far from Paradise. "Mama sneezes, baby gets knocked 'unconscious.' Mama runs lightly and blithely into a table and baby gets its head stoved in. Mama has constipation and baby, in the anxious effort, gets squashed. Papa becomes passionate and baby has the sensation of being put into a running washing machine. Mama gets hysterical, baby gets an engram. Papa hits Mama, baby gets an engram. Junior bounces on Mama's lap, baby gets an engram. And so it goes.''

It is also very noisy in the uterus. "Intestinal squeaks and groans, flowing water, belches, flatulation and other body activities of the mother produce a continual sound. . . . When mother takes quinine a high ringing noise may come into being in the foetal ears as well as her own—a ringing which will carry through a person's whole life.'' Moreover, the uterus is very tight in later prenatal life. If Mama has high blood pressure, "it is extremely horrible in the womb."

In addition to being knocked out by blows, coughs, sneezes, vomiting, and so on, the poor embryo can also be rendered unconscious by the violent pressures of the sex act, and—understandably—by attempted abortions. Throughout his book, Hubbard reveals a deep-seated hatred of women, but this hatred is most clearly indicated by his obsession with what dianeticians call "AA"—attempted abortion. When Hubbard's Mamas are not getting kicked in the stomach by their husbands or having affairs with lovers, they are preoccupied with AA—usually by means of knitting needles. "Twenty or thirty abortion attempts are not uncommon in the aberee," Hubbard writes, "and in every attempt the child could have been pierced through the body or brain." These experiences naturally produce the worst engrams because they are usually accompanied by verbal expressions charged with emotion. Since all these remarks are recorded literally by the embryo, they create engrams capable of causing great damage in later life when they are fed as data to the conscious mind.

To cite an example from Hubbard: Papa beats Mama on the stomach, knocking baby unconscious. At the same time Papa yells, "Take that! Take it, I tell you. You've got to take it!" Later in life, these sentences are interpreted literally, and the person becomes a kleptomaniac or thief. "Oh, this language of ours," Hubbard exclaims sadly, "which says everything it doesn't mean! Put into the hands of the moronic reactive mind, what havoc it wreaks! Literal interpretation of everything!"

Before a prenatal engram can cause damage, however, it must be "keyed in." This occurs when the person has a painful experience which closely resembles, in some respect, the dormant engram. Hubbard illustrates this by citing another mother, struck in the abdomen by her husband. The husband shouts, "God damn you, you filthy whore: you're no good!" This engram contains a headache, a falling body, the grating of teeth, and the mother's intestinal sounds. Several years later, the child is slapped by the father who says, "God damn you: you're no good." The child cries, and that night has a headache. The engram has been "keyed in." "Now the sound of a falling body or grating teeth or any trace of anger of any kind in the father's voice will make the child nervous. His physical health will suffer. He will begin to have headaches."

Here are a few additional samples from Hubbard of how prenatal engrams cause later difficulties. A pregnant mother is straining for a bowel movement. This compresses the baby into painful unconsciousness. The mother talks to herself and says, "Oh, this is hell. I am all jammed up inside. I feel so stuffy I can't think. This is too terrible to be borne." Later in life the child has frequent colds ("I feel so stuffy. . . ."). An inferiority complex develops because he feels he is "too terrible to be *born.*" (Puns of this sort turn up frequently in dianetic therapy. An auditor reported recently that a psychosomatic rash on the backside of a lady patient was caused by prenatal recordings of her mother's frequent requests for aspirin. The literal reactive mind had been feeding this to her analytical mind in the form of "ass burn.")

Another of Hubbard's patients was a morose young man whose attitude toward life was expressed by Hamlet's famous line, "To be or not to be, that is the question." Hubbard's therapy revealed that the man's mother, when pregnant, had been beaten by an actor husband who then proceeded to recite from *Hamlet.* And so, Hubbard writes, the young man "would sit for hours in a morose apathy wondering about life."

Some of the most horrible engrams arise from the fact that a child is named after his father. If the pregnant mother is committing adultery, as so many of Hubbard's Mamas do, she is likely to make unkind remarks about George—meaning her husband. These remarks are recorded, of course, by the innocent embryo who is being knocked unconscious by the sex act. If the child is also named George, one can imagine the awful consequences. Since engrams are taken literally, Junior assumes that all these remarks are about *him!* "It is customary," writes Hubbard, "to shudder, in dianetics, at the thought of taking on a Junior case."

The technique of dianetic therapy—known as auditing—is designed to "erase" the patient's engrams. The process begins by having the patient relax on a comfortable chair, or lie on a couch, in a darkened room. He closes his eyes and keeps them closed throughout the session. The auditor sits beside him, and by talking to him, places him in a "dianetic reverie." This is indicated by fluttering eyelids, and is similar to the early stages of hypnosis. The patient remains in full possession of his will, however, and after the session (usually two hours long) will recall everything.

Prompted by the auditor, the patient "goes back" along his "time track," returning to early engram-forming experiences. As he recounts these experiences, the engrams slowly lose their evil power. Eventually they are totally "erased." This means they have been taken from the "reactive bank" and refiled in the "standard memory bank" where they can be recalled by the conscious mind.

The auditor tries to send the patient back to his earliest engram, known as the "basic-basic." The reason for this is that once the BB (basic-basic) has been erased, later engrams erase more easily. The BB is usually formed a few weeks after conception, though it may trace back as far as the zygote (fertilized ovum). Eventually, almost every patient experiences a "sperm dream," in which he imagines himself swimming up a channel, or (as the egg) waiting to be met by the sperm. At first, Hubbard thought this dream had little meaning as far as engrams are concerned, but more recently he has found cases of engrams formed in the sperm and ovum before fertilization occurs.

While an engram is being "reduced" by recounting, the patient tends to yawn and stretch. The yawn is regarded as a significant sign of successful therapy, and must not be misinterpreted as meaning that the patient is bored or drowsy. Curious aches and pains appear in various parts of the body, then vanish mysteriously. These are the ghosts of psychosomatic ills which he will never have again. When the engram is finally erased, the patient experiences sudden relief and pleasure, and usually laughs wildly. One patient, Hubbard reports, was so relieved when an engram was erased that he laughed for two days without stopping. This also must not be misinterpreted. One might wrongly suppose that the patient was laughing at how preposterous the whole procedure was.

It takes about twenty hours of auditing to turn an aberrated patient into a "release." A release is a person free of all *major* neuroses and ills. According to Hubbard, it "is a state superior to any produced by several years of psychoanalysis, since the release will not relapse." As the auditing continues, the release becomes a "pre-clear," and finally a "clear." The clear is, literally, a superman—an evolutionary step toward a new species. He is a person completely free of engrams. All of them have been erased and refiled. He has no neuroses or psychosomatic ills. "Clears do not get colds," Hubbard informs us. If he is wounded, the wounds heal faster. His eyesight is better. His I.Q. is raised markedly. As Hubbard expresses it, "The dianetic clear is to the current normal individual as the current normal is to the severely insane."

Hubbard points out that the length of time required to "process" a clear varies widely with the patient, and although he intimates that a few people have been cleared, exactly who they are is considerably less than clear. In 1950, speaking to an audience of 6,000 in the Shrine Auditorium, Los Angeles, Hubbard introduced a coed named Sonya Bianca as a clear who had attained perfect recall of all "perceptics" (sense perceptions) for every moment of her past. In the demonstration which followed, however, she failed to remember a single formula in physics (the subject in which she was majoring), or the color of Hubbard's tie when his back was turned. At this point, a large part of the audience got up and left. Hubbard later produced a neat dianetic explanation for the fiasco. He had called her from the wings by saying, "Will you come out here *now*, Sonya?" The "now" got her stuck in present time. As for Hubbard himself, he freely admits *he* is not a clear. He decided, he says, to devote all his energies to giving dianetics to the world rather than spend more time having himself processed.

One of the most important of the many branches of dianetics is what Hubbard calls "preventive dianetics." This consists in exercising great care to prevent the formation of engrams while a person is unconscious or when there is a possibility an embryo may become unconscious. It means, for example, maintaining absolute silence while helping people severely injured or ill. "Say nothing and make no sound around an 'unconscious' or injured person," Hubbard writes. "To speak, no matter what is said, is to threaten his sanity. Say nothing while a person is being operated upon. Say nothing when there is a street accident. Don't talk!"

Again: "Maintain silence in the presence of birth to save both the sanity of the mother *and* the child and safeguard the home to which they will go. And the maintaining of silence does not mean a volley of 'sh's,' for those make stammerers."

A mother, of course, must be exceedingly careful during pregnancy. She must not talk while having a bowel movement, coughing, sneezing, having intercourse, being beaten by her husband, or punched in the abdomen by a doctor seeking to determine whether she is pregnant. Nor should anyone else talk in her presence during these events. "If the husband uses language during coitus," writes Hubbard, "every word of it is going to be engramic. If the mother is beaten by him, that beating and everything he says and that *she* says will become part of the engram."

By a combination of dianetic therapy and preventive dianetics, the world may now move forward toward a superior culture. Hubbard closes his book by picturing two plateaus, one higher than the other, and separated by a chasm. An engineer builds a bridge across the canyon. People start to cross over from the lower plateau to the higher. "What sort of an opinion would you have of the society on the lower plateau?" Hubbard asks, "if they but moaned and wept and argued and gave no hand at all in the matter of widening the bridge or making new bridges?" The answer is clear. Dianetics is the first crude bridge, but it must be improved. Hubbard closes his book with these ringing words: "For God's sake, get busy and build a better bridge!"

A more revealing picture of the coming dianetic order is given by Hubbard in a lengthy letter in *Astounding Science Fiction*, August, 1950. Since clears are superior persons with higher I.Q.'s, he writes, they will naturally become the aristocracy of the new culture—a wide gulf separating them from all others. " . . . One sees with some sadness that more than three-quarters of the world's population will become subject to the remaining quarter as a natural consequence and about which we can do exactly nothing." Fortunately, Hubbard adds, the clears will be free from evil motives (Hubbard's conviction being that human nature, without engrams, is basically good), and this "will inhibit their exploitation of the less fortunate."

Science of Survival, a new book covering simplified and speedier processing techniques, was published by Hubbard in 1951. If *Dianetics* was written in three weeks, this book, almost as big, appears to have been written in three days. It introduces dozens of new terms such as MEST (the initial letters of matter, energy, space, and time), theta (life energy), entheta, and enMEST—and goes in heavily for metaphysics and reincarnation.

A staff-written book, *Child Dianetics,* for processing children of the ages five to thirteen, has also appeared, as well as a *Handbook for Pre-Clears.* Recent circulars from Hubbard advertise additional works, all by himself. They include *Symbological Processing; What to Audit; How to Live and Still Be an Executive* (guaranteed to eliminate "management ulcers"); *Original Thesis* (the first written version of *Dianetics*—a manuscript Hubbard tried unsuccessfully to sell to numerous publishers, including Shasta, a Chicago science-fiction press); and *Excalibur.*

The amazing story behind *Excalibur* was revealed by Arthur J. Cox in the July, 1952, issue of *Science-Fiction Advertiser,* a magazine published by science-fantasy fans in Glendale, California. In 1948, Hubbard told the California fans that during an operation performed on him for injuries received while in the Navy, he was actually dead for eight minutes. As Cox tells it, "Hubbard realized that while he was dead, he had received a tremendous inspiration, a great Message which he must impart to others. He sat at his typewriter for six days and nights and nothing came out—then, *Excalibur* emerged. *Excalibur* contains the basic metaphysical secrets of the universe. He sent it around to some publishers; they all hastily rejected it. . . . He locked it away in a bank vault. But then, later, he informed us that he would try publishing a 'diluted' version of it. . . . *Dianetics,* I was recently told by a friend of Hubbard's, is based upon one chapter of *Excalibur.*"

On Hubbard's advertising sheets, the blurb for *Excalibur* is worth quoting. "Mr. Hubbard wrote this work in 1938. When four of the first fifteen people who read it went insane, Mr. Hubbard withdrew it and placed it in a vault where it remained until now. Copies to selected readers only and then on signature. Released only on sworn statement not to permit other readers to read it. Contains data not to be released during Mr. Hubbard's stay on earth. The complete fast formula of clearing. The secret not even *Dianetics* disclosed. Facsimile of original, individually typed for manuscript buyer. Gold bound and locked. Signed by author. Very limited. Per copy . . . $1,500.00."

Another recent Hubbard work, called *Self Analysis* (published in 1951 by the International Library of Arts and Sciences, whatever that is), carries even further Hubbard's intrepid attempts to produce parodies of his original ideas. This book enables the reader to give himself a "light processing." The author's claims, as usual, are quite modest. "Self analysis cannot revive the dead," he says in his opening sentence. "Self analysis will not empty insane asylums or stop wars. These are the tasks of the dianetic auditor and the group dianetic technician." The book is written only for stable readers who want to improve their health, happiness, and efficiency. If you are stable enough, there is no danger. Otherwise? "I will not mislead you," Hubbard confesses. "A man could go mad simply reading this book."

Upon inspection, the book seems harmless enough. It consists mainly of page after page of questions which the reader asks himself, such as "Can you recall a time when somebody you liked was asleep?" Or "Can you recall a time when you skipped rope?" To aid the reader in meditating on these episodes, Hubbard provides a cardboard disk with slots cut in it. The disk is placed on the page so that a question shows through one of the slots. If the top of the disk says "sight," you try to "see" the incident. On the next question you rotate the disk so another "sense" appears on top—say "smell." You now try to recall the "smell" of the episode. As you can imagine, many curious combinations of senses and memories result from this ingenious process. "Without using the disk," Hubbard warns, "the benefit of processing is cut more than eighty per cent." Two disks are provided, one green and one white. "Use the one you like best," Hubbard says.

At the back of the book the "editor"—probably Hubbard—steals some of Wilhelm Reich's current thunder. While the atom bomb was being developed, the editor says, Hubbard was quietly working on a constructive use of atomic energy. "In 1947 he had found how this unruly energy could be smoothed out and rearranged in a mind so that thought would be sane, not insane. He had found how this energy governed the body functions." A touching footnote informs the reader of Hubbard's present financial plight. "He carries on the advance line of dianetic research without even the assistance of a secretary. He does not even own a car and he writes on a second hand Remington he bought years ago. A few volunteer contributions from friends and people whom his work has helped are his chief support. He has refused to take advantage of any part of the money made by the Foundation on the grounds that he would rather it helped others. Any contribution that you might care to make to him would help a man who is giving everything he has to help you—the Editor."

The most prominent convert to dianetics from the ranks of medical men has been Dr. Joseph Augustus Winter. He was a general practitioner in St. Joe, Michigan, when John Campbell, Jr. introduced him in 1949, by correspondence, to Hubbard. Winter had previously been interested in Count Alfred Korzybski's methods of treating neurotics by teaching them general semantics, and like so many other members of the semantics movement, he found dianetics even more intriguing. His correspondence with Hubbard induced him to visit the Master in Elizabeth, New Jersey, where he underwent a dramatic auditing. Back in Michigan, he tried dianetics on his six-year-old son who had developed a fear of ghosts. The fear vanished when his son recalled his delivery by an obstetrician in a white apron, with white gauze over his mouth.

Dr. Winter's enthusiasm knew no bounds. He moved to New Jersey, and became the first medical director of the newly formed Dianetics Research Foundation. In less than a year, however, disenchantment set in. By October, 1950, he resigned. He is now practicing his own modified version of dianetics in a swanky Manhattan office off Park Avenue.

In his book, *A Doctor's Report on Dianetics,* published in 1951, Dr. Winter pays tribute to what he thinks is a solid core of truth in dianetics, then cites the points on which he now disagrees. For example, although he is convinced that prenatal engrams can be formed, he suspects (though he is not sure) that the "sperm dream" is something imagined by the patient rather than a true memory. He also objects to the therapeutic value of having a patient recall his deaths in previous incarnations (now a standard Hubbard procedure). Hubbard's authoritarian attitude and the foundation's utter disregard for scientific method, he found appalling. For instance, the "Guk" program. Guk, Winter explains, was the name for a "haphazard mixture of vitamins and glutamic acid, which was taken in huge doses in the belief that it made the patient 'run better.' There were no adequate controls set up for this experiment, and it was a dismal, expensive failure."

Dr. Winter also disagrees with Hubbard's view that anybody can be an auditor. "Any person who is intelligent and possessed of average persistency," Hubbard wrote in *Dianetics,* "and who is willing to read this book thoroughly should be able to become a dianetic auditor." Moreover, Hubbard insisted that even a bad auditor was better than none at all and that no possible harm could be caused by clumsy auditing. Dr. Winter thinks otherwise. His book cites several cases of patients who seemed to be sane until they underwent dianetic therapy, after which they had to be institutionalized as psychotics.

And last but not least, Dr. Winter was puzzled by the conspicuous absence of any clears. "I have yet to see a 'clear' before and after dianetic therapy," he writes. "I have not reached that state myself nor have I been able to produce that state in any of my patients. I have seen some individuals who are supposed to be 'clear,' but their behavior does not conform to the definition of the state. Moreover, an individual supposed to have been 'clear' has undergone a relapse into conduct which suggests an incipient psychosis.''

Perhaps the most revealing parts of Dr. Winter's book are the records of his own dianetic sessions—revealing because they indicate with unmistakable starkness the manner in which the auditor suggests to a patient what sort of things he is supposed to recall. The patient, it must be remembered, in the vast majority of cases, is already familiar with dianetic theory. With this in mind, let us examine one of Dr. Winter's cases.

(Therapist) What sensation do you have now?
(Patient) My eyes feel as if I want to rub them.
(T) What do you suppose could cause that feeling?
(P) Having something in my eye—a cinder maybe.
(T) Anything else?
(P) Having "pink eye."
(T) Anything else?
(P) I can't think of anything.
(T) There are some possibilities I can think of; you don't have to accept them, of course. Could your eyes feel like this if you were crying?
(P) Yes, I guess so.
(T) Could your eyes feel like this if someone puts drops in them?
(P) Certainly.
(T) All right, let's try to recall the first time your eyes felt this way.

Notice the way in which the cinder and pink eye explanations are ignored. After the patient is unable to think of anything else, the therapist suggests crying and eye drops. In a few moments the patient will be back along his time track to the time of his birth, imagining the delivery scene and connecting it with the present sensation in his eyes.

Here is another of Winter's cases. The patient has reported a headache and stuffy nose:

(T) What else do you suppose that you'd feel?
(P) I don't know. Say, I can't take this much longer.
(T) Do you suppose that someone might have used the phrase, "Take this," during your birth?
(P) Yes, I suppose that the doctor might have said it.
(T) What might he be doing at the time?
(P) I guess that he'd be handing me over to the nurse.
(T) And what would the doctor be saying?

(P) "Here, you can take this now." No, that doesn't seem quite right.

(T) Change the words to suit yourself.

(P) "Here, you take *him* now." That's it.

(T) Repeat the phrase, please, and notice how your head feels.

(P) (Repeats phrase 5 or 6 times.)

(T) Notice how your nose feels as you go over these words. Repeat them again.

(P) (More repetitions.)

(T) How's the headache now?

(P) It's getting worse. (Rubs his eyes.)

(T) How about your eyes—what sensation do you suppose they'd have?

(P) They're stinging; it must be those damn drops he put in.

(T) How do you feel about the doctor putting drops in your eyes?

(P) I'm mad at him; that's a dirty trick. ·

(T) Supposing that you could get even with the doctor; what would you like to do to him?

(P) I'd like to hit him. (Words are spoken in a resentful tone.)

(T) All right—imagine that the doctor's face is on the couch beside you. Now hit it!

(P) (Clenches jaws and strikes at the couch with closed fist; makes about ten blows.)

(T) Go ahead—get good and mad at him. Hit him again!

(P) (Laughs.) I can't—it's too silly.

(T) How's the headache now?

(P) Better.

(T) Now let's put all these associations together in a pattern. Notice your headache . . . notice how your eyes feel . . . your nose . . . the feeling of anger. Anything else?

(P) (Scratches at ribs along left axillary line.) Funny—I was just thinking about the way my sister used to tickle me. I haven't thought about that in years.

(T) What sensation might you have had in birth that would remind you of being tickled?

(P) I don't know. (Scratches chest again.)

(T) How do you suppose that the doctor picked you up?

(P) He could have picked me up with his hand under my chest there.

(T) Imagine how it would feel to have someone pick you up. What would the temperature of his hand be?

(P) Warm, I guess.

(T) And what does he say?

(P) "Here, you can take him now."

(T) Where is "now"?

(P) Why, *now*—present time.

(T) Are you being born in 1951?

(P) No—of course not.

(T) You can differentiate between "now," if it was said at the time of your birth, and "now" in 1951, can't you?

(P) Sure.

(T) Supposing that your headache obeyed the command, "Take him now." What might happen?

(P) I don't know—I can't seem to figure that one out.

(T) What does "take" mean?

(P) It means to carry . . . to steal . . . to grasp . . . to attract.

(T) And where is now?

(P) Oh, I see—that could mean that my headache would be taken to present time.

(T) Do you have to bring your birth-headache up to present time just because the doctor said, "Take him now"?

(P) No, that's silly.

(T) How's the headache now?

(P) Much better—practically gone.

Nothing could be clearer from the above dialogue than the fact that the dianetic explanation for the headache existed only in the mind of the therapist, and that it was with considerable difficulty that the patient was maneuvered into accepting it. The therapist's questions are of such a "leading" character that even Dr. Winter admits they "encourage fantasy." In fact, the doctor says, it does not matter much whether such memories are real or imaginary! This is a startling admission. If there is no evidence such memories are real, then the whole Hubbardian notion of prenatal and birth engrams must be discarded. Perhaps that is exactly what the doctor has since done with it. His latest book, *Are Your Troubles Psychosomatic?*, 1952, contains not a single reference to dianetics.

Hubbard himself admits that many patients indulge in fantasies about their uterine experiences. "The patient tells about father and mother," he writes, "and where they are sitting and what the bedroom looks like, and yet there he is in the womb." Hubbard rejects the theory "that the tortured foetus develops extrasensory perception in order to see what is coming next." This is a good theory, he admits, but must be rejected in view of the fact that the foetus has no mind and therefore lacks clairvoyant powers.

Actually, the notion that neuroses and psychosomatic ills trace back to experiences when the mind was unconscious—whether in or out of the womb—is so completely unsupported by anything faintly resembling controlled research that not a single psychiatrist of standing has given it a second thought. More than one psychoanalyst has pointed out that the practice of blaming one's ills on events that occurred when one was an embryo, is an extremely convenient device for avoiding any real understanding of the roots of a neurosis. Even Dr. Winter speaks of the strong feeling of escape from guilt which accompanies fantasies of womb sensations. With this in mind, the entire structure of dianetics appears to be one vast attempt on Hubbard's part to dodge a genuine understanding of his own compulsions.

Of all the defenses which can be made of dianetics, the defense that "it works" is the most irrelevant. It is irrelevant because in the cure of neurotic symptoms, *anything* in which a patient has faith will work. Such "cures" are a dime a dozen. The case histories of dianetics are not one whit more impressive than the hundreds of testimonials to be found in young Perkins' book on the curative power of his father's metallic tractors. They prove that dianetics can operate on some patients as a form of faith healing. They prove nothing more.

Hubbard is prepared, of course, to expect this sort of opposition to his views. "Should the pre-clear discover that anyone is attempting to prevent him from starting or continuing dianetic therapy," he writes, "the fact should be communicated immediately to the auditor. . . . Anyone attempting to stop an individual from entering dianetic therapy either has a use for the aberrations of that individual . . . or has something to hide."

At the time of writing, the dianetics craze seems to have burned itself out as quickly as it caught fire, and Hubbard himself has become embroiled in a welter of personal troubles. In 1951, his third wife, twenty-five-year-old Sara Northrup Hubbard, sued him for divorce. She called him a "paranoid schizophrenic," accused him of torturing her while she was pregnant, and stated that medical advisers had concluded Hubbard was "hopelessly insane."

In February, 1952, the Dianetic Foundation in Wichita went bankrupt. It was later purchased from the bankruptcy court by a Wichita businessman who refuses to have anything to do with Hubbard. At the moment, the founder of dianetics is living in Phoenix, Arizona. From there the Hubbard Association of Scientologists ("scientology" is a new Hubbardian term, meaning the "science of knowledge") is mailing out literature fulminating against the Wichita group, hawking Hubbard's latest books, publishing a periodical called *Scientology,* and selling a *Summary Course in Dianetics and Scientology,* complete with tape recordings, for $382.50. The Hubbard College Graduate School, in Phoenix, charges a registration fee of $25.00 and offers a degree of Bachelor of Scientology.

For $98.50 Hubbard will send you an electropsychometer, which "registers relative degrees of dynamic psychophysical stress from moment to moment during the dianetic session." It also "indicates the approximate Hubbardian tone-scale of the preclear from 1.0 to *infinitely high ranges!*," and "immediately discloses points of entry into 'armored' or 'shut off' cases. . . ." On one leaflet, Hubbard states, "Bluntly, auditing can't be at optimum without an electropsychometer. An auditor auditing without a machine reminds one of a hunter hunting ducks at pitch black midnight, firing his gun off in all directions." A manual by Hubbard on *Electropsychometric Auditing* comes free with the device. For $48.50 you can obtain a smaller model called the "minemeter."

A recent letter from Hubbard asked for donations of $25 to help pay his living expenses, establish free dianetic schools "across America," and a few other little projects he has in mind. In return, donors are to be given membership in a new dianetic organization called "The Golds."

John Campbell, Jr., who had been introduced to dianetics many years earlier when Hubbard began treating him for sinusitis, and who in turn introduced dianetics to the world, has likewise been divorced. He married Dr. Winter's sister.

And he still has his sinusitis.

CHAPTER 23

GENERAL SEMANTICS, ETC.

AFTER DISCUSSING orgonomy and dianetics, a description of any other cult in which psychiatric techniques are prominent is certain to be anticlimactic. Nevertheless, our survey would be incomplete if it did not touch upon the "general semantics" of Polish-born Count Alfred Habdank Skarbek Korzybski, and the "psychodrama" of the Rumanian-born psychiatrist, Jacob L. Moreno. Neither movement, it should be stated, approaches the absurdity of the two previously considered cults. For this reason, general semantics and psychodrama must be regarded as controversial, borderline examples, which may or may not have considerable scientific merit.

Korzybski was born in 1879 in Warsaw. He had little formal education. During World War I, he served as a major in Russia's Polish Army, was badly wounded, and later sent to the United States as an artillery expert. He remained in the States, and for the next ten years drew on his personal fortune to write *Science and Sanity,* the 800-page Bible of general semantics. The book was published in 1933 by the Count's International Non-Aristotelian Library Publishing Company. It is a poorly organized, verbose, philosophically naive, repetitious mish-mash of sound ideas borrowed from abler scientists and philosophers, mixed with neologisms, confused ideas, unconscious metaphysics, and highly dubious speculations about neurology and psychiatric therapy.

Allen Walker Read, in two scholarly articles on the history and various meanings of the word "semantics" (*Trans/formation,* Vol. 1, Numbers 1 and 2, 1950, 1951), disclosed that the word had not been used in the Count's original draft of *Science and Sanity.* Before the book was published, however, the word had been adopted by several Polish philosophers, and it was from them that Korzybski borrowed it.

Most contemporary philosophers who use the word "semantics," restrict it to the study of the meaning of words and other symbols. In contrast, the Count used the word so broadly that it became almost meaningless. As Read points out, Korzybski considered a plant tropism, such as growing up instead of down, a "semantic reaction." In *Science and Sanity* he discusses a baby who vomited to get a second nursing, and writes, "Vomiting became her semantic way of controlling 'reality.' " Modern followers of the Count tend to equate "semantic" with "evaluative," defining "general semantics" as "the study and improvement of human evaluative processes."

Korzybski never tired of knocking over "Aristotelian" habits of thought, in spite of the fact that what he called Aristotelian was a straw structure which bore almost no resemblance to the Greek philosopher's manner of thinking. Actually, the Count had considerable respect for Aristotle (one of the many thinkers to whom his book is dedicated). But he believed that the Greek philosopher's reasoning was badly distorted by verbal habits which were bound up with the Indo-European language structure,[1] especially the subject-predicate form with its emphasis on the word "is." "Isness," the Count once said, "is insanity," apparently without realizing that such concepts as "isomorphic," which he used constantly, cannot be defined without assuming the identity of mathematical structures.

Another "Aristotelian" habit against which the Count inveighed is that of thinking in terms of a "two-valued logic" in which statements must be either true or false. No one would deny that many errors of reasoning spring from an attempt to apply an "either/or" logic to situations where it is not applicable, as all logicians from Aristotle onward have recognized. But many of the Count's followers have failed to realize that there is a sense in which the two-valued orientation is inescapable. In all the "multi-valued logics" which have been devised, a deduction within the system is still "true" or "false." To give a simple illustration, let us assume that a man owns a mechanical pencil of a type which comes in only three colors—red, blue, and green. If we are told that his pencil is neither blue nor green, we then conclude that it is red. This would be a "true" deduction within a three-valued system.[2] It would be "false" to deduce that the pencil was blue, since this would contradict one of the premises. No one has yet succeeded in creating a logic in which the two-valued orientation of true and false could be dispensed with, though of course the dichotomy can be given other names. There is no reason to be ashamed of this fact, and once it is understood, a great deal of general semantic tilting at two-valued logic is seen to be a tilting at a harmless windmill.

One finds in *Science and Sanity* almost no recognition of the fact that the battle against bad linguistic habits of thought had been waged for centuries by philosophers of many schools. The book makes no mention, for example, of John Dewey (except in bibliographies added to later editions), although few modern philosophers fought harder or longer against most of what the Count calls "Aristotelian." In fact, the book casts sly aspersions on almost every contemporary major philosopher except Bertrand Russell.

Korzybski's strong ego drives were obvious to anyone who knew him or read his works carefully. He believed himself one of the world's greatest living thinkers, and regarded *Science and Sanity* as the third book of an immortal trilogy. The first two were Aristotle's *Organon* and Bacon's *Novum Organum*. Like Hubbard, he was convinced that his therapy would benefit almost every type of neurotic, and was capable of raising the intelligence of most individuals to the level of a genius like himself. He thought that all professions, from law to dentistry, should be placed on a general semantic basis, and that only the spread of his ideas could save the world from destruction. In the preface to the second edition of *Science and Sanity,* he appealed to readers to urge their respective governments to put into practice the principles of general semantics, and in the text proper (unchanged in all editions) expressed his belief that ultimately his society would become part of the League of Nations.

The Count's institute of General Semantics, near the University of Chicago, was established in 1938 with funds provided by a wealthy Chicago manufacturer of bathroom equipment, Cornelius Crane. Its street number, formerly 1232, was changed to 1234 so that when it was followed by "East Fifty-Sixth Street" there would be six numbers in serial order. The Count—a stocky, bald, deep-voiced man who always wore Army-type khaki pants and shirt—conducted his classes in a manner similar to Kay Kyser's TV program. Throughout a lecture, he would pause at dramatic moments and his students would shout in unison, "No!" or "Yes!" or some general semantic term like "Et cetera!" (meaning there are an infinite number of other factors which need not be specified). Frequently he would remark in his thick Polish accent, "I speak facts," or "Bah—I speak baby stuff." He enjoyed immensely his role of orator and cult leader. So, likewise, did his students. In many ways the spread of general semantics resembled the Count's description, on page 800 of *Science and Sanity,* of "paranoiac-like semantic epidemics" in which followers fall under the spell of a dynamic leader.

According to the Count, people are "unsane" when their mental maps of reality are slightly out of correspondence with the real world. If the inner world is too much askew, they become "insane." A principal cause of all this is the Aristotelian mental orientation, which distorts reality. It assumes, for example, that an object is either a chair or not a chair, when clearly there are all kinds of objects which may or may not be called chairs depending on how you define "chair." But a precise definition is impossible. "Chair" is simply a word we apply to a group of things more or less alike, but which fade off in all directions, along continuums, into other objects which are not called chairs. As H. G. Wells expressed it, in his delightful essay on metaphysics in *First and Last Things:*

> . . . Think of armchairs and reading-chairs and dining-room chairs, and kitchen chairs, chairs that pass into benches, chairs that cross the boundary and become settees, dentist's chairs, thrones, opera stalls, seats of all sorts, those miraculous fungoid growths that cumber the floor of the Arts and Crafts Exhibition, and you will perceive what a lax bundle in fact is this simple straightforward term. In cooperation with an intelligent joiner I would undertake to defeat any definition of chair or chairishness that you gave me.

The non-Aristotelian mental attitude is, in essence, a recognition of the above elementary fact. There is so such thing as pure "chairishness." There are only chair 1, chair 2, chair 3, et cetera! This assigning of numbers is a process Korzybski called "indexing." In similar fashion, the same chair changes constantly in time. Because of weathering, use, and so forth, it is not the same chair from one moment to the next. We recognize this by the process of "dating." We speak of chair 1952, chair 1953, et cetera! The Count was convinced that the unsane, and many insane, could be helped back to sanity by teaching them to think in these and similar non-Aristotelian ways. For example, a neurotic may hate all mothers. The reason may be that a childhood situation caused him to hate his own mother. Not having broken free of Aristotelian habits, he thinks all mothers are alike because they are all called by the same word. But the word, as Korzybski was fond of repeating, is not the thing. When a man learns to index mothers—that is, call them mother 1, mother 2, mother 3, et cetera—he then perceives that other mothers are not identical with his own mother. In addition, even *his* mother is not the same mother she was when he was a child. Instead there are mother 1910, mother 1911, mother 1912, et cetera. Understanding all this, the neurotic's hatred for mothers is supposed to diminish greatly.

Of course there is more to the non-Aristotelian orientation than just indexing and dating. To understand levels of abstraction, for example, the Count invented a pedagogical device called the "structural differential." It is a series of small plates with holes punched in them, connected in various ways by strings and pegs. The "Semantic Rosary," as it was

called by *Time* magazine, is impressive to anyone encountering epistemology for the first time.

Obviously there is nothing "unsane" about the various general semantic devices for teaching good thinking habits. In psychiatry, they may
even be useful to doctors of any school when they try to communicate
with, or instruct, a patient. But Korzybski and his followers magnified
their therapeutic value out of all sane proportions. At conventions, general semanticists have testified to semantic cures of alcoholism, homosexuality, kleptomania, bad reading habits, stuttering, migraine, nymphomania,
impotence, and innumerable varieties of other neurotic and psychosomatic ailments. At one conference a dentist reported that teaching general
semantics to his patients had given them more emotional stability, which
lessened the amount of acid in their mouths. As a consequence, fillings
stayed in their teeth longer.

Korzybski's explanation of why non-Aristotelian thinking has therapeutic body effects, was bound up with a theory now discarded by his
followers as neurologically unsound. It concerned the cortex and the
thalamus. The cortex was supposed to function when rational thought was
taking place, and the thalamus when emotional reflexes were involved.
Before acting under the impulse of an emotional response, Korzybski
recommended a "semantic pause," a kind of counting-to-ten which gave
the cortex time to arrive at an integrated, sane decision. For a person who
developed these habits of self-control, there was a "neuro-semantic
relaxation" of his nervous system, resulting in normal blood pressure,
and improved body health.

It is interesting to note in this connection that a special muscular
relaxation technique also was developed by the Count after he observed
how often he could ease a student's worried tenseness by such gestures as
a friendly grasp of the student's arm. The technique involves gripping
various muscles of one's body and shaking them in ways prescribed in
The Technique of Semantic Relaxation, by Charlotte Schuchardt, issued
by the Institute of General Semantics in 1943.

Modern works of scientific philosophy and psychiatry contain almost
no references to the Count's theories. In Russell's technical books, for
instance, which deal with topics about which Korzybski considered himself a great authority, you will not find even a passing mention of the
Count. This is not because of stubborn prejudice and orthodoxy. The
simple reason is that Korzybski made no contributions of significance to
any of the fields about which he wrote with such seeming erudition.[3]
Most of the Count's followers admit this, but insist that the value of his
work lies in the fact that it was the first great synthesis of modern
scientific philosophy and psychiatry.

But is it? Few philosophers or professional psychiatrists think so. On matters relating to logic, mathematics, science, and epistemology, *Science and Sanity* is far less successful as a synthesis than scores of modern works. It is more like a haphazard collection of notions drawn from various sources accessible to the Count at the time, and bound together in one volume. Many of the Count's ideas give a false illusion of freshness merely because he invented new terms for them. For example, his earlier book, *The Manhood of Humanity*, 1921, describes plants as "energy binders," animals as "space binders," and men as "time binders."[4] When this is translated, it means that plants use energy in growing; animals, unlike plants, are able to move about spatially to meet their needs; and man makes progress in time by building on past experince. All of which would have been regarded by Aristotle as a set of platitudes.

It is true that Korzybski made a valiant attempt to integrate a philosophy of science with neurology and psychiatry. It is precisely here, however, that his work moves into the realm of cultism and pseudoscience. Teaching a patient general semantics simply does not have, in the opinion of the majority of psychiatrists, the therapeutic value which followers of the Count think it has. Where the Count was sound, he was unoriginal. And where he was original, there are good reasons for thinking him "unsane."

Samuel I. Hayakawa, in many ways a saner and sounder man than the Count, is still waving the banners of general semantics in Chicago, even though he made a break with Korzybski shortly before the Count moved his headquarters to Lakeville, Connecticut, in 1946. Hayakawa continues to edit his lively little magazine, *Etc.*, and work with the International Society of General Semantics, founded in Chicago in 1942 and not connected with the Lakeville group. His *Language in Action*, 1941 (revised in 1949 as *Language in Thought and Action)* remains the best of several popular introductions to Korzybski's views. One night in a Chicago jazz spot—Hayakawa is an authority on hot jazz—he was asked what he and the Count had disagreed about. Hayakawa paused a few moments (perhaps to permit a neurological integration of reason and emotion), then said, "Words."

Since the Count's death in 1950, the cult seems to be diminishing in influence. An increasing number of members, including Hayakawa himself, are discovering that almost everything of value in Korzybski's pretentious work can be found better formulated in the writings of others. Then too, many of its recruits from the ranks of science fiction enthusiasts, especially in California, have deserted general semantics for the more exciting cult of dianetics.

The case of A. E. van Vogt of Los Angeles suggests the new trend. Van Vogt is the author of many popular science-fiction novels of the superman type, including one called *The World of AĀ*, the action of which involves a future society that has adopted AĀ, or Korzybski's non-Aristotelian orientation. A few years ago, van Vogt was proposing that general semantics go underground on a cellular basis. The United States might have another great depression, he feared, and fall into the hands of the Communists, who do not care for Korzybski's views. He even toyed with the notion of a General Semantics Church, with its own sacred literature, but this idea proved abortive and nothing came of it. At the moment, van Vogt has lost his former enthusiasm for semantics and Dr. Bates' eye exercises. He is head of the California branch of the dianetics movement.

The psychodrama movement of Jacob L. Moreno also seems to have passed the peak of its popularity, though the cult was never very large. This is a form of therapy which places the patient into impromptu dramatic scenes related to his neuroses or psychoses. It is closely allied to a technique known as "play therapy," widely used in the diagnosis and treatment of neurotic children.

Psychodramatic skits take place on a relatively bare circular stage with three concentric levels, the levels having various symbolic meanings. There are no scenery or curtains—only two pillars in the background, a table, and some chairs. About eighty people are accommodated in the orchestra, and there is a high balcony for patients in the audience who have delusions of grandeur. The first theater of this sort was founded by Dr. Moreno in Vienna in 1922. At present, headquarters for the Psychodrama Institute are at Beacon, N. Y. though there are similar theaters in Manhattan's Bellevue Hospital and St. Elizabeth Hospital, Washington, D.C.

The patient may play a wide variety of roles—his father, mother, himself as a small child, Hamlet, God, and so on. If his neurotic situation involves a love triangle, an effort may be made to get the three real life participants on to the stage at one time. Their spontaneous reactions, as one can easily imagine, are often dramatic. Colored-lights are sometimes used to provide mood atmosphere. Thus if a patient feels a need to portray the role of Satan, only a crimson light is cast on the stage. This gives him the feeling that he is surrounded by the lurid flames of hell, and increases the effectiveness of the therapy.

Usually the patient takes part in the skits, with other roles acted either by other patients or by trained actors and actresses. In the jargon of psychodrama, persons taking these other roles are called "auxiliary egos." If the patient is shy or refuses to participate for some other reason, an auxiliary ego may substitute for him while the patient watches from the audience. This is known as "mirror technique."

The dramatic scenes are used both for diagnosis and therapy. Sometimes the patient achieves a Freudian catharsis (purging of a neurotic drive) while he is acting. At other times the catharsis comes later when the scene is being reviewed and analyzed by the therapist. Sound and movie recordings often are made so the scene can be reviewed more accurately. Members of the audience also experience a therapeutic benefit from watching the acting, thus making possible an inexpensive kind of group therapy.

Naturally this is far too brief an account of psychodrama to give much insight into its theory which is almost as intricate as dianetics. There are many neologisms—like tele, warming-up process, sociodrama, audience constellations, psychomusic, physiodrama, and *statu nascendi*—but we lack space to go into them here. Explanations for them may be found in the *Psychodrama Collected Papers,* printed by Beacon House in 1945.

Of Moreno's published works, none is more baffling than *The Words of the Father,* issued by Beacon House in 1941. The book purports to be a new revelation from God. Moreno's preface and commentary state that all previous revelations have been only partial expressions of divine truth. This is a "final and total expression." For the first time in history, God is speaking in the "first person." The words came, of course, through a human being—"an anonymous, isolated man, somewhere on the continent." His name does not appear on the title page, Moreno writes, any more than a tree would bear on its trunk the signature of the gardener. The reason is that the message, like the tree, is from God.

With this buildup, one turns the pages with trembling fingers, in expectation of imperishable thoughts. Alas, they consist only of stale religious platitudes—brief sentences printed at the top of each page in capital letters, and surrounded by a great expanse of blank space. God's opening statement is:

I AM GOD
THE FATHER
THE CREATOR OF THE UNIVERSE

THESE ARE MY WORDS
THE WORDS OF THE FATHER

A sample of the Father's words—the entire content of page 128—is as follows:

YOU CANNOT SERVE
TWO MASTERS
SERVE ME

Moreno's methods, like a good part of the Reichian therapy, are part of a recent trend toward having patients play active, dynamic roles rather than lie passively on the couch. Hundreds of new therapeutic gimmicks are making their appearance in the more eccentric Freudian fringes. Dr. Francis I. Regardie, for example, a psychotherapist in Los Angeles, is now inducing patients to vomit as part of their treatment. "Initiating a gag reflex" is his way of putting it. Here is the doctor's description of this valuable new technique, taken from his article on "Active Psychotherapy" in the Winter 1952 issue of *Complex*.

> The second of these somatic procedures is to ask the patient to regurgitate by using a tongue depressor and a kidney pan. Usually, the patient is puzzled and resists with some vigor. If a brief and simplified explanation is given, or if the therapist states unequivocally that this is no time for intellectual discussion which must wait for a later occasion, the patient as a rule will comply. My procedure is to let him gag anywhere up to a dozen times, depending on the type of response. In itself, the *style* of gagging is an admirable index to the magnitude of the inhibitory apparatus. Some gag with finesse, with delicacy, without noise. These are, categorically, the most difficult patients to handle. Their character armor is almost impenetrable, and their personalities rigid almost to the point of petrifaction. They require to be encouraged to regurgitate with noise, without concealment of their discomfort and disgust, and with some fullness. Others will cough and spit, yet still remain unproductive. Still others sneer and find the whole procedure a source of cynical amusement. Yet another group will retch with hideous completeness.

When one of Dr. Regardie's patients develops hostility, he is encouraged to "ventilate it overtly in a variety of different ways. One effectual way is to permit him frankly, by direct instruction, to employ all the so-called filthy language and obscenities he has acquired in the course of living. . . . Some people in this situation evince an astonishing familiarity with these visceral linguistics."

Other means of ventilating rage, used by the doctor, are to let the patient buffet a pillow, tear up a phone book, or punch one of those inflated rubber clown toys that have sandbags at the base. ". . . Replacement is required at the rate of two or three per month," he writes, in reference to the rubber clowns. "But the emotional discharges that occur by using these devices are altogether remarkable. In all of the dozen or more years of practicing psychotherapy I can say in all honesty and humility that I have never witnessed such awe-inspiring demonstrations."

CHAPTER 24

FROM BUMPS TO HANDWRITING

THERE ARE MANY "sciences" by which a person's inner character may be determined from such outer manifestations as the shape of his skull, nose, hands, and so forth, or the manner in which he crosses his "t's" and dots his "i's." Each of these alleged sciences has a literature of its own, running to many thousands of volumes, in some cases extending back to the ancient Greeks. Not until recent years, however, have attempts been made to place such studies on a sound scientific basis. A few pseudo-scientists, comparable to some of the men we have been discussing, have put in their appearance, especially in Europe, but their work does not lend itself to amusing or dramatic treatment. We shall, therefore, take only a fleeting glance at four of these dubious arts—phrenology, physiognomy, palmistry, and graphology.

Phrenology, when it was first advanced around 1800 by the Austrian anatomist, Francis Joseph Gall, aroused enormous public interest. Hundreds of phrenological societies sprang up here and there in Europe—later in England and America—with devotees who defended it with religious passion. Numerous periodicals flourished and died, of which the *Phrenological Journal,* edited in Edinburgh, was the most influential. And of course there were tons of books and pamphlets.

In essence, Gall and his disciples argued that human personality consisted of a number of independent, inborn mental "faculties," each of which was localized in a part of the brain. The larger the size of each region, the stronger the faculty. Consequently, an examination of skull bumps would reveal a person's character. Eventually, the Austrian government prohibited Gall from lecturing (on the grounds that his belief in the close connection between brain and personality violated religious dogma), but he continued to spread his doctrine in Germany and France. When he died in Paris, in 1828, a post-mortem revealed, curiously, that his skull was twice as thick as the normal—a fact which provoked many unkind witticisms.

During the nineteenth century phrenology was widely accepted by intelligent people, particularly those inclined toward the occult. When Sherlock Holmes deduced from the large size of a hat that the wearer was "highly intellectual," he was basing his deduction on a phrenological dictum widely current at the time. Alfred Russel Wallace, like Conan Doyle, became a spiritualist, and also a believer in phrenology. On one occasion, when he had a subject in a deep trance, he touched various regions on a phrenological model head. With each touch the subject responded by making facial expressions appropriate to the bump. It was not telepathy, Wallace writes, because once he thought he was touching one bump when actually his finger was on another. Nevertheless the subject made the correct response.

Walt Whitman was so proud of the fact that phrenology showed him well developed in *all* faculties, that on five occasions he published a chart of his own head bumps. *Leaves of Grass* is filled with phrenological terms. For years critics puzzled over such lines as, "O adhesiveness—O pulse of my life," until they discovered that "adhesiveness" was a phrenological faculty having to do with the attraction of one soul for another. In some of his personal notes Whitman referred occasionally to "16" and "164." For instance: "Always preserve a kind spirit and demeanor to 16. But pursue her no more." Early critics who couldn't believe that Whitman was homosexual used these references to prove his interest in women. It later turned out that 16 and 164 were sections in a popular book on phrenology, and had reference to the faculties of hope and acquisitiveness.

Modern research on the brain has, as most everyone knows, completely demolished the old "faculty psychology." Only sensory centers are localized. The base of the brain at the back of the head, for example, is related to sight, and not, as Gall taught, to "philoprogenitiveness" —the love of children. As a pseudo-science, phrenology is almost completely defunct except for an occasional carnival fortuneteller.[1]

Physiognomy, the art of judging character from facial features, goes all the way back to a treatise ascribed (falsely) to Aristotle. From the Greeks down to the present there have been divided opinions about whether the countenance expresses personality traits or, as Shakespeare wrote, "There's no art to find the mind's construction in the face." The literature on physiognomy is vast and contradictory, with the most impressive works appearing in the Renaissance. A number of recent investigations, by competent psychologists, have yielded negative results. The shape of the nose, ears, lips, the color of hair and eyes, texture of skin, and other features seem to show no correlation whatever with mental traits.

Certain exceptions, however, must be made. These are cases where a persistent mental attitude, with characteristic facial expressions, has left an impress in the form of lines or set muscular features. A person who has *felt* gloomy for thirty years is likely to *look* gloomy, for example, or a person who laughs a great deal may develop strong mirth wrinkles about the eyes. Sociologists have pointed out that certain folkway beliefs—like the notion that red-haired people have fiery tempers, or that a jutting chin is a sign of determination—may tend to bring out those traits in children who have the appropriate physical features. If everyone expects a carrot-topped child to have temper tantrums, this may help produce them. As yet, however, very little attempt has been made at scientific studies along such lines.

The theory that criminals have characteristic "stigmata"—facial and bodily features which distinguish them from other men—was the theme of an elaborate quasi-scientific study by an Italian occultist, Cesare Lombroso (1836–1909). His statistical methods were so shoddy that the work soon passed into deserved limbo, but in recent years its underlying notions have been revived by Professor Earnest A. Hooton, of the Harvard anthropology faculty. In a study made in the thirties, Dr. Hooton found all kinds of body correlations with certain types of criminality. For example, robbers tend to have heavy beards, diffused pigment in the iris, attached ear lobes, and six other body traits. Hooton must not be regarded as a crank, however—his work is too carefully done to fall into that category—but his conclusions have not been accepted by most of his colleagues, who think his research lacked adequate controls.

The work of Dr. William H. Sheldon in correlating mental traits with body types (endomorph, ectomorph, and mesomorph) should also be mentioned in this connection. His books are highly controversial, but clearly not the sort suitable for discussion here. Nor are we competent to discuss the "Szondi test," devised by Leopold Szondi, a Zurich psychiatrist. In this test, a patient is given a set of photographs of persons who represent eight major psychotic types. He picks out those he dislikes most and those he dislikes least, then his choices are carefully analyzed. The theory behind the test is that patients tend to like pictures of psychotics similar to themselves in facial expression, muscular tensions, and so on. If the reader is interested, he will find it all explained in *Introduction to the Theory and Practice of the Szondi Test,* 1949, by Susan K. Deri, who teaches the technique at Manhattan's New School for Social Research.

Palmistry is another ancient art with a literature of unbelievable immensity. It is usually divided into two branches—chiromancy, which has to do with telling the future from the lines in the palm; and chirosophy, the art of reading character from the shapes of fingers, size of the "mounts," and other features of the hand. In China there is a related art called pedomancy—fortunetelling and character reading from the lines and mounds of the feet—but so far this has made little progress in the tea rooms of the West. A few recent books have attempted to place chirosophy on a scientific basis—notably *The Human Hand,* 1943, by Charlotte Wolff, of the University of London, and several popular books by Josef Ranald. Their studies are interesting, but based on extremely shaky experimental methods.

Graphology, the art of reading character from handwriting, seems to have started in Italy in the early seventeenth century. Its modern popularity began in the mid-nineteenth, when the Abbé Jean-Hippolyte Michon, of Paris, worked out an elaborate set of "signs"—the shape of loops, form of "*t*" crosses, position of "*i*" dots, and more—which he related to specific mental traits. A great deal of subsequent research, chiefly European, has invalidated most of the Abbé's "signs," but there is some evidence that more general, or "gestalt," features of handwriting may have statistical correlations with certain traits. It is reasonable to suppose, for instance, that a person who is lazy and easygoing is not likely to become suddenly animated when he picks up a pen, or that a neat and orderly person will, when writing a formal letter, produce an extremely disorderly looking page. Nor would you expect a thoroughly conventional type of person to develop highly eccentric ways of forming capital letters.

One cannot even positively rule out in advance, particularly in the light of depth psychology, the possibility that there may be subtle unconscious connections between certain mental attitudes and their symbolic analogues in handwriting. For example, a religious mystic might unconsciously express spiritual yearning by terminating words with flourishes which rise upward to an unusual height. From this point of view, handwriting is looked upon as a form of "expressive behavior," like talking, walking, facial expressions, laughter, or shaking hands—and as such would be expected to have *some* connection with personality.[2] Exactly how much connection however, or what sort, is still far from satisfactorily demonstrated in spite of the many impressive studies which have appeared in recent years.

If the reader is interested in exploring this psychological borderland, a good reference is Klara G. Roman's *Handwriting: a Key to Personality,* 1952. Mrs. Roman, a product of the Hungarian school of graphology, has been teaching the subject for several years at the New School for Social Research. Her book, like most books on graphology, has little to offer in the way of careful experimental verification, but it contains a good summary of recent studies, and is an excellent introduction to modern graphological theory. Attention should also be called to *Diagrams of the Unconscious,* 1948, a book stressing a Freudian approach to graphology, by Prof. Werner Wolff, of Bard College.[3]

One of the major difficulties in all forms of character reading research is that no really precise methods have yet been devised for determining whether an analysis fits the person or not. Wide margins on a written letter, for example, are supposed to indicate "generosity." Is there anyone who would not feel that such a trait applied to himself? People are generous in some ways and not in others. It is too vague a trait to be tested by any empirical method, and even good friends may disagree widely on whether it applies to a given individual. The same is true of most of the graphological traits. If you are told you have them, you can always look deep enough and find them—especially if you are convinced that the graphologist who made the analysis is an expert who is seldom wrong.

Here is a simple test to try on any amateur or professional handwriting analyst. Obtain sample scripts from twenty friends, all of the same sex and approximate age, and turn them over to the graphologist. When you receive his twenty reports, give them (without the handwriting samples) to someone who is personally acquainted with all twenty writers. Ask this person to pair each report with a name on a list of the twenty friends. If he matches them at random, the laws of chance would allow him one correct pair. Consequently, if his score is high—say ten or more correct pairs—you have an objective basis for assuming there is something to the graphologist's ability. If the score is low, it strongly suggests the contrary.

This is the sort of testing which has been done in the past by a few psychologists, and usually with poor showings for the graphologist. Until a character analyst can consistently score high on tests of this sort, his work will remain on the fringes of orthodox psychology. The fact that millions of people were profoundly impressed by the accuracy of phrenological readings suggests how easy it is to imagine that a character analysis fits the person analyzed—provided you know exactly who the person is!

Closely related to modern graphology are a number of expressive behavior tests which have been developed by psychoanalysts. Jung and his followers place considerable emphasis on the study of pencil doodling, and analytical literature is filled with reports along such lines. Somewhat similar is the "draw a person test." A patient is given a sheet of blank paper and told simply to "draw a person." When the sketch is properly analyzed, it is supposed to provide valuable clues about the patient's personality disorders. Karen Machover, of Kings County Hospital, Brooklyn, has done most of the pioneer work with this test, and teaches a course on it at the New School for Social Research. The "drawing completion test," invented by Dr. Ehrig Wartegg of Germany, is along similar lines. The patient is given a set of squares, each containing a few meaningless symbols, and told to make eight drawings which incorporate the symbols in some way. *The Drawing Completion Test,* by G. Marian Kinget, of the University of Chicago, is a recent book explaining the technique. Drawing and painting have, of course, long been used by psychiatrists and analysts for diagnostic purposes. The literature dealing with the analysis of art produced by neurotic children is particularly extensive.

All these forms of expressive behavior testing—and new ones spring up every year[4]—are still in the experimental stage. It may be decades before adequate tests are devised for testing these tests. Until then, it is unwise to be dogmatic in labeling them either science or pseudo-science.

CHAPTER 25

ESP AND PK

THE BELIEF IN psychic phenomena—or "parapsychology" as its more dignified proponents like to call it—is as old as humanity. Not until the last century, however, were attempts made to give it a scientific, laboratory-tested foundation. These studies range all the way from obviously crackpot research to the work of sane, reputable psychologists like Dr. Joseph Banks Rhine of Duke University, and Gardner Murphy at the City College of New York. In this chapter we shall focus our attention on the work of Rhine, who has done more than any one man in history to give scientific respectability to the investigation of psychic forces.

It should be stated immediately that Rhine is clearly not a pseudo-scientist to a degree even remotely comparable to that of most of the men discussed in this book. He is an intensely sincere man, whose work has been undertaken with a care and competence that cannot be dismissed easily, and which deserves a far more serious treatment than this cursory study permits. He is discussed here only because of the great interest that centers around his findings as a challenging new "unorthodoxy" in modern psychology, and also because he is an excellent example of a borderline scientist whose work cannot be called crank, yet who is far on the outskirts of orthodox science.

There is obviously an enormous, irrational prejudice on the part of most American psychologists—much greater than in England, for example—against even the possibility of extra-sensory mental powers. It is a prejudice which I myself, to a certain degree, share. Just as Rhine's own strong beliefs must be taken into account when you read his highly persuasive books, so also must my own prejudice be taken into account when you read what follows.

Dr. Rhine was born in 1895 at Waterloo, Pa., but spent most of his childhood in a small Ohio town. As a young man he served two years in the Marines, then entered the University of Chicago where he was graduated in 1922. Later he received a doctorate in botany from Chicago, and for a time taught that subject at West Virginia University. As a youth he had intended to become a Protestant minister. But over the years his orthodox convictions evaporated and he began to look elsewhere for something to support the broad religious views he still retained.

It was in this frame of mind that he and his wife, in the early twenties, attended a lecture on spiritualism by Sir Arthur Conan Doyle. The lecture made a deep impression on them both. ". . . clearly if there was a measure of truth in what he [Doyle] believed," Rhine has written, "misguided though Sir Arthur might be in details, it would be of transcendental importance. This mere possibility was the most exhilarating thought I had had for years."

These emotions led to a long period of immersion, on the part of Rhine and his wife, in the literature and practice of spiritualism. "Psychic adventures" is a term Rhine has used for this early exploratory work. In 1927, he became a research assistant at Duke University, working on psychic forces under Professor William McDougall, formerly of Oxford and Harvard. He joined the Duke faculty in 1928, and since 1940 has been director of the school's Parapsychology Laboratory.

Rhine's first report on his experiments was published in 1934 under the title *Extra-Sensory Perception*. This was followed *by New Frontiers of the Mind*, 1937; *Extra-Sensory Perception After Sixty Years*, 1940 (of which he was a co-author); and *The Reach of the Mind*, 1947.[1] Since 1937, he has edited the *Journal of Parapsychology*, probably the most important journal in the history of scientific psychic investigation, and written numerous articles for popular magazines.

In his books and articles, Rhine puts forth the claim that ESP ("extra-sensory perception," a term including telepathy and clairvoyance) has been demonstrated beyond all reasonable doubt by means of several million tests with ESP cards. These are cards bearing five easily distinguishable symbols—a square, circle, cross, star, and wavy lines. They are usually used in decks of twenty-five cards each, five cards for each symbol. In more recent years Rhine has turned his attention toward another type of "psi" phenomena (his term for "psychic") which he calls PK, an abbreviation of "psychokinesis." This is the ability of the mind to control matter, as exemplified in mediumistic levitations, faith healing, haunted house phenomena, and so on. By having subjects concentrate on certain faces of dice, which are shaken by hand or thrown by a machine, he claims to have found that these faces show up more often than laws of chance allow.

Both ESP and PK, Rhine reports, are curiously free of space and time restrictions. For example, ESP works just as well when the subject and cards are separated by a considerable distance. They also work just as well when the subject is calling the order of a deck *before* it is shuffled. The latter ability is called "pre-cognition." As Rhine has written, ". . . there was no appreciable difference in the scoring, whether the subjects were calling the present or the future order of the cards." The phenomenon of pre-cognition has been a great headache to Rhine, because it makes it extremely difficult to devise a test for pure telepathy. Instead of reading a sender's mind, the subject may be "seeing" by pre-cognitive clairvoyance the final tabulated results of the test! Since every test must eventually be tabulated in some manner, it becomes an understandably difficult task to rule out pre-cognition as the explanation.

The independence of psi phenomena from space and time makes it impossible to explain them by any known physical theory. This leads to the view, Rhine argues, that at least part of the mind is detached from the physical world—a fact which lends support to beliefs in the soul, free will, and survival after death. In addition, if ESP and PK exist, then a solid core of truth is established behind all the obvious fraud and flummery of spiritualism. Rhine believes, for instance, that mediums, Ouija board operators, and automatic writers often give out information picked up by telepathy, clairvoyance and pre-cognition.

Rhine's work also lends support to common folk beliefs about the psychic powers of animals. In *American Magazine,* June 1951, he published an article called "Can Your Pet Read Your Mind?" in which he presents the case for these views. Rhine is convinced that "Lady," a professional mind-reading horse in Richmond, Virginia, had psychic powers.[2] He also tells about a dog in California that howled all day and stopped just as an earthquake occurred at Long Beach, and a collie that whimpered under a bed until an explosion at a nearby plant killed his master. In both cases, Rhine thinks, the dogs had pre-cognition of coming events. (See also his article, "The Mystery of the Animal Mind," in *The American Weekly,* March 30, 1952.)

Exactly what is one to make of such startling claims? Have Dr. Rhine and his associates given psi phenomena a sound empirical basis, or is his research highly suspect by scientific standards? Since the case for his work is so readily obtainable in his books and articles, and the case against him buried in academic publications, we will summarize the most important criticisms which have been made by sceptical psychologists.

The most damaging fact against Rhine is that, with very few exceptions, the only experimenters who have confirmed his findings are men who share his strong belief in psychic phenomena. Hundreds of tests by doubting psychologists have been made, and yielded negative results. Rhine attributes this to the fact that the attitude of the experimenter has a marked influence on the subject. If the scientist is a disbeliever it will upset the delicate operation of the subject's psi abilities. Critics of Rhine counter by accusing him of having performed his own experiments under a loose system of laboratory controls, and of having selected for publication only a small portion of the total number of tests made. H. L. Mencken (in an article on Rhine, Baltimore *Evening Sun,* Dec. 6, 1937) summed up the second criticism as follows: "In plain language, Professor Rhine segregates all those persons who, in guessing the cards, enjoy noteworthy runs of luck, and then adduces those noteworthy runs of luck as proof that they must possess mysterious powers. . . ."

This alleged "selection" is not a deliberate process, but something which operates subtly and unconsciously. To give one example, let us suppose an experimenter tests 100 students in a classroom to determine who should be given additional testing. By the laws of chance, about fifty of these students will score above average and fifty below (group tests, according to Rhine, nearly always show such over-all chance results). The experimenter decides that high scorers are most likely to be psychic, so they are called in for further testing. The low scorers in this second test are again dropped, and work continued with the high ones. Eventually, one individual will remain who has scored above average on six or seven successive tests. As an isolated case, the odds against such a run are high, but in view of the selective process just described, such a run would be expected.

A competent experimenter would not, of course, be guilty of anything as crude as this, but the illustration suggests how tricky the matter of selection is. To give a better illustration, let us imagine that one hundred professors of psychology throughout the country read of Rhine's work and decide to test a subject. The fifty who fail to find ESP in their first preliminary test are likely to be discouraged and quit, but the other fifty will be encouraged to continue. Of this fifty, more will stop work after the second test, while some will continue because they obtained good results. Eventually, one experimenter remains whose subject has made high scores for six or seven successive sessions. *Neither experimenter nor subject is aware of the other ninety-nine projects, and so both have a strong delusion that ESP is operating.* The odds are, in fact, much against the run. But in the total (and unknown) context, the run is quite probable. (The odds against winning the Irish sweepstakes are even higher. But someone does win it.) So the experimenter writes an enthusiastic paper, sends it to Rhine who publishes it in his magazine, and the readers are greatly impressed.

At this point one may ask, "Would not this experimenter be disappointed if he continues testing his subject?" The answer is yes, but as Rhine tells us, subjects almost always show a marked decline in ability after their initial successes. In addition, he writes, " . . . experimenters who were once successful may even lose their gift. There are cases . . . of research workers who found evidence of psi capacities in one or more experimental series and became less successful in later ones even with the same experimental conditions. This failure is understandable . . . in view of the loss of original curiosity and initial enthusiasm, but it shows clearly the extreme elusiveness of psi." The failure is also understandable, one must add, in terms of the laws of chance.

In testing individuals, when a score falls to chance or below, Rhine has a great many "outs" which make use of that score to support ESP rather than count against it. Thus the subject may be scoring not on the correct card (known as the "target") but on the card ahead. This phenomenon is called "forward displacement." Or he may be scoring on the card behind ("backward displacement"). Such displacement of ESP may even be two or three cards ahead or behind! Clearly if one can choose between all these possible variations, there is a strong likelihood *one* of them will show scores above average. If no displacement is found, however, a chance score may be attributed to some disturbance of the subject's mental state. He may be worried about his studies, or bored, or distracted by visitors, displeased with the experimenter, ill, tired, skeptical of the work, low in I.Q., neurotic, or in a state of emotional crisis. Even the experimenter, if *he* is in any of these regrettable states, may inhibit the subject by unconscious telepathy. All these factors are specifically cited by Rhine as contributory to low scoring. In his dice experiments, a low score may even be due, he says, to the fact that a person "dislikes" a certain set of dice! (We are not told whether the person expresses his dislike before or after the test.) As Rhine understates it, "the subtlest influences seem to disturb the operation of these abilities."

Of course when scores are high, no one is likely to look for "subtle influences." But if scores drop low, the search begins. Naturally they will not be hard to find. Usually if low scores continue, the tests are discontinued. If the scores are *extremely* low, they are regarded as a *negative* form of ESP. This is called "avoidance of the target." In our previous chapter on dowsing, we reported Henry Gross' poor scoring when tested by Rhine. Recently, however (in *The American Weekly,* March 23, 1952), Rhine has disclosed that Henry didn't fail after all! He was "unconsciously rebelling," Rhine writes, and adds, "He could not have made so many misses by mere coincidence."

"Avoidance of the target" may occur even in the dice work. "There is a common characteristic of shiftiness of aim in both ESP and PK" Rhine writes, " . . . a tendency under certain conditions not only to avoid the target, but sometimes to fix upon the target nearest in some respect to the one intended." The phrase "in some respect" deserves attention. Let us suppose the subject is trying to make the dice show 3's. The results are negative. But when they are examined more closely, they may show a high number of 4's. Now 4 is "in some respect" close to 3, since it comes after 3. Likewise 2 would be close, because it comes before 3. Or perhaps 6's show up, a number the subject had tried to make in the previous test. This would be a "lag effect." And of course pre-cognition might cause the subject to throw the *next* number to be chosen for testing. With this sort of jugglery possible, how can one lose? I myself have often noticed, in playing the "26 game" in Chicago, an exasperating tendency the cubes have of showing extremely high hits on numbers other than the one desired.

Obviously, unless all the "displaced" scores are averaged into the total picture, a statistical distortion results. If one experiment is published to show how the subject "hit the target," and the next experiment is published to show how the subject hit some other target than the one intended, then nothing at all is proved. That the number of hits falls close to chance when *all* results are lumped together is strongly indicated by the following statement. "This rejection of the target depresses the scoring rate below the chance average, and if it continues long enough, as it has in a number of researches, the total negative deviation reaches the point where it cannot be attributed to chance." A gambler would put it more simply. He would say that he often has runs of extremely good luck, but usually interspersed with runs of very bad luck.

"Sometimes the subject will score on the negative side of the mean chance line at one session," Rhine continues, "and on the other side at the next. He may not even know what is causing the swing. Worse still, he may begin his run on one side of the mean and swing to the other side just as far by the time he ends the run; or he may go below in the middle of the run and above at both ends. The two trends of deviation may cancel each other and the series as a whole average close to 'chance.' "

Here is the description of an ESP card test. " . . . the displacement was both forward and backward when one of the senders was looking at the card, and only in the forward direction with another individual as sender; and whether the displacement shifted to the first or the second card away from the target depended upon the speed of the test." No wonder Rhine speaks of ESP as "incredibly elusive"—having a "fickle-ness," "skittishness," and "shiftiness" which makes it the "most vari-able ability on record."

In addition to selection of data, the possibility of "recording errors" must also be taken into account in evaluating the Duke experiments. A number of tests have been made in recent years at other universities which have demonstrated dramatically the fact that believers in ESP are prone to make mistakes in recording calls, and such mistakes almost always favor ESP. At Stanford University, for instance, a test was made with 1,000 ESP cards. The calls were recorded by a person who believed strongly in ESP. According to chance, 200 cards should have been correctly guessed. The final score showed 229 guesses. Unknown to the person, however, a sound recording was made of the experiment. When this was checked against the records, 46 spurious hits were found. This reduced the score to slightly below chance level. When the experiment was repeated, and the person knew recordings were being made, only two errors were found.

A more recent test, made in 1952 by Richard S. Kaufman at Yale, involved a PK experiment with 96 dice. Each of eight persons kept records of forty successive tosses, without being aware that a hidden camera was recording everything. The four persons who believed in PK made errors in favor of PK, while the four disbelievers made errors of the opposite sort. According to the camera, the actual results conformed to chance. But the records of two believers had enough favorable errors to indicate PK. Mistakes of this sort need not be conscious fraud. Under the excitement and tenseness of such work, the mind can play strange tricks, and it is easy to understand how mistakes would be guided by one's bias. Incidentally, these favorable recording errors would be more likely at the beginning of a session, when interest and expectancy of success is at a pitch, then decrease as the clerks become tired and bored and start making *random* mistakes. This would explain the consistent decline in success which is shown in most of Rhine's dice tests, and which he regards as the strongest evidence in favor of PK. Dice tests also have a tendency to decline, then pick up again toward the close. Again, this is understandable. The knowledge that the experiment will soon be over could easily cause clerks to perk up a bit, and start making favorable errors once more.

The possibility of recording error runs through all of Rhine's work. His tests have been made under hundreds of widely different conditions, the descriptions of which are usually vague.[3] You seldom are told exactly who recorded, how it was recorded, or what the beliefs were of clerks who kept the records. Only in later years did Rhine tighten his controls to prevent such mistakes, and it is significant that the more rigid these controls became, the less ESP was found. ". . . Elaborate precautions take their toll," he writes. "Experimenters who have worked long in this field have observed that the scoring rate is hampered as the experiment is made complicated, heavy, and slow-moving. Precautionary measures are usually distracting in themselves."

In short, elaborate precautions disturb the subject's psi ability. ESP and PK can be found only when the experiments are relatively careless, and supervised by experimenters who are firm believers. Although Rhine has found many subjects whom he regarded as psychic to a high degree, he has not found a single subject capable of demonstrating ESP to sceptical scientists at other universities. Naturally he has explanations for this, but is it surprising that "orthodox" psychology hesitates to accept the existence of psi phenomena as proven?

In connection with Rhine's dice experiments, Clayton Rawson has pointed out (in *Scarne on Dice*, 1945) that a considerable PK push of some sort must be required to make a rolling die shift to another side (the size and weight of the dice have no effect on the results, Rhine has stated). Such a force could easily be demonstrated, Rawson writes, by a delicately balanced arrow, under a vacuum jar, which the subject would cause to rotate by concentrating. If mediums are capable of lifting heavy tables by PK, surely a medium should be able to set in motion such a simple laboratory device. Why, Rawson and Scarne want to know, does Rhine neglect such an unambiguous test and turn to experiments with dice which are subject to the same pitfalls of statistical error and unconscious selection that are involved in card testing?[4]

Another disturbing question comes to mind. For decades Chicagoans have played the "26 game" in their bars and cabarets. Ten dice are shaken from a cup, the player betting a certain number will show up at least 26 times in 13 rolls. Obviously the tired and bored dice-girl, who tallies each roll, doesn't care one way or another. Obviously the player is doing his damnedest to roll the number. How does it happen that these tally sheets, year after year, show precisely the percentage of house take allowed by the laws of chance? One would expect PK to operate strongly under such conditions.

Aristotle wrote that it was probable the improbable would sometimes take place; or as Charlie Chan once expressed it, "Strange events permit themselves the luxury of occurring." Considering the fact that Rhine has now supervised several million calls of cards and dice, most of which were made without adequate safeguards against recording errors, it would be unusual indeed if he had not encountered runs of incredibly high scores. In some early cases, there may even have been deliberate cheating on the part of subjects, since they are often paid for their services and only a subject showing psi ability would be repeatedly tested. (Three cards in an ESP deck are all that need be marked, by a fingernail scraped on the edge for example, to enable a subject who sees the deck to make average scores of above 7.) By treating high runs as evidence of ESP and PK, and finding plausible excuses for chance and low scores, is it possible Rhine has become the victim of an enormous self-deception?

When proponents of ESP accuse orthodox psychologists of having ignored psi phenomena, the answer is that it isn't true. Many careful experiments have been made, and with negative results.[5] To mention one outstanding example, Prof. John E. Coover, of Stanford University, made extensive and carefully controlled ESP tests which were published in detail in 1917, in a 600-page work, *Experiments in Psychical Research.* Recently Rhine and others have gone over Coover's tables, looking for forward and negative displacement, etc. They insist that ESP is concealed in his figures. But other statisticians regard this as a process akin to the marvelous cipher messages which Ignatius Donnelly and other Bacon Shakespeare scholars managed to dig out of Shakespeare's plays. You can always find patterns in tables of chance figures if you look deep enough.

In spite of the far from conclusive quality of the evidence, however, many intelligent and prominent moderns have accepted ESP. Among philosophers and psychologists—William James, Henry Sidgwick, William McDougall, Henri Bergson, and Hans Driesch are outstanding. Among writers—Conan Doyle, Aldous Huxley,[6] Gerald Heard (whose book *Preface to Prayer* explores the role of ESP in prayer fulfillment), Jules Romains,[7] H. G. Wells, Maurice Maeterlinck, Sir Gilbert Murray, Upton Sinclair, Arthur Koestler, and many others may be cited. In Koestler's recent book, *Insight and Outlook,* he speaks of Rhine's work as having ushered in a new Copernican Revolution.

It may surprise many to learn that there is a strong tendency to accept ESP on the part of psychoanalysts.[8] Freud himself wrote several papers on the subject, pointing out that dreams received telepathically would be subject to the usual distortions, that fortune-tellers might pick up repressed wishes from clients, and that a patient engaged in free-associating on the couch may have his thoughts confused by telepathic contact with the analyst. Jung and Stekel were two other prominent analysts who accepted telepathy. A few contemporary Freudians, like New York City analyst Nandor Fodor, have managed to combine psychoanalysis with a thoroughgoing occultism. The results are as weird as you might imagine. Fodor's *Search for the Beloved,* 1949, is a good example. The book discusses the telepathic influence of a mother's thoughts on the mind and body of her unborn child, and even speculates on the possibility that the embryo's mental state may be communicated by ESP to the mother. The old superstition about birthmarks is defended by Fodor, who thinks the mother's shock experiences may be sent by telepathy to the child. He cites the case of a ''coffee-colored splash on the neck of a beautiful woman whose mother had spilt hot coffee over herself while carrying the child.'' In this case, the woman with the birthmark happened to have a

twin sister without a mark, but Fodor is untroubled by such a fact. ". . .
We know nothing about the relative sensitivity of twins," he points out.

The book was published by the same house which later published
Dianetics, and it is interesting to note that Fodor anticipates much of
Hubbard's work by his stress on the harm that can be done to an unborn
child's mind by attempted abortion. In Fodor's view, however, most of
the harm is telepathically caused by the mother's wish to have the child
done away with. A chapter titled "The Love Life of the Unborn," in
which he traces the origin of nymphomania, satyriasis, and other aberrations to the influence of the mother's sex life on her unborn child, is
surely one of the most curious in all the annals of Freudian literature.
Nevertheless Fodor contributes to many respectable psychoanalytical
journals.

Upton Sinclair's *Mental Radio* deserves comment because it is the best
known and most impressive record of clairvoyance in recent psychic
literature. The book was published in 1930, with an introduction by
Einstein to the German edition. It is a record of a series of clairvoyant
experiments by Sinclair's second (and present) wife, in which she attempted to duplicate 290 simple drawings, most of them by her husband,
made at a variety of times and under varying conditions. The tests took
place while she was passing through a period of profound depression and
with a heightened interest in occult phenomena. The interest was occa-
sioned by the Sinclairs' friendship with Roman Ostoja, a professional
mind reader identified in the book only as "Jan." This young psychic
had such a standard and conventional repertory of stunts that it is hard to
understand how the Sinclairs could have taken him seriously. He would
remain buried several hours, for example, in an air-tight coffin. He
performed what is known in the magic trade as "muscle reading." On
one occasion, in a seance, he levitated a 34-pound table eight feet over
Upton's head![9] The Sinclairs accepted all these powers as genuine.

As *Mental Radio* stands, it is a highly unsatisfactory account of
conditions surrounding the clairvoyancy tests. Throughout his entire life,
Sinclair has been a gullible victim of mediums and psychics. His first
article, written at the age of twenty, was contributed to a theosophical
magazine. His latest writings, the ten novels of the Lanny Budd series,
are marred throughout by frequent intrusions of psychic matters. The last
volume, *O Shepherd Speak!,* is filled with eulogies of Dr. Rhine. It also
describes a faith cure performed by Lanny, and seances in which Lanny
speaks to the departed spirits of Bergson and Franklin D. Roosevelt. The
title of the book comes from the FDR seance, at which Roosevelt
apologizes for his agreement with Stalin at Yalta—surely the most embarrassing scene in all of Sinclair's novels.

Of the 290 pictures used in the tests with his wife, Sinclair estimates that she successfully duplicated 65 of them, with 155 "partial successes," and 70 failures. The paired drawings reproduced in the book, showing Mrs. Sinclair's sketch next to her husband's, are strikingly similar, though on a few the similarity requires that one drawing be turned upside down. It is not necessary, however, to assume clairvoyance in order to explain these results. In the first place, an intuitive wife, who knows her husband intimately, may be able to guess with a fair degree of accuracy what he is likely to draw—particularly if the picture is related to some freshly recalled event the two experienced in common. At first, simple pictures like chairs and tables would likely predominate, but as these are exhausted, the field of choice narrows and pictures are more likely to be suggested by recent experiences. It is also possible that Sinclair may have given conversational hints during some of the tests—hints which in his strong will to believe, he would promptly forget about. Also, one must not rule out the possibility that in many tests, made across the width of a room, Mrs. Sinclair may have seen the wiggling of the top of a pencil, or arm movements, which would convey to her unconscious a rough notion of the drawing. (Many professional mentalists are highly adept at this art.)

Very few of the "partial successes" are shown. Are they as successful as Sinclair says they are? Again, we are told that 290 drawings were used in the tests. Can we be sure this is an accurate *total* count of all drawings tested? It is easy to find excuses for "not counting" an attempt—a headache, an upsetting event, or others. There would be little incentive to save such attempts, if they were failures, and Sinclair is certainly capable of losing them and forgetting about them. Successes, of course, would be carefully preserved. In many such ways an unconscious selective process may have operated.

One can only wish that the tests had been supervised by competent psychologists familiar with the subtle ways experiments of this sort can go astray. Like Conan Doyle, Sinclair is a persuasive writer on psychic subjects, but we must remember that like Doyle, he is also exceedingly naive about the safeguards necessary to insure a controlled experiment. As a consequence, it is as impossible to rely on Sinclair's memory of the crucial details as to rely on Doyle's account, say, of his investigations of spirit photography. Later, when Mrs. Sinclair's health was better, she was tested by Prof. William McDougall, under better precautions, and the results were far less satisfactory. Since Sinclair ran for Governor of California in 1934, she has made no more experiments with ESP, though occasionally she startles her husband by some spontaneous manifestation of apparently psychic insight.

Thomas Edison is another good example of the self-styled psychic expert. He not only believed in telepathy, but on one occasion was completely taken in by a professional mountebank, Dr. Bert Reese, who had no more genuine mind-reading ability than modern stage mentalists like Dunninger, whose methods are well-known to magicians.[10] Edison was so impressed, however, that he wrote a letter to *The Evening Graphic,* in New York City, in which he stated his conviction Reese was a genuine psychic. The recently published *Diary and Sundry Observations of Thomas Alva Edison,* reveals a theory of the human mind surprisingly similar to the menorgs and disorgs of Alfred Lawson! Edison expressed his belief that in the human brain are millions of submicroscopic intelligences whom he called "little peoples." They rush about performing the desired mental functions, and are under the control of "master entities" who "live in the fold of Broca." Before he died in 1931, Edison was working on a sensitive piece of apparatus for communicating with departed spirits.

Edison's friend, Luther Burbank, is another example of a man who held curious occult views even though he achieved greatness in one field of practical science. Burbank communicated often with his sister by telepathy, had visions of his departed mother, and was firmly persuaded that plants had a sensitive nervous system capable of responding to love and hate. The Indian Yogi Paramhansa Yogananda, in an autobiography, quotes his friend Burbank as saying, "Yes, I have often talked to my plants in order to create a vibration of love. . . . One person will plant a flower, attend it carefully, and it will wither. But under identical physical care, a second person may develop that same flower into a healthy thriving plant. The secret . . . is love."

A book worth noting in this connection is *Secret Science Behind Miracles,* 1948, by Max Freedom Long, issued by the Kosmon Press of Los Angeles. Long is the director of the Huna Research Foundation, devoted to the study of Polynesian and Hawaiian Kahuna mystic lore. The book deals with the influence of the mind's vital force (which the Kahunas call "mana") over matter. To test the power of mana, Long recently has been experimenting with growing plants. He has discovered that if you talk to plants, as Burbank did, giving them love and attention, they grow faster and more luxuriantly than neglected plants. Similar work with corn and soy beans has been reported by Edgar Block, a Huna associate in Indianola, Indiana, and Dr. Rhine recently disclosed in *The American Weekly* (April 20, 1952) that a French doctor and his wife were able to stimulate plant growth by mental efforts. Dr. Littlefield's work with evaporating salt solutions, described in Chapter 10, may be regarded as still another variety of PK phenomena.

Charles Fort accepted both ESP and PK, in spite of his own private experiments which were not too successful. One month he made a thousand attempts to "see" clairvoyantly the contents of a store window before he had walked close enough to inspect it. Only three of his attempts met with success, the most striking being the first. He was walking along West Forty-Second Street when he concentrated on a window ahead and immediately thought "turkey tracks in red snow." (Red snow was one of Fort's special interests at the time.) The window proved, he reports in *Wild Talents,* to have in it track-like lines of black fountain pens against a background of pink cardboard.

On another occasion Fort experimented with PK. "The one great ambition of my life," he writes, "for which I would abandon my typewriter at any time—well, not if I were joyously setting down some particularly nasty little swipe at priests or scientists—is to say to chairs and tables, 'Fall in! forward! march!' and have them obey me. I have tried this. . . . But a more unmilitary lot of furniture than mine, nobody has."

Nevertheless Fort regarded PK as one of the "wild talents." "Teleportation" was his term for it—a term now common in science fiction. Objects, people, ships—anything could be teleported (moved from place to place) by individuals who possessed the power. Cargoes might be teleported commercially. Space ships might be teleported from planet to planet. It was Fort's belief that intelligences in other worlds frequently picked up human beings, perhaps for research purposes, and teleported them away. "I think we're fished for," he wrote. His books are filled with press accounts of men and women who vanished mysteriously and were never seen again. In 1910 a girl disappeared from Central Park. On the same day, Fort discovered, a swan appeared on the Park's lake near 79th Street. Could there be a connection? Ambrose Bierce vanished from Texas. An Ambrose Small vanished from Canada. "Was somebody collecting Ambroses?" Fort asked.

Not only are people teleported off the earth, Fort thought, but perhaps inhabitants of other worlds are teleported here. They would be throwbacks—men with affinities for our barbarisms. "They would join our churches . . . They'd lose all sense of decency and become college professors. Let a fall start, and the decline is swift. They'd end up as members of Congress."

"Poltergeists" are mischievous spirits who are supposed to cause the disturbances in bewitched houses. *(Haunted People,* a Freudian study of poltergeist and related phenomena, by Nandor Fodor and Hereward Carrington, was published by Dutton in 1952.) One of Fort's "acceptances" was that poltergeists are children with the powers of teleporting objects. Perhaps such children could be used in time of war. Not even Dr. Rhine, who in his latest book speculates on how increased psi powers might make war obsolete (since no military secrets could be kept), has equaled this Fortean vision:

Girls at the front—and they are discussing their usual not very profound subjects. The alarm—the enemy is advancing. Command to the poltergeist girls to concentrate—and under their chairs they stick their wads of chewing gum.

A regiment bursts into flames, and the soldiers are torches. Horses snort smoke from the combustion of their entrails. Reinforcements are smashed under cliffs that are teleported from the Rocky Mountains. The snatch of Niagara Falls—it pours upon the battlefield.

The little poltergeist girls reach for their wads of chewing gum.

CHAPTER 26

BRIDEY MURPHY AND OTHER MATTERS

THE SEARCH FOR Bridey Murphy began in Pueblo, Colorado, in 1952. Under hypnosis, a brown-haired, trim-figured little housewife, Mrs. Virginia Tighe, began to talk in Irish brogue about her previous incarnation as a red-headed colleen named Bridey Murphy. William J. Barker, assistant editor of the *Denver Post*'s Sunday supplement, *Empire,* serialized the story in 1954 (Sept. 12, 19, and 26) under the title, "The Strange Search for Bridey Murphy." The reader response suggested that here was material for a national best-seller. Morey Bernstein, the Pueblo businessman who had hypnotized Virginia, decided to write a book about it. Barker helped on the manuscript and Doubleday printed it in 1956 as *The Search for Bridey Murphy.*

For many weeks the book topped the country's best-seller lists. It was translated into five other languages. A tape recording of one of Mrs. Tighe's trance sessions was placed on a long-playing record and tens of thousands of copies were sold at $5.95 each. *True* magazine condensed the book. More than forty newspapers syndicated it. Movie rights were sold. Hostesses gave "Come as you were" parties. Juke boxes blared *Do You Believe in Reincarnation?, The Love of Bridey Murphy,* and the *Bridey Murphy Rock and Roll.* Night club hypnotists who hadn't worked for years suddenly found themselves in great demand. All over the country, and especially in California, amateur hypnotists began sending parlor subjects back to previous lives. One lady described her existence in 1800 as a horse. In Shawnee, Oklahoma, a teen-age boy shot himself, leaving a note saying that he was curious about the Bridey theory and would now investigate it in person. Two studies of Edgar Cayce (see p. 188) were rushed back into print simply because Bernstein mentioned them favorably. A rash of new books on hypnotism, reincarnation, and related occult topics broke out on publishers' lists.[1] In the words of a Houston bookdealer, Bridey was "the hottest thing since Norman Vincent Peale."[2]

One would be hard put to find a choicer sample of an utterly worthless book designed to exploit a mass hunger for scientific evidence of life after death, or a better example of the power of modern huckstering to swindle the gullible, simple folk who take such books seriously. The book is not even well written. All the miserable tricks of a low-grade pulp thriller are there to punch up the narrative (e.g., the single sentence paragraph: "Reincarnation—oh, no!," "It was," "I went," "I did.") As tasteless as the text is, Virginia's remarks under hypnosis are even drabber. W. B. Ready, purchase librarian of Leland Stanford University, expresses it this way: "It is not the misstatements, the vagueness, the ignorance that are so depressing; it is the downright dullness. . . . There is not a single line. . . . to brighten the eye, to quicken the breath, to cause anything but a feeling of discomfort."[3] Mr. Ready rests his case on Virginia's answers to some leading questions about her courtship:

> Brian came to your house?
> Uh-huh.
> When you were seventeen?
> Uh-huh.

"Now there is many an answer that an Irish girl would give to a query about her young and burgeoning love," comments Ready, "but by all the powers that be it would never be *Uh-huh.*"

Obviously the book did not sell on its literary merits. It sold because readers of little faith thought that here at last was some sort of tangible "proof" of life after death even though it be the proof of a Christian heresy. As usual, prominent leaders of science fiction were properly impressed. Robert Heinlein, writing in *Amazing Stories,* April, 1956, predicted that before the year 2001 the survival of the soul after death would be demonstrated with "scientific rigor" by following the path broken by Bernstein. And John Campbell Jr., in *Astounding Science Fiction,* Sept., 1956, wrote, "Having had some personal experience with the profound professional ignorance of the nature of the hypnotic phenomena, I am all for Morey Bernstein's highly successful effort to call attention to the lack of real understanding."[4]

If Virginia had been in the hands of a trained, well-informed psychologist, what would his reaction have been? In the first place he would have immediately recalled the classical cases exactly like Virginia's. Almost any hypnotic subject capable of going into a deep trance will babble about a previous incarnation if the hypnotist asks him to. He will babble just as freely about his future incarnations. Usually what he has to say is dreary and uninspired. At times, however, he spins such a detailed story that it becomes a matter of special interest. In every case of this sort where there has been adequate checking on the subject's past, it has been found that the subject's unconscious mind was weaving together long forgotten bits of information acquired during his early years. There was nothing in the least exceptional about Virginia's case. The obvious first step, therefore, of a trained psychologist would have been an investigation of Virginia's childhood. Did this occur to Mr. Bernstein? If it did, there is no indication that he thought about it twice. Instead, he began a search for Bridey in Ireland! Irish librarians, an Irish legal firm, and others were asked to check on place names mentioned by Virginia and to look for evidence of a Bridey Murphy who once lived in Cork. Nothing of significance turned up. The *Chicago Daily News* ordered its London correspondent to Ireland. The results again were negative. The *Denver Post* then sent their reporter Barker, who began it all, to Ireland, printing his report as a special supplement to their March 11, 1956 issue. Barker summarizes this report in a chapter added to the Pocket Book edition of Bernstein's book. Although Barker was firmly convinced of Bridey's nineteenth century existence, the best he could discover on his expensive jaunt was that Virginia at some time had been exposed to a certain amount of Irish lore.

And then, in a wonderfully hilarious and crushing climax, Hearst's *Chicago American,* whose Sunday supplement *The American Weekly* was once the nation's outstanding purveyor of pseudo-science, actually found the missing Bridey![5] With admirable scientific acumen, *American* reporters began to prowl about Virginia's old home town—Chicago. With the help of Rev. Wally White, pastor of the Chicago Gospel Tabernacle where Virginia once attended Sunday School, it did not take them long to locate Mrs. Anthony Corkell. Now a widow with seven children, she was still living in the old frame house where she had lived when Virginia was in her teens. For five years Virginia lived in a basement apartment across the street. Mrs. Corkell's Irish background had fascinated the little girl. One of her old friends recalled that Virginia even had a "mad crush" on John, one of the Corkell boys. Another Corkell boy was named Kevin, the name of one of the imaginary Bridey's friends. Note also the similarity of Corkell and Cork, the city where Bridey was supposed to have lived. And what was Mrs. Corkell's maiden name? Bridie (with an "ie',) Murphy!

The more reporters talked to relatives and friends of Virginia, the more it became obvious that in her trances she had simply dredged up childhood memories. Bernstein had been careful to give Virginia a fictitious name in his book in order to conceal her identity. But so striking were the parallels between Virginia's early life and the life of Bridey Murphy that several of her childhood chums, who had no idea she was living in Pueblo, recognized her when they read the book!

Virginia was born in 1922 in a white frame house in Madison, Wisconsin, exactly like the house that Bridey described. Her mother's name was Katherine Pauline (Bridey's mother's name was Kathleen). Virginia's sister once suffered a bad fall down the stairs just like the fall Bridey said had caused her own death.

When Virginia was four, her parents separated and she was taken to Chicago to live with an uncle and aunt. Her uncle recalls how she used to dance jigs on the street for pennies. (One of the few impressive incidents in Bernstein's book is when Virginia, in a trance, dances an Irish jig.) Bridey's brother died when Bridey was four. Virginia had a brother who died when she was five. At the age of six or seven she had been soundly whipped for scratching fresh paint off her metal bed, another incident that figures prominently in the book and which greatly impressed the naive Bernstein. It is odd that Virginia did not tell this to her hypnotist because she certainly had not forgotten it. One of her friends recalled her laughing about it some dozen years after it had happened.

A friend of Virginia's foster parents was a man whose first name was Plezz, but whom Virginia called "Uncle Plazz." Uncle Plazz told reporters that he remembered Virginia well. Bernstein had been enormously excited by the "authenticity" of Bridey's "Uncle Plazz." One of his "Irish investigators" told him that this was "phonetic spelling" of the "very rare Christian name Blaize," after the Irish Saint Blaize. Evidently the Irish "investigator" was confusing "Blaize" with Saint Blaise, a fourth century *Armenian* saint. "I had been unable to find anyone who had even heard of such a name," Bernstein writes. It never occurred to him to ask Virginia if *she* had ever known an Uncle Plazz.

Attending high school on Chicago's north side, Virginia was both active and talented in school dramatics. Her teacher recalled that she memorized several Irish monologs which she delivered with a heavy brogue. In one monolog she was Bridget Mahon; in another, Maggie McCarthy. Her uncle told a reporter, "She could put on a brogue so well you would swear there was an Irishman in the room with you." Moreover, Virginia had an Irish aunt (no longer living) of whom she was very fond and who used to tell her Irish tales about the old country. There is some evidence that it was from this aunt that the lonely Virginia, rejected by her parents, received the most affection. If so, it would explain her lifelong love for things Irish. One of young Virginia's favorite songs was the *Londonderry Air (Danny Boy)* which, oddly enough, was also one of Bridey's favorites. Virginia liked potato pancakes. So did Bridey. Even Bridey's flaming red hair had a parallel, for Virginia had so admired red hair that she once dyed her own hair red!

Bridey's Irish husband was named Sean Joseph Brian MacCarthy. Sean is Gaelic for John, the name of the Corkell boy on whom Virginia had her childhood crush. Brian happens to be (though Bernstein never mentions it) the middle name of Virginia's husband! And we all know who Joe McCarthy is, or rather was back in 1952 when Virginia first slipped into her trances. Then, too, Virginia was a "McCarthy" in one of her Irish monologs.[6]

"Virginia had a good imagination," one of her childhood friends told a reporter. "I always thought she could write a book." That of course, is exactly what she did, only Bernstein's name is on the title page and the royalty checks from Doubleday. It would have been the simplest thing in the world for Bernstein to have checked on Virginia's past, but then of course he wouldn't have had his book. Fortunately, it looks as though the *Chicago American* articles gave the book its much needed *coup de grace.* At any rate it dropped quickly from best-seller lists.

Bernstein ends his book with a cliff-hanger that obviously hints at plans for a dramatic sequel. He speaks of an "idea for expanding the Bridey experiment—an idea so fascinating that I can hardly wait to set up the experiment. It looks as though I'm about to take another step on the long bridge."

One would guess that Bernstein's next step, if he takes it at all, will be neither as long nor as profitable as his first one. But one should never underestimate the skill of Manhattan's copy writers or the credulity of the suckers for whom the copy is so carefully written.

Dr. Joseph B. Rhine, in his brief comments on the Bridey mania (*Tomorrow,* Summer, 1956), agreed that there was nothing in the book that a student of science would want to take seriously. On the other hand, he felt that the wave of Bridey excitement had done more good than harm by stimulating public curiosity concerning matters on the fringe of orthodox psychology and so paving the way for serious study of such phenomena. To the extent that Bernstein's book did this, I would agree that some good may come of it. But I am more inclined to think that the book has encouraged only crank research. There already is an enormous popular interest in the occult, as the astrology magazines testify, and no doubt Bridey Murphy stimulated this interest. But the total absence of scientific insight on Bernstein's part has resulted in a book more likely to repel the serious student of psychology than fascinate him.

Of course *some* good may result from even the wildest pseudo-scientific work. In previous chapters I made no effort to pause and point out what such good results might be, so let me summarize a few of them here. In the first and obvious place, it is good to have off-trail work constantly going on because there is always the possibility that the eccentric scientist, however incompetent, may stumble upon something worth-while. In fact most pseudo-scientific cults, especially those that attract a mass following, contain many praiseworthy elements. The homeopathic movement, for example, arose in pre-scientific days of medicine when doctors were fond of giving patients heavy doses of drugs that were little understood and often harmful. By diluting their doses to an infinitesimal amount, the homeopaths developed a *materia medica* that caused no harm because it had no effect whatever. As someone once put it, homeopathic patients died only of the disease whereas patients of the orthodox doctors died of the cure as well. The success of the homeopathic movement thus not only called attention to the evils of large and indiscriminate doses, but it also pointed up the psychosomatic value of the placebo.

Similarly, the naturopathic movement helped publicize the value of sunlight, fresh air, exercise, fresh foods, low-heeled shoes for women, and a host of other good things. Orthodox doctors were not in opposition to these views, but it is true that they made little effort to stress them. With all his crank and at times dangerous medical opinions, one must grant that Bernarr Macfadden's publications did a certain amount of good in bringing a few sound ideas to the attention of persons who might not otherwise have encountered them.

Even when a pseudo-scientific theory is completely worthless there is a certain educational value in refuting it. "False facts are highly injurious to the progress of science," Darwin wrote, "for they often endure long; but false views, if supported by some evidence, do little harm, for everyone takes a salutary pleasure in proving their falseness; and when this is done, one path toward error is closed and the road to truths often at the same time-opened." Darwin had in mind the theories of scientists more competent than most of the men discussed in this book. It is not likely that any new roads to truth will be opened by refuting Velikovsky. Nonetheless, Darwin's remark is applicable in a way to almost any fringe scientist. Anyone who refutes Velikovsky, for example, can hardly avoid learning a great deal of physics and astronomy. The *Chicago American* exposé of Bridey Murphy has, let us hope, made it more difficult for pseudo-psychologists of the future to be misled by hypnotically induced memories of previous lives. This "path toward error" has been closed so many times before, however, that one hesitates to be optimistic.

The spectacular recent successes of pseudo-science have a value also in publicizing aspects of our culture that are much in need of improvement. We need better science education in our schools. We need more and better popularizers of science. We need better channels of communication between working scientists and the public. And so on.

Finally, we must not forget that these Don Quixotes of the scientific world, with their mad antics and fantastic rationalizations, are often vastly amusing fellows. Alfred Lawson, for instance, is one of the nation's great unintentionally comic figures. If you read the works of these men in the right spirit, you will find much high and refreshing humor.

In view of these merits, should one then conclude that publishers do us all a favor by bringing out in hard covers a worthless scientific work? To answer this question we shall have to return to one of the continuums cited in Chapter 1. Off-trail science is a spectrum that ranges from the obviously crackpot views of a Voliva to the reputable views of a Rhine. We can all agree that a distinguished publishing house should not lower its standards by printing a book proving the earth to be flat, just as we can all agree that the same house should not hesitate to publish a book by Dr. Rhine. Somewhere near the middle of this spectrum is a vague area where a manuscript will be in doubt; where there will be equally good arguments on both sides. To ask, therefore, whether a given book should or should not be published is to ask whether it lies so far below this area of doubt that nothing will be gained (aside from financial profits) by publishing it. And how is the location of a manuscript on this spectrum to be determined? Obviously, it can only be decided by scientists themselves. A reputable house should no more think of submitting a scientific work to a literary critic for his opinion than it would think of submitting an unpublished novel to a research scientist. Actually, a scientist's opinion of a novel would likely be of more value than a literary critic's opinion of a work of science. Most publishing houses recognize this, of course, and when a scientific manuscript falls into their hands it is immediately routed to an appropriate expert for his opinion.

In many recent cases, however, this commendable practice was violated and pseudo-scientific works far below the area of reasonable doubt were published and heavily promoted by houses that either did not seek expert opinion or disregarded it. It is in these cases that the publication of a book begins to smack of fraud, for the public has grown to expect the larger houses to weed out worthless manuscripts. The very fact that their imprimatur is on a book suggests to the man in the bookstore that here is something worthy of his consideration. And if the promotion of the book portrays it as a revolutionary new scientific hypothesis to compare with Darwin's, then the fraud is compounded.

Eric Larrabee, an editor of *Harper's* magazine who introduced Velikovsky's views to the public by way of a *Harper's* article, has, like a man with an uneasy conscience, repeatedly defended his action. In a letter to *Scientific American*, May, 1956 (to which physicist Harrison Brown makes a long and good reply), he states his opinion that Velikovsky has behaved better than his detractors, and that when a few scientists threatened to boycott Macmillan's textbooks unless it abandoned *Worlds in Collision* their behavior was a "disgrace to American science." He is shocked at how slender is the faith of scientists in the "open testing of ideas."

In my opinion Mr. Larrabee never comes to grips with the central problem—whether he, as a non-scientist, did a good thing when he trusted his own judgment concerning Velikovsky as opposed to the judgment of astronomers and physicists. To the professional scientist, Velikovsky's manuscripts no more deserved the dignified treatment Larrabee accorded them than scores of similar manuscripts that are rejected every week by editors. Larrabee has persistently ignored the fact that scientific societies provide a highly efficient means for the "open testing" of ideas. He has only to glance at back issues of the *Bulletin of the American Physical Society* to see how often scientists sit and listen with patience and interest to a weird and unorthodox paper. Obviously there are still lower levels of the spectrum where views are so preposterous that the only way the scientist can get a hearing is to publish the views himself. This of course is precisely what he does. It would be foolish to expect the leading publishing houses and magazines like *Harper's* to take on the staggering burden of proclaiming these off-trail theories.

Let us put it this way. If Velikovsky deserves to be praised in the pages of *Harper's* then so do dozens of other men mentioned in this book. Let's have an article by Larrabee on the great work that Wilhelm Reich, certainly a man with a far more distinguished scientific background than Velikovsky, is doing with orgone energy. Let's have a *Harper's* article on the virtues of Krebiozen as a cancer cure, so that the research of Dr. Andrew Ivy can be tested in the market place of public opinion. One sees at once that editors have to be ruthlessly selective. They can't publicize every eccentric theory that comes along. And this snaps us quickly back to the basic question of who should do the selecting. Editors untrained in science?

By all means let the Don Quixotes of science be heard. But let them be heard in a manner befitting their position on the spectrum of unorthodoxy, and let that position be determined by those who alone are qualified to do so.

Appendix and Notes

CHAPTER 1

Notes

1. This point is emphasized in I. Bernard Cohen's excellent paper, "Orthodoxy and Scientific Progress," *Proceedings of the American Philosophical Society,* Oct., 1952. The same issue also contains Edwin G. Boring's wise and witty lecture, "The Validation of Scientific Belief," which opened the society's 1952 symposium on scientific unorthodoxies. See also L. Sprague de Camp's informative article, "Orthodoxy and Science," *Astounding Science Fiction,* May, 1954.

2. In 1952 a science crank's paranoia led to a senseless murder. A young war veteran and self-styled genius named Bayard P. Peakes walked one morning into the offices of the American Physical Society, Columbia University, whipped a pistol from his pocket and killed the 18-year-old stenographer whom he had never seen before. After his capture he gave his reasons. The society had refused to publish his book *How to Live Forever,* explaining his theory of prolonging life by electronics. He believed that if he killed someone connected with the society the publicity would win recognition for this work. In 1949 he had issued a pamphlet titled *So You Love Physics,* distributing some 6,000 copies to scientists, including ten to Einstein. "Einstein didn't answer me," Peakes told a reporter. "I think he's crazy."

3. One of the most delightful of these "inverted" theories is the "granular universe" of Osborne Reynolds (1842–1912), professor of Engineer-

ing at Owens University, Manchester, England. On the basis of experiments with wet sand, Reynolds concluded that space was made up of solidly packed spheres, each with a diameter of one seven hundred thousand millionth part of the wave length of light. Material particles are simply bubbles of nothing, moving about in this dense, elastic, granular medium. The larger the "hole" in the medium the stronger the distortion in the otherwise normal "piling" of surrounding grains. Gravity is a pressure that results from this distortion. See his *On an Inversion of Ideas as to the Structure of the Universe,* 1902, and *The Sub-Mechanics of the Universe,* 1903, both published by Cambridge University Press.

4. Bertrand Russell, in his article, "In the Company of Cranks," *Saturday Review,* Aug. 11, 1956, writes:

> Experience has taught me a technique for dealing with such people. Nowadays when I meet the Ephraim-and-Manasseh devotees I say, "I don't think you've got it quite right. I think the English are Ephraim and the Scotch are Manasseh." On this basis a pleasant and inconclusive argument becomes possible. In like manner, I counter the devotees of the Great Pyramid by adoration of the Sphinx; and the devotee of nuts by pointing out that hazelnuts and walnuts are just as deleterious as other foods and only Brazil nuts should be tolerated by the faithful. But when I was younger I had not yet acquired this technique, with the result that my contacts with cranks were sometimes alarming.

CHAPTER 2

Readers interested in learning more about Voliva may consult the following two articles: "Croesus at the Altar," by Alfred Prowitt, *American Mercury,* April, 1930, and "They Call Me a Flathead," by Walter Davenport, *Colliers,* May 14, 1927.

For historical background on Symmes' hollow earth see "Symmes' Theory," by John W. Peck, *Ohio Archeological Historical Publications,* Vol. 18, 1909, p. 28; "The Theory of Concentric Spheres," by William M. Miller, *Isis,* Vol. 33, 1941, p. 507; and "The Theory of Concentric Spheres," by Conway Zirkle, *Isis,* Vol. 34, July, 1947.

It is interesting to note that Teed was the author of a novel about the future. It was published posthumously by his followers in 1909 under the title, *The Great Red Dragon; or, The Flaming Devil of the Orient,* and bearing the pseudonym of Lord Chester.

Teed's hollow earth, or a theory very similar to it, has been taken up by Duran Navarro, a Buenos Aires lawyer. According to a story in *Time,* July 14, 1947, Navarro contends that gravity is really centrifugal force generated by a rotating hollow earth inhabited on the inside. The force naturally diminishes as you move away from the surface toward the central point where protons and electrons come together to form "fotons" that in turn produce the sun. Simultaneously with Navarro's announcement, *Time* adds, comes news from Berlin that the earth does not rotate from west to east. An accountant named Valentin Herz has proved that it really rotates the opposite way.

It was also in Germany that pseudo-science recently took a drubbing. A West German patent attorney, Godfried Bueren, boldly offered 25,000 marks (about $6,000) to anyone who could disprove his hollow sun theory. According to Herr Bueren, the sun's flaming outer shell surrounds a cool inner sphere. Covered with vegetation, the dark core can be glimpsed occasionally through sunspots which are nothing more than temporary rents in the blazing shell. The German Astronomical Society carefully ripped the theory apart and when Bueren refused to pay, the society took legal action. Incredible as it may seem, the court decided in favor of the astronomers. Herr Bueren was ordered to pay the sum he had offered, plus court costs and interest. See *Time*, Feb. 23, 1953.

Notes

1. For further details concerning the German cult and other hollow earth theories see Willy Ley's "The Hollow Earth," *Galaxy*, March, 1956.

CHAPTER 3

Velikovsky's second book, *Ages in Chaos*, Vol. I, 1952, is a drastic revision of ancient Hebrew and Egyptian history to make it conform to the author's interpretation of the *Old Testament*. Velikovsky's historical method, as reviewer William Albright observed (*N.Y. Herald Tribune Book Review*, April 20, 1952), is on a level with that of the professor who identified Moses with Middlebury by dropping the "-oses" and adding "-iddlebury."

In 1955 Velikovsky's third work, *Earth in Upheaval*, appeared. It defends his theory that evolution has proceeded by a series of catastrophic leaps, the catastrophes caused by comets. The book is markedly inferior in ingenuity to George Price's *The New Geology* (discussed in Chapter 11). Several prominent non-scientists continued to lend the dignity of their names to the doctor's fantasies. "I am impressed with how Dr. Velikovsky has piled up the evidence," stated Horace Kallen, "and built a case for the catastrophic theory of evolution which no self-respecting man of science can disregard." And Clifton Fadiman, in the *Book-of-the-*

Month Club News, Dec., 1955, likened Velikovsky to Leonardo DaVinci and asserted that to charge the doctor with "eccentricity" was the "merest obscurantism." Fadiman admitted that he himself possessed "virtually no scientific training," nevertheless he thought Velikovsky's thesis carried "a great deal of persuasion."

One of the best of the many attacks on Velikovsky's twaddle is a paper by Harvard astronomer Cecilia Payne-Gaposchkin in the *Proceedings of the American Philosophical Society,* Vol. 96, Oct., 1952. This paper was read at a session on scientific unorthodoxies during the 1952 meeting of the society. Velikovsky attended the session, and after the paper was read, the chairman permitted the doctor to make a rebuttal. As one present on this occasion I can report that Velikovsky—tall, distinguished looking, and completely at ease—gave a magnificent performance as the misunderstood genius, patiently resigned to the stubbornness of his orthodox critics. His chief fear, he said, was that in the future his theories might become a new dogma, as difficult to modify as the theories of present-day astronomy. Everyone applauded politely when he finished.

CHAPTER 4

Of the many articles that have been written about Fort, the following are recommended: "The Fortean Fantasy," by Benjamin De Casseres, *The Thinker,* April, 1931; "The Mad Genius of the Bronx," by H. Allen Smith, Chapter 5 of *Low Man on a Totem Pole,* 1941; "Charles Fort," by Robert Johnson, *If,* July, 1952; and "Charles Fort: Enfant Terrible of Science," by Miriam Allen DeFord, *Fantasy and Science Fiction,* Jan., 1954.

Tiffany Thayer is still editing *Doubt,* the back cover of which continues to plug Chakotin's *Rape of the Masses,* and a 425-page work by Iktomi titied *America Needs Indians.* I was startled to find a favorable review of *In the Name of Science* in a 1953 issue. "The author gets a little sore at YS [your secretary]," writes Thayer, "and took a few snipes at him, but we don't hold that against him."

It is not generally known that Mr. Thayer is one of the country's top-flight advertising copy writers, working six months of each year for a Manhattan agency where he turns out radio jingles for Pall Mall cigarettes. In 1956 he published the first three volumes of his projected 21-volume novel, *Mona Lisa.* At the end of this trilogy Mona is not yet born, but plenty of things have been going on in the streets and bedrooms of Thayer's Renaissance Italy.

CHAPTER 5

Since this chapter was written several dozen hard cover books on the saucers have enjoyed profitable sales. They range in quality from the works of George Adamski, that out-Scully Scully, to two new books by Keyhoe. Keyhoe does not, like Adamski, claim to have ridden in a saucer where he conversed with a voluptuous golden-sandaled Venusian. This restraint has led many reviewers who should know better to take Keyhoe's speculations seriously. Groff Conklin, writing in *Galaxy,* April, 1954, found the data in Keyhoe's *Flying Saucers from Outer Space* "ominously persuasive" and "unassailably factual." The editors of *Fantasy and Science Fiction,* Feb., 1954, found the same book "uncontrovertible" and thought it likely the most important work ever mentioned in their review column.

In spite of the flood of zany books, articles, lectures, documentary films, and special magazines trumpeting the extraterrestrial saucer theory, the mania seems to have slowly abated as far as the general public is concerned, leaving saucer speculation in the hands of the occultists. Some dozen or so cults of a theosophical type have now integrated the saucers with their other beliefs, the general approach being that the space people are here to prepare the earth for a New Age. In most cases a leader or prominent member of the cult is in touch with the space people by extra-sensory perception, receiving detailed instructions which are then passed on to the neurotic middle-aged ladies who make up the bulk of the cult's membership.

In the fall of 1954 the saucer mania struck France, then fanned out over Europe. The French press outdid even the United States in unbridled reports of fantastic little men observed here and there stepping out of the saucers (see *Time,* Oct. 25, 1954; *Life,* Nov. 1, 1954; and the *New York Herald Tribune,* Dec. 10, 1954).

The wire services in this country have made two valiant but unsuccessful Fortean attempts to replace the worn out saucer craze with something new and less boring. In 1954 it was a "glasspox" epidemic—a mysterious pitting of automobile windshields. The epidemic started on the west coast then rapidly moved eastward. Even England suffered a mild attack. Then in the hot summer months of 1955 garden hoses began to burrow their way into the soil. Nobody mentioned Shaver's "deros," obviously responsible for this mischief.

The mania of past centuries for seeing sea serpents has many obvious parallels with the flying saucer craze and should be just about due for a revival. One or two sensational eye witness accounts, a solemn article in a mass circulation magazine, and soon half the sailors on the high seas will be bringing back reports about monsters of the deep. The recent flurry of interest in the "abominable snowman" of the Himalayas, an elusive beast who leaves giant footprints but manages to evade being captured or photographed, suggests that public interest in monsters "unknown to science" is still a lively one. (See Ralph Izzard's book, *The Abominable Snowman*, 1955.)

Notes

1. The best discussion to date of the Shaver hoax will be found in two articles by Thomas S. Gardner—"Calling All Crackpots!" and "Crackpot Heaven"—in the science fiction magazine, *Fantasy Commentator*, spring and summer issues, 1945.
2. An enthusiastic Fortean, Arnold once sought to have the Fortean Society sponsor him on a lecture tour. This is disclosed by Tiffany Thayer in *Doubt,* No. 40, 1953, an issue devoted entirely to a chronological report on saucer sightings.
3. Father Connell provides an appendix on extra-terrestrial theology to Aimé Michel's *The Truth About Flying Saucers,* 1956, a translation of a French roundup of American and European saucer data.
4. Frank Scully's latest work is *Blessed Mother Goose,* a rewriting of the familiar nursery rhymes so that, as an advertisement has it, they "echo the Catholic way of life."
5. "Dr. Gee" turned out to be Leo GeBauer, proprietor of a radio and television supply house in Phoenix, Arizona. It was he and his friend Silas Newton, Denver geophysicist, who provided Scully with the data on which his book was based. Newton and GeBauer were arrested in 1952 and later found guilty by a Denver court of swindling a Denver businessman out of some $250,000. The two men had sold their victim an electronic "doodlebug" for finding oil, and he had sunk a small fortune in worthless oil leases as a result. The machine proved to be a radio frequency changer that could be bought as war-surplus for about $3.50.
6. For a recent article that takes seriously the many "maps" which have been sketched of Martian canals, see Wells Alan Webb's "Correlation of the Martian Canal Network," in *Astounding Science Fiction,* March, 1956. Webb subjects these maps to an elementary topological analysis, finding them similar to networks that are manmade *(e.g.,* airline routes), thus leading him to conclude that they have an "animal origin." His theory has one simple fallacy. If the "maps" are merely the doodlings of imaginative astronomers, as most astronomers think they are, the topological analysis naturally still applies. A footnote credits editor John Campbell Jr. with the theory that the canals may be pathways beaten out by migrating herds of animals.

CHAPTER 6

Notes

1. Cy Q. Faunce (probably a Lawson pseudonym) is also the author of *The Airliner and Its Inventor,* 1921, a biography of Lawson.
2. *Time,* March 24, 1952. After a 1954 investigation for tax dodging by a Congressional House committee, Lawson sold the University to a Detroit businessman for $250,000. It is to be turned into a large shopping center *(New York Times,* Nov. 21, 1954, p. 81).
3. This incident took place in 1945. Edwin A. Baker, of Alexandria, La., won a court fight to "free" his daughter, age 12, from the university. The girl described the school as "nightmarish" and "full of generals."
4. According to a story in the *New York Herald Tribune,* Aug. 25, 1954, the school tore down its large smokestack early in the year but continued to buy tons of coal. "It was rumored that Mr. Lawson had put his theory of 'penetrability' to work and built a maze of tile pipes leading into a big tunnel in the ground where the smoke settled. Engineers said it was impossible, but could find no trace of smoke escaping anywhere."
5. See also *Lawson's Mighty Sermons,* published by Lawson in 1948.

CHAPTER 7

Notes

1. Two classic essays in English deal with the historic Goethe-Newton battle: one by Helmholtz, in his *Popular Lectures,* first series, and one by Tyndall, in *New Fragments.* The battle was so decisively won by Newton that it is hard to conceive of any twentieth century thinker taking Goethe's speculations seriously. But the pseudo-scientist is a man of boundless courage. Ernst Lehrs, in his book *Man or Matter,* combines Goethe's metaphysical physics with the anthroposophical poppycock of Rudolf Steiner (see pp. 169, 224f). Lehrs' most amusing feat is his revival of the ancient notion of "levity," a force opposed to gravity. The book was published by *Harpers* in 1951.
2. It is surprising to find Prof. Michael Polanyi, writing on "From Copernicus to Einstein," *Encounter,* Sept., 1955, still taking Miller's work seriously. "The experience of D. C. Miller," he declares, "demonstrates quite plainly the hollowness of the assertion that science is simply based on experiments which anybody can repeat at will." Dr. Polanyi forgets that the "repeatability" of a complicated experiment does not, and could not, demand that every single person who tried it would get

identical results. "Repeatability" is a matter of degree. It is always possible to find *someone* unable to perform an experiment.

3. Augustus de Morgan's *Budget of Paradoxes,* recently reissued by Dover Publications, is a mine of information on early angle trisecters, circle squarers, cube duplicaters, and parallel postulate provers. The most fantastic of them all was James Smith (1805–1872), the Liverpool merchant who wrote book after book to prove that *pi* was exactly 3⅛. See de Morgan's work, Vol. II, p. 103f, for a hilarious account of Smith's labors and personality.

CHAPTER 8

Notes

1. Inventor of the Birds Eye frozen food process, now owned by General Foods.

CHAPTER 9

In 1953 Mr. Roberts brought out *The Seventh Sense,* a sequel to his first book on dowsing. It is more of the same, its scientific value confined solely to the psychological insight it provides into the working of Mr. Roberts' mind. An enormous number of popular magazine articles appeared in the wake of publicity attending Roberts' two books, but they differ in no essential respect from similar articles of past centuries.

Notes

1. Sir Arthur Conan Doyle's introduction to an article on "The Sideric Pendulum," *Strand* magazine, Vol. 60, 1920, p. 180, speaks of his own pendulum tests as having "never failed." Doyle attributes the phenomenon to the same occult forces that make dowsing possible.

2. See also Riddick's excellent article, "Dowsing—An Unorthodox Method of Locating Underground Water Supplies or an Interesting Facet of the Human Mind," *Proceedings of the American Philosophical Society,* Vol. 96, Oct., 1952.

3. And sometimes the gullible investor. See footnote 5 of chapter 5, concerning a recent doodlebug swindle in Denver.

4. One of the few references known to me on oil doodlebugs is L. W. Blau's entertaining article, "Black Magic and Geophysical Prospecting," *Geophysics,* Jan., 1936.

CHAPTER 10

CHAPTER 11

Notes

1. From a letter of Price's quoted in the booklet *Miscellaneous Documents,* by W. C. White and D. E. Robinson, 1933, Elmshaven Office, St. Helena, Calif.
2. *The Phantom of Organic Evolution,* 1924, p. 210.
3. *Ibid,* p. 8.
4. *Ibid,* p. 20.
5. *Catholic World,* Oct., 1930.
6. An English biologist, Mivart's outstanding work was a monumental 557-page treatise, *The Cat,* 1881. Some magazine articles advocating liberal theological views led to his excommunication in 1900 and he died a few months thereafter. Years later, after his friends convinced the church that his heretical opinions were not due to willful disobedience but to the diabetic condition which caused his death, his body was given a Christian burial.
7. From *Feet of Clay,* 1949, a booklet issued by the Christian Evidence League, Malverne, N. Y.

CHAPTER 12

Lysenko's subsequent downfall may be chronicled as follows:

In 1953, a few months after Stalin's death, *Pravda* published Lysenko's "Eulogy" of Stalin in which he revealed that the dictator had helped prepare his famous speech of 1948. "Comrade Stalin found time even for detailed examination of the most important problems of biology . . . ," Lysenko declared. "He directly edited the plan of my paper, 'On the Situation in Biological Science,' in detail explained to me his corrections, provided me with directions as to how to write certain passages in the paper."

In 1954 Lysenko was severely rebuked in a speech of Khrushchev's and later by several official party organs. He was branded a "scientific monopolist" and "academic schemer" who stifled all theories opposed to his own. He was accused of failing to make practical contributions to Soviet agriculture.

Lysenko made one last futile attempt to regain prestige. He dramatically announced a sensational new agricultural discovery by Soviet agronomist Terenty Maltsev. This "new" discovery proved to be identical with the view advocated by Edward H. Faulkner in two books published by the University of Oklahoma Press: *Plowman's Folly,* 1943, and *A Second Look,* 1947. According to Faulkner's highly questionable thesis, crop yields can be greatly increased by loosening the soil with a disk harrow instead of turning it over with a conventional plow. Nothing much came of Lysenko's announcement, and in 1956 he resigned as head of the All-Union Academy of Agricultural Sciences.

Shortly before his resignation, the Academy ordered the republication of the works of Lysenko's old enemy, the great Vavilov who died in Siberian exile after Lysenko ousted him as head of the Academy. Although this heralds a return to Mendelianism, there is little likelihood it will be called by that name. So far, all criticism of Lysenko's views, though identical with western criticism, has been in terms of his "perversions" of Michurinism! "If Mendelian genetics ever comes to its own again," Abraham Brumberg prophecies in his article on the fall of Lysenko, *New Leader,* Aug. 9, 1954, "it will be done stealthily, unostentatiously, through the back door of pristine Michurinism."

Nevertheless, at the moment a fresh breeze may be blowing through Soviet biology. Whether it will grow stronger or weaker in the years ahead is a question about which one hesitates to hazard an opinion.

Notes

1. For an hilarious documentation of Nazi party "misorganization" of science during World War II, see Samuel Goudsmit's remarkable book, *ALSOS,* 1947.
2. A hope rapidly fading. For a time there were strong political pressures in the Soviet Union against Bohr's principle of complementarity and other widely accepted concepts of modern nuclear theory, but even these pressures seem to have lessened since Stalin's death. On the level of technical achievement, Soviet war research now appears to be only a step or two behind the United States.

CHAPTER 13

CHAPTER 14

Notes

1. For a magnificent compendium of information on the literature dealing with Atlantis and other sunken cultures see L. Sprague de Camp's *Lost Continents,* Gnome Press, 1954.
2. Lewis Spence died in Edinburgh, 1955.

CHAPTER 15

Notes

1. Most distinguished recent convert to Jehovah's Witnesses is Mickey Spillane. He was baptized into the sect in 1950 and has since been preaching from house to house, selling *Watch Towers* on street corners, and feeling pangs of remorse over the wickedness in his best-selling sex-plus-sadism crime thrillers. See Marion Hargrove's "The Secret Life of Mickey Spillane," *Redbook*, June, 1955.
2. As the fatal day passed without visible signs of Armageddon, the Anglo-Israelites were not dismayed. The new age, they now argue, began imperceptibly and it may be many years until it becomes clear to everyone that we are in it. The best that *Destiny* could do with the Aug. 20, 1953 date was to note that on that day the Russians *announced* they had exploded (eight days earlier) a hydrogen bomb. (See *Destiny*, Nov., 1953.)
3. The latest occult work on the pyramid is *Secret: The Gizeh Pyramids*, by someone who calls himself Thothnu Tastmona. The publisher, Thothmona Book Company, Manhattan, took a full page ad in the *New York Times Book Review*, Oct. 31, 1954, to huckster his 144-page book at $15.00 a copy.

CHAPTER 16

Notes

1. Likewise discarded are Hahnemann's views on mesmerism. In his *Organon* he recommends it highly as a curative agent. "It acts in part homeopathically," he writes, "by exciting symptoms similar to those of the disease to be cured, and is applied for this purpose by a single pass or stroke of the hands held flatwise over the body, and carried, during moderate exertions of the will, from the crown to the tips of the toes; this process is efficacious in uterine haemorrhages, even when death is imminent."
2. John Kellogg's most massive work was *Rational Hydrotherapy*, 1900, a volume of more than 1200 pages. It was his younger brother, W. K. Kellogg, who became the country's corn flake king.
3. There is no question that Macfadden's death from jaundice in 1955 was hastened by a three-day fast, self-imposed in an effort to cure himself without medical aid. *Dumbbells and Carrot Strips*, an uninhibited biogra-

phy of Macfadden by his third wife, Mary Macfadden, was published in 1953.

4. Medical follies, like religious sects, never completely die. The German picture magazine *Quick,* in its May 30, 1954 issue, featured an article on iridiagnostician Emil Stramke, of Hamburg; and in June, 1954, police in Tulsa, Okla., arrested "iriologist" J. D. Levine, to the annoyance of a crowd assembled to hear him lecture at the Alvin Hotel.

5. The best references on Shaw's crank medical views are the lengthy preface of his play *The Doctor's Dilemma,* and the papers collected under the title *Doctor's Delusions,* Vol. 13 of the *Standard Edition of the Works of G.B.S.*

6. Dewey also penned an introduction to a work much stranger than any of Alexander's books: *Universe,* 1921, privately printed by the author, Scudder Klyce, a retired Navy officer of Winchester, Mass. According to Klyce's own estimate, "This book unifies or qualitatively solves science, religion, and philosophy—basing everything on experimental, verifiable evidence. . . . All the qualitative problems set forth by the race—by 'religion, science, and philosophy'—are herein positively, definitely, and verifiably solved."

Although they were friends at first, Klyce later developed a violent hostility toward Dewey's views, resulting in his collection of angry letters, *Dewey's Suppressed Psychology,* 1928. *Sins of Science,* 1925, is Klyce's 432-page attack on all leading scientists and thinkers.

CHAPTER 17

At the moment, the largest cancer clinic in the United States operated by a man without a medical license is the Hoxsey Cancer Clinic, Dallas, Texas. In 1953 the Food and Drug Administration obtained an injunction on interstate shipments of Hoxsey medicines, but the case is still limping through the courts. The FDA issued an unprecedented public warning in 1956 against these medicines, pointing out that some contain potassium iodide, believed to *accelerate* certain types of cancer.

Harry Mathias Hoxsey never got beyond the eighth grade. He began his medical career in Illinois in the twenties by peddling a cancer tonic and salve inherited from his father, a veterinarian. He was thrice convicted in Illinois for illegal medical practice and he himself boasts of having been arrested a hundred times. Eventually Hoxsey became a licensed naturopath, establishing his Dallas clinic in 1936. Most of his staff doctors are osteopaths and it is largely through osteopaths that residents of other cities obtain the Hoxsey treatment. In 1956 five of his osteopaths were suspended from practice by the Texas State Board of Medical Examiners for practicing with a layman.

Hoxsey is the author of a book called *You Don't Have to Die,* published by guess who? His motto, engraved on a plaque in his office: "The world is made up of two kinds of people—dem that takes and dem that gets took." See *Time,* Aug. 9, 1954, and *Life,* April 16, 1956.

Hoxsey's most influential convert, Pennsylvania state senator John J. Haluska, promoted the recent founding of another Hoxsey clinic, in Portage, Pa. The Oct., 1956 issue of *Search* features the senator's picture on the cover and an article by him titled "Hoxsey Does Cure Cancer." A similar piece by him appeared in the May issue of the same year. *Search* is an occult pulp magazine published by Ray Palmer to compete with his other occult magazine, *Fate* (see p. 60).

More fantastic than the story of Hoxsey is the story of the secret cancer drug Krebiozen—fantastic because it involves a widely respected medical scientist, Dr. Andrew C. Ivy. The story begins in Argentina in the late forties. There two Yugoslav refugees, Dr. Stevan Durovic and his brother Marko, obtained a culture from a cattle disease called "lumpy jaw" and injected it into what must have been an extraordinarily large number of horses. From thousands of gallons of horse blood they extracted, they claim, two grams of a mysterious whitish powder which they named Krebiozen or "K." The Durovic brothers eventually settled near Chicago, dissolved the "K" in mineral oil, and began selling it as a cancer drug.

Physiologist Ivy, then a vice president of the University of Illinois and head of its department of clinical science, became an ardent booster of the drug. George D. Stoddard, the university's president, ordered Ivy to stop using "K" in university clinics after Ivy and the Durovic brothers refused to produce samples of the drug for analysis. Dr. Ivy found strong support among the trustees (including ex-football hero Red Grange) and they eventually succeeded in having Dr. Stoddard dismissed from his post. He is now dean of New York University's School of Education. In 1955 he published *Krebiozen: The Great Cancer Mystery,* a book about his sad and incredible experience.

Seven careful investigations of *"K,"* including one by the American Medical Association, have resulted in the charge of "worthless." To date, no one except Dr. Durovic and his associates have been permitted to analyze the powder. The reader who wishes to read the pro-"K" side of the controversy may consult *K: Krebiozen—Key to Cancer,* by Chicago newspaperman Herbert Bailey, Hermitage House (publishers of *Dianetics,* see Chapter 22), 1955; and *Observations on Krebiozen in the Management of Cancer,* by Dr. Ivy and two associates, Regnery, 1956.

Notes

1. The Electronics Medical Foundation, founded by Abrams in 1922, is still operating. With headquarters in San Francisco, it publishes *The Electronic Medical Digest,* leases out oscilloclasts and some dozen other curious contraptions, and diagnoses blood samples shipped to the headquarters.

In 1954 the Food and Drug Administration obtained an injunction against interstate shipment of the devices. The FDA estimated that about 5,000 of the machines were in use throughout the country by osteopaths, chiropractors, naturopaths, and other fringe practitioners. In making its investigation, government scientists submitted a spot of coal-tar dye as a sample of a woman's blood and were told that the patient suffered from "systematic toxemia." The therapeutic machines contain nothing but low powered short wave radio transmitters and coils capable of producing a weak magnetic effect.

Fred J. Hart, president of the foundation, raised the usual howls of persecution by the medical trusts and vowed he would continue his great work in Germany and Mexico if necessary. The foundation's director, Thomas Colson, B.S., Ll.B., and D.O. (doctor of osteopathy), is the author of *Molecular Radiation,* a learned book published by himself in 1953.

2. For an amusing chapter on Sinclair's faith in Abrams, see H. L. Mencken's *Prejudices,* Vol. 6. Mencken's interesting thesis: the same rebellious impulses that make a political radical too often find similar outlets in quack medical opinions.

3. *Time,* April 10, 1950.

4. Prof. Estep is still stepping lively to avoid the law. In 1955 the Food and Drug Administration confiscated a number of his "automotrones" that were being shipped from Texas to Modesto, California, for distribution to purchasers. The automotrone, a cabinet with a sun lamp, short-wave unit, and colored slides, is used to irradiate water which is then swallowed to cure 87 different ailments. Estep is currently a fugitive from Illinois where he was sentenced to prison for three to five years for medical malpractice.

5. The frightening story of Ghadiali's trial is told by Rita Halle Kleeman in her informative article, "Beware of the Medical Frauds!" *Saturday Evening Post,* Nov. 22, 1947.

6. For fascinating details on the healing power of both red and blue light, see Chapters 14 and 15 of Dr. Margaret A. Cleaves' 827-page pseudo-scientific opus, *Light Energy,* published in New York in 1904. There is also a chapter on the physiological effects of N Rays (see Note 1, Chapter 21).

7. Color therapy also plays a prominent role in the "anthropotherapy" of cults stemming from the work of Edmund Székely. It is practiced along with phonotherapy (sound healing), aromotherapy (odor healing), masticotherapy (Fletcherism—see next chapter), cellulotherapy (fasting), and dozens of other therapies. Of Székely's many books, perhaps the best introduction to his occult brand of nature healing is the 1948 revised edition of L. Purcell Weaver's English translation of *Cosmos, Man, and Society: A Paneubiotic Synthesis*. Typographical errors and other mistakes in the earlier edition were kindly corrected by Gerald Heard.

8. Sugrue died in 1953 at the age of 45. Since 1937 he had been confined to a wheel chair with a painful and rare form of arthritis.

9. *Many Mansions*, 1950, by Gina Cerminara, is a detailed study of Cayce's teachings on reincarnation and allied topics. See also Sherwood Eddy's *You Will Survive After Death*, 1950, and Morey Bernstein's *The Search for Bridey Murphy*, 1956, both of which take Cayce's visions seriously.

10. For ten dollars Ray Palmer (see p. 51) will sell you, through his Venture Bookshop in Evanston, Ill., a pair of "Aura Goggles" for seeing auras of "the human body, animals, and inanimate things." The goggles come complete with "pinacyanole bromide" filters.

CHAPTER 18

Notes

1. A striking recent example of how the public will welcome a book by a layman in spite of all protests by the medical profession is the spectacular success of *Arthritis and Common Sense*, by Dan Dale Alexander, of Newington, Conn. The book was privately printed by the author in 1951, but sold so well that a Hartford publisher took over publication. In the opinion of Dr. W. D. Robinson, president of the American Rheumatism Association, Alexander's remedy (cod liver oil mixed with orange juice) belongs to "the era of snake-oil, bear grease, and the torch-lighted medicine show."

2. This illustrates one of the most elementary of statistical fallacies. People in Wisconsin tend to be long-lived and since cancer is a disease of middle and elderly years, it is a more frequent cause of death in Wisconsin than in many other states. An area low in cancer deaths is likely to be an area of poor health where inhabitants tend to die young.

3. One of the nation's most colorful vegetarians was Sylvester Graham (1794–1851), the New England Presbyterian minister for whom the Graham cracker is named. In lectures and books (his most important was

Science of Human Life, 1836) he railed against meat, liquor, tea, coffee, tobacco, featherbeds, and corsets. *A Graham Journal of Health and Longevity* was published weekly in Boston.

4. For an entertaining picture of Hauser's personality and background, see "You Can Live to be a Hundred, He Says," by Noel Busch, *Saturday Evening Post,* Aug. 11, 1951.

5. Actually, Hauser's "system" varies widely from year to year. One of his funniest books, *Types and Temperaments, with a Key to Foods,* second edition, 1930, classifies everyone into basic chemical types depending on their personalities and facial characteristics. Norma Shearer and Dolores Costello are "sulphur" types ("fiery, spontaneous and talented . . . as explosive as sulphur"); Billy Sunday and Douglas Fairbanks are "sodium"; and so on. A photograph of Annie Besant, the theosophical leader, reveals that she is a "striking example of the negative body and positive head," as well as a "harmonious balance of calcium, carbon, and phosphorus." Each type naturally should follow a special type of diet prescribed by Hauser.

6. Blackstrap is far from the rich source of vitamin B that Hauser claims it is. To obtain minimum daily requirements of this vitamin one would have to drink a gallon of the stuff. Its calcium content does not come from nature at all, but from limewater used in the refining of sugar, and its iron and copper content come largely from contact with factory machinery. See A. D. Morse's authoritative article, "Don't Fall for Food Fads," *Woman's Home Companion,* Dec., 1951.

CHAPTER 19

I am happy to report that since this chapter was written, a detailed, authoritative, completely unanswerable book exposing the absurdities of the Bates system finally has been written. It is titled *The Truth about Eye Exercises,* published 1956 by the Chilton Co., Philadelphia. The author, Philip Pollack, is a Manhattan optometrist who brings to his task a comprehensive knowledge of his subject. It is a rare occasion indeed when anyone so well informed troubles to take apart a pseudo-scientific cult in such a thorough and painstaking manner. The book will not put an end to the Bates movement, but let us hope that it will at least have a dampening effect on editors and publishers who do not realize the harm that results from such pro-Bates pieces as a recent one in *Coronet* (Oct., 1955).

Notes

1. He is called "Dr." Peppard on the title page, leading the reader to suppose he is either a medical doctor or eye doctor. Actually, as Pollack discloses in his book cited above, Peppard is an osteopath who worked for a time with Bates.
2. See Pollack's book, opposite page 25, for recent photographic evidence of the change in the lens' shape during accommodation. The proof is so simple that even a Bates practitioner can understand it.
3. Prentice Hall published her *Help Yourself to Better Sight* in 1949, and Crown issued her *How to Improve Your Sight* in 1953. Mrs. Cobett has never bothered to learn even the most elementary facts about the eyes. She thinks that proper breathing has a great effect on eye ailments and reports the case of one man whose cataracts temporarily vanished when he took a deep breath!
4. Miss Hackett's book, *Relax and See,* was published by Harper in 1955. The foreword is by William Gutman, Manhattan *homeopath* (see p. 164).

CHAPTER 20

Notes

1. Dr. Montagu expanded this article to book length, publishing it in 1953 under the title, *The Natural Superiority of Women*.
2. The notion that right or left testicles determine sex is as old as ancient Greece. One of the few experiments proposed by Aristotle is that of tying or removing one testis in order to disprove this theory.
3. The recently published *Origins of Psychoanalysis: Letters to Wilhelm Fliess,* by Sigmund Freud, Basic Books, 1954, discloses that Fliess, a Berlin physician and good friend of Freud, was the holder of many amusing biological views. He thought, for example, that in psychosomatic illnesses there is a close connection between the sex organ and certain disorders of the nose, and he worked out a fantastic number theory about cycles of 23 and 28 days in the health of men and women. For a time Freud was enormously impressed by these views.
4. Dr. Stopes' latest book, *Sleep,* 1956, recommends sleeping in a north-south position. "It is comparatively unimportant whether the head or feet are at the north end of the bed," she writes, "but it is very important that . . . the body should lie . . . south-north or north-south."
5. Cf. the strange method of love-making proposed by W. J. Chidley of

Australia in his little book, *The Answer, or the World as Joy,* 1915. Chidley believed that all the crime, madness, and misery of the race were due to a false method of coitus and that he had discovered the correct technique that would restore happiness to mankind. Havelock Ellis discusses Chidley's theories briefly in Chapter 3 of his *The Mechanism of Detumescence.* See also Norman Douglas' autobiography, *Looking Back,* 1934, p. 451f., for his opinion of Chidley, who called upon him one day in London. "I would have given a good deal," Douglas writes, "if Mr. Chidley had obliged me with an oracular demonstration of his 'correct method,' which I hold to be physiologically impossible or else, if practicable, a sight worth seeing."

6. See David Riesman's *Faces in the Crowd,* 1952, for the sad case of "Henry Friend," the name Riesman uses for a 15-year old Los Angeles boy undergoing Reichian analysis, and whose views are a crude mishmash of Communism and Reichian theory.

CHAPTER 21

Reich's most spectacular recent invention is a rain-making device, one of the first of his C.O.R.E. (Cosmic Orgone Engineering) projects. Irwin Ross, in a long and amusing article on Reich *(N. Y. Post, Sunday Magazine Section,* Sept. 5, 1954) describes the device as follows:

". . . a bank of long hollow pipes tilting at the sky and sections of hollow cable, all of which are mounted on a metal box; it resembles a stylized version of an anti-aircraft gun, and works with surpassing ease. The clouds are not sprayed with any substance; the hollow pipes merely draw orgone out of them—thus weakening their cohesive power and eventually causing them to break up."

At that time three of Reich's cloudbusters were operating in different sections of Orgonon, two others in North Carolina. It was raining furiously when Irwin Ross visited Orgonon, and when Ross asked Reich if his devices were responsible, the scientist modestly assured him that they were. "But did you have to produce so *much* rain?" Ross asked. "Well," Reich replied, "you know, we haven't yet learned to control it completely."

In 1954 the Food and Drug Administration brought suit against Reich, his wife Ilse Ollendorff, and the Wilhelm Reich Foundation, to prevent interstate shipping of orgone energy accumulators and all literature mentioning orgone energy. The FDA estimated that more than a thousand of the accumulators had been rented or sold. After a series of carefully conducted tests, research scientists for the FDA concluded that "there is no such energy as orgone and that Orgone Energy Accumulator devices are worthless in the treatment of any disease or disease condition of man. Irreparable harm may result to persons who abandon or postpone rational medical treatment while pinning their faith on worthless devices such as these."

Reich chose to ignore the injunction and an Orgone Legal Fund, headed by the cartoonist William Steig, began raising money for the trial. It took place in May, 1956, at Portland, Maine, with Reich acting as his own attorney. After deliberating twenty minutes the federal jury returned a verdict of guilty. Reich was given a two-year prison sentence, his foundation was fined $10,000, and his associate, Dr. Michael Silvert, received a year and a day in jail. Reich is appealing of course, and each man is out on $15,000 bail.

Because the federal injunction ordered the destruction of many of Reich's books that contain little mention of orgone energy, the case soon acquired a civil rights aspect that had nothing to do with a scientific evaluation of Reich's work. The American Civil Liberties Union issued a rebuke and protesting letters appeared in various liberal journals *(e.g. New Leader,* July 30, 1956). See Steig's letter in *Time,* June 25, 1956, in which he states, "Reich's great findings are factual, demonstrable, irrefutable, as were those of Galileo. How much longer will it be before officials, the press, the public shake off their apathy, accept the largesse of orgonomy, and fight to defend it?"

Reich has expressed his opinion that there is a "Red Fascist" *(i.e.,* Communist) group within the FDA, seeking access to his unpublished papers in order to learn the secret "Y" factor of his orgone energy motor. This motor presumably runs on orgone energy and offers promise of immense power. See *A Report on the Jailing of a Great Scientist in the USA,* 1956, a 20-page booklet published in 1956 by Raymond R. Rees and Lois Wyvell.

Notes

1. The history of modern physics is spotted with reports of nonexistent radiations, and it is not unusual for the discoverer to attribute dowsing and similar occult phenomena to them. A good example is the nineteenth century discovery of a force called "Od" by German physicist Baron Karl von Reichenbach. Oddly enough, other scientists were unable to duplicate the baron's experiments.

In 1903 Prosper Blondlot, a reputable French physicist at the University of Nancy, detected what he called *"N"* (for Nancy) rays. Scores of papers describing the curious properties of N rays appeared in French journals and the French Academy actually awarded Blondlot a prize for his discovery. The *coup de grace* was deftly executed by American physicist Robert W. Wood (best known to laymen as the author of *How to Tell the Birds from the Flowers)* when he called upon Blondlot at his laboratory. While Blondlot was observing and describing an N-ray spectrum, Wood slyly removed an essential prism from the apparatus. This had no effect on what poor Blondlot fancied he was seeing!

There is no question but that the French scientist was sincere. The explanation of his behavior lies in the realm of psychology—of self-induced visual hallucinations. Wood's exposure led to Blondlot's madness and death. See Wood's letter on the episode, Nature, Vol. 70, 1904, p. 530, and Chapter 17 of William Seabrook's biography, *Dr. Wood,* 1941.

CHAPTER 22

I have made no effort to keep up with Hubbard's views since he plunged dianetics into occultism, but the following quotations suggest the atmosphere in which he has been doing his recent research:

Issue 3-G of his journal *Scientology* opens with the headline, "Source of Life Energy Found. Scientology enters third echelon far ahead of schedule; revival of dead or near-dead may become possible." The article beneath states that "The Greek gods . . . probably existed, and the energy glow and potential of Jesus Christ and early saints are common knowledge to every school boy. . . . The recovery of this energy potential and the ability to use it has become suddenly a matter of two to 25 hours of competent practice." The same issue contains a spine-chilling piece by Hubbard on Black Dianetics, warning against misuse of the science by unscrupulous groups.

Dr. Hubbard (he has awarded himself a degree of doctor of scientology) was running his patients back into previous incarnations long before the search for Bridey Murphy began. Each individual, according to Hubbard, has a "theta being" that has been reappearing in MEST bodies for about 74 trillion years. (See *Time,* Dec. 22, 1952.)

A. E. Van Vogt, in an incredible article on dianetics in *Spaceway Science Fiction,* Feb., 1955, takes a cautious attitude toward this work, but states that in his opinion "this is the first time that anyone has investigated this territory [i.e., the human soul isolated from the body] in a manner that can be scientifically acceptable." Van Vogt was the first head of the Hubbard Dianetic Research Foundation of California, Inc. "I was reluctant to become involved," he writes, "but since doing so I have done more than 5,000 hours of Dianetic processing on other people, and have taken upward of 800 hours of training in the methods."

Dianetics has not been mentioned in *Astounding Science Fiction* for many years. Editor Campbell has found something even more revolutionary—"psionics," a combination of electronics with psi (psychic) phenomena. Campbell first wrote about it in an editorial, "The Science of Psionics,"

Astounding, Feb., 1956. The editorial asked his readers if they would like to see a series of articles about psychic electronic machines. Campbell described psionics as "honest non-scientific research," pointing out that "Buddha, Jesus, and President Eisenhower" also are excluded from the category of "honest scientific research" since they "use methods other than those used by physicists in their laboratory work."

After a resounding "Yea!" from his readers, Campbell ran the first article, "Psionic Machine—Type One" in his June, 1956 issue. The article was written by himself. It tells how to build a Hieronymous machine, patented in 1949 by one Thomas G. Hieronymous, at that time a resident of Kansas City, Mo., and tested with positive results by "nuclear physicist" (see p. 265) Campbell. The machine was designed by the inventor to analyze the "eloptic radiation" of minerals, a new type of radiation discovered by Hieronymous. Among electronic engineers, Hieronymous patent (No. 2,482,773) is passed around for laughs, and considered in a class with Socrates Scholfield's famous patent of 1914 (No. 1,087,186), consisting of two intertwined helices for demonstrating the existence of God.

Campbell thinks Hieronymous' theory is "cockeyed" and he has also made several basic changes in the construction of the machine. Hieronymous claimed that his detector worked on *photographs* of minerals. Campbell hasn't bothered to test that. Nevertheless, the machine Campbell built did detect something "not detectable by any standard form of meter," and he knows there is no "jiggery-poker" because he constructed the thing himself.

In a lecture on psionics at the New York Science Fiction Convention, 1956, Campbell displayed his second and "more precise" version of the Hieronymous machine. It works just as well, he claimed, without the electric power supply. But it won't work, he added, if there is a burned-out vacuum tube! The device is beautifully subjective. You turn a dial with one hand while you stroke a plastic plate with the other. The plate is supposed to feel "sticky" when the dial reaches a certain setting, the setting varying with each individual. Some people get the proper tactile sensation the first time they try it. Willy Ley and others at the convention couldn't feel a thing. Campbell solemnly informed his audience that the machine does not work well with either scientists or mystics. Five mystics tried it, he stated, and got only random responses. His own personal "hunch" is that the machine is detecting something "beyond space and time." Or as he expresses it in the October, 1956 issue of his magazine, "there is a reality-field other than, and different in nature from, that we know as Science."

Another Campbell "hunch" is that the device operates because of certain "relations" between its parts. Someone at the lecture stood up and asked the obvious question: had Campbell tried varying the circuit or even removing it altogether to see if the device still worked? No, Campbell hadn't tried that. He was just an amateur, he explained, "having fun" with psionics, and he felt no obligation to try all the experiments that are possible, particularly without getting paid for it.

No one of course expects a researcher to perform all possible experiments with a device before he publishes results. But one does expect at least a minimum of experimentation to insure fairly adequate controls. As it is, psionics promises to be even funnier than dianetics or Ray Palmer's Shaver stories. It suggests once more how far from accurate is the stereotype of the science fiction fan as a bright, well-informed, scientifically literate fellow. Judging by the number of Campbell's readers who are impressed by this nonsense, the average fan may very well be a chap in his teens, with a smattering of scientific knowledge culled mostly from science fiction, enormously gullible, with a strong bent toward occultism, no understanding of scientific method, and a basic insecurity for which he compensates by fantasies of scientific power.

Notes

1. Prof. Schuman opens this letter by quoting Oliver Cromwell's "I beseech ye, in the bowels of Christ, to consider whether ye may not be mistaken." He goes on to say that the *New Republic,* by printing such an irresponsible review, has made itself "the laughing stock of the rapidly growing throng of people who know what dianetics is all about. Not the book, but the review, is 'complete nonsense,' a 'paranoiac system' and a 'fantastic absurdity.' There are no authorities on dianetics save those who have tested it. All who have done so are in no doubt whatever as to who is here mistaken."

CHAPTER 23

Notes

1. A point of view held chiefly by philologists and cultural anthropologists who like to imagine that *their* subject-matter (words or culture) underlies logic and mathematics. See "Words, Logic, and Grammar," by H. Sweet, *Transactions of the Philological Society,* 1876. Because Aristotelian logic rests upon grammatical rules peculiar to the Aryan language, Sweet argues, the "whole fabric of formal logic falls to the ground."

2. Strictly, this is a three-valued logic with two-valued functions. But even in the more exciting multi-valued logics that have multi-valued functions, deductions remain two-valued in the sense that they must be either valid or invalid in terms of the rules of the system.

3. Dr. Ernest Nagel, in a letter to the *New Republic*, Dec. 26, 1934 (replying to protests against his unfavorable review of *Science and Sanity* in the Oct. 24 issue), expresses this point as follows:

> ". . . it is my considered opinion that *Science and Sanity* has no merit whatever, and is not worth the serious attention of readers of the *New Republic*. Its main thesis rests on a misunderstanding of recent work on the foundation of logic. The few interesting suggestions on technical problems, to which I referred in my note, have not been systematically developed by Count Korzybski, and they play only a very inconsiderable role in his book."

See also Paul Kecskemeti's penetrating "Review of General Semantics," *New Leader,* April 25, 1955.

4. Max Eastman, in his amusing piece "Showing up Semantics," *The Freeman,* May 31, 1954, quotes the following pompous passage from *The Mankind of Humanity:* "This mighty term—time-binding—when comprehended, will be found to embrace the whole of the natural laws, the natural economics, the natural governance, to be brought into the education of time-binders; then really peaceful and progressive civilization, without periodical collapses and violent readjustments, will commence; not before."

CHAPTER 24

Notes

1. For a lively account of the history of phrenology see *Phrenology: Fad and Science,* by John D. Davis, 1955.

2. If character can influence handwriting then why not vice versa? In Paris, graphologist Raymond Trillat asked himself this question, then developed what he calls "grapho-therapy"—the science of treating neurotics by teaching them how to write differently. Hundreds of French children, he claims, have been benefited by this novel therapy. See *Time,* April 23, 1956 for the grotesque details.

3. Dr. Wolfe also is convinced that the two composite faces that can be formed by placing the edge of a mirror vertically on the center of a person's front-view photograph indicate two basic sides of that person's

personality. See his *The Expression of Personality,* 1943, and *The Threshold of the Abnormal,* 1950.

4. Two recent tests: the Swiss "draw a tree" test, and the test developed by Prof. David L. Cole, Occidental College, Los Angeles ("What kind of an animal would you like to be if you had to be one, and why?")

CHAPTER 25

The most recent attack on Rhine comes from Dr. George R. Price, of the University of Minnesota's department of medicine. In a lengthy article, "Science and the Supernatural," *Science,* Aug. 26, 1955, he reaches the conclusion that many of the findings of parapsychology are "dependent on clerical and statistical errors and unintentional use of sensory clues, and that all extrachance results not so explicable are dependent on deliberate fraud or mildly abnormal mental conditions."

Dr. Price's article stirred up a hornet's nest of controversy. In its Jan. 6, 1956 issue, *Science* published replies by Rhine, S. G. Soal, and others, with a rejoinder by Price and a further rejoinder by Rhine. Physicist P. W. Bridgman also contributed a brief article suggesting that probability theory may be more complicated than scientists think it is.

It was Price's suggestion of "deliberate fraud" that most annoyed his opponents. Yet it is hard to suppose that out of the thousands of individuals tested by Rhine, there would not be a few with the incentive and skill to cheat. This particularly applies to Rhine's naive early work when no attempt was made to shield the backs of the cards and when some of his most sensational results were achieved. No two card backs are exactly alike, especially after the cards have been handled for a while. If a sharp-eyed subject happened to notice a faint smudge of dirt, say, at one corner of a card, he would hardly consider it "cheating" if he spotted the same smudge on later tests. Already firmly convinced of his own psychic ability, what harm would there be, he might reason, in raising his score a trifle, particularly on days when he felt that his psi ability was at a low ebb? His fondness and admiration for Dr. Rhine would forever prevent him from mentioning this later, to say nothing of the personal humiliation of confessing fraud. The history of occultism swarms with personalities possessing precisely this combination of sincerity and duplicity.

Notes

1. Rhine's latest book, *New Worlds of the Mind,* appeared in 1953. The book differs from earlier ones mainly in its emphasis on religious specu-

lation and its suggestions for empirically testing such beliefs as the existence of God and the efficacy of prayer. Putting spiritual reality on scientific foundations would, Rhine declares, "do for religion something like what the germ theory did for medicine. It would open the range of religious exploration to horizons beyond all present conceptions." In the meantime, Rhine's latest research program has to do with "anpsi" (animal psi). He cites experiments which already have shown that cats are both telepathic and clairvoyant, and hints of exciting news to come.

2. Rhine's first testing of Lady Wonder, then a three-year-old filly, took place in the winter of 1927–28. He and his wife described the tests in their article, "An Investigation of a 'Mind-Reading' Horse," *Journal of Abnormal and Social Psychology,* Vol. 23, 1929, p. 449. The horse spelled out answers to questions by touching her nose to lettered and numbered blocks. Rhine was shrewd enough to perceive that the horse could read his mind only when her owner, Mrs. Claudia Fonda, was nearby. But instead of concluding that Mrs. Fonda was signaling the horse in the standard manner of all "talking horse" and "talking dog" acts, Rhine decided that Lady was getting her cues telepathically. His reason? Lady was successful on many tests in which Mrs. Fonda was kept "ignorant of the number." Instead of sending Mrs. Fonda out of the room, however, which would have enormously strengthened the telepathic theory, she was permitted to remain at all times to aid in controlling the unruly colt!

With his usual vagueness in describing such experiments, Rhine nowhere states exactly what he means by keeping Mrs. Fonda "ignorant of the number." It is not until we read his "Second Report on Lady, the 'Mind-Reading' Horse," in the same journal, Vol. 24, 1929, p. 287, that he unintentionally lets the cat out of the bag. "When he [Rhine]," he writes, "stood behind F [Mrs. Fonda] and wrote the number on a pad as he had once done with excellent results, there was now complete failure." In other words, the numbers were *written down,* but since Mrs. Fonda was not permitted to see this information, Rhine concluded that telepathy was operating.

Now it so happens there are some fifty different ways a clever medium or mentalist can secretly obtain information that has been written down. Unless Rhine was aware of these methods, and there is no indication he knew any of them, his testing of the horse was valueless. There is even the very strong possibility that Mr. Fonda at times played the role of a confederate, for Rhine tells us in his first article that Mr. Fonda was present "part of the time," but since he played a very inconspicuous role in the proceedings, "we leave him out of the account for the sake of brevity.'' In brief, Rhine's ''controls'' were laughably inadequate.

In 1956 my friend Milbourne Christopher, a professional magician, attended a performance of Lady Wonder without telling Mrs. Fonda what his profession was. She gave him a long pencil and a pad, told him to stand across the room and write a number. He moved the pencil in the path of a figure eight, but touched the paper only at such points that it wrote the figure three. Lady Wonder guessed the number to be eight, a clear indication that Mrs. Fonda was ''pencil reading.'' This is the term mentalists use for the art of guessing what a person writes by observing the motions of the pencil. Dr. Rhine could easily have used such tricks as this to determine exactly what part Mrs. Fonda played in Lady's demonstrations, but there is no evidence that he made the slightest effort along such lines.

Rhine's belief today is that Lady Wonder *used to have* genuine telepathic powers, but that by December, 1928, she had lost these powers and Mrs. Fonda had taken up the practice of signaling. For a picture story on the aged Lady, see *Life,* Dec. 22, 1952. Milbourne Christopher contributed an informative article on famous mind-reading animals of the past to the April, 1955 issue of *M-U-M,* official organ of the Society of American Magicians.

3. One of Rhine's vaguest descriptions is his account in *New Frontiers of the Mind,* p. 94, of the historic occasion on which one of his subjects scored 25 correct ''hits'' in a row. This was the most sensational test Rhine ever personally observed and one that he returns to again and again in his later lectures and writings. Here is a partial list of important questions that one would wish to have answered before evaluating the test. (1) How many cards were in the pack used? (2) What exactly does Rhine mean when he says that each card was ''returned to the pack and a cut made?'' (If this means that each card was placed on top, then cut to the center, this procedure would tend to keep bringing the same few cards back to the top each time.) (3) Did Rhine look at each card before it was named, or after? (4) Did Rhine show each card to the subject after it was named? (5) Was it a pack the subject could have handled on previous occasions? (6) Were the cards examined later by card gambling experts to determine if they were marked in any way? (7) Was a careful check made of the room to insure against possible reflections of the cards in shiny surfaces? (8) Exactly how many cards were incorrectly called before the run of 25 correct hits? (It is clear from the text that there were at least seven misses, possibly more.)

4. There is an obvious and suggestive analogy between parapsychology's preoccupation with purely statistical evidence, with all its murky aspects,

and the preoccupation of mediums with phenomena that for some odd reason take place only in darkness.

5. For an excellent recent example see "A Methodological Refinement in the Study of 'ESP,' and Negative Findings," by Kendon Smith and Harry J. Canon, *Science,* July 23, 1954.

6. See Aldous Huxley's two articles, "A Case for ESP, PK, and Psi," *Life,* Jan. 11, 1954, and "Facts and Fetishes," *Esquire,* Sept., 1956.

7. See his *Eyeless Sight,* translated by C. K. Ogden, Putnam, 1924. This ridiculous work is devoted to Romains' extensive experiments proving that microscopic rudimentary organs of vision are present everywhere on the body in cells of the skin. The skin is thus able to "see" both form and color, providing a physical explanation of certain types of clairvoyance.

8. An anthology of psychoanalytic papers dealing favorably with telepathy, *Psychoanalysis and the Occult,* edited by George Devereux (see Note 6 to the next chapter), was published in 1953. See also *New Dimensions of Deep Analysis,* 1954, by Jan Ehrenwald; *The Use of the Telepathy Hypothesis in Psychotherapy,* 1952, by Jule Eisenbud; and two articles by Eisenbud, "Psychiatric Contributions to Parapsychology: A Review," *Journal of Parapsychology,* Dec., 1949; and "Telepathy in Psychoanalytical Treatment," *Tomorrow,* Winter, 1952–53.

9. In reminiscing about Einstein, *Saturday Review,* April 14, 1956, Sinclair tells once more the story of this table tipping. The levitation occurred in a darkened room under familiar seance circle conditions. A later attempt was made to repeat the performance, Sinclair reveals, with Einstein and several other physicists present, but owing to a "hostile influence" in the room, nothing happened.

10. The following amusing excerpt from Frank Joglar's column in a magic trade journal *(Hugard's Magic Monthly,* Aug., 1953), should give the astute reader some insight into Dunninger's methods:

> One of his [Dunninger's] best recent stunts was a twist on his familiar sealed prophecy test. In this one he wrote in advance the name and address of an envelope later to be selected, also the sender's name and address and the postmark. He sealed the cardboard in an envelope and passed it to three newsmen. *If you didn't recognize Bob Dunn, the method may have puzzled you.* On the screen the postmaster of the N.Y. Post Office was seen by a conveyor belt. Sacks of mail went past. He selected one. Opened it, spilled out the letters. Then he chose one. The camera took a closeup. *The three newsmen made notes of what was written on it, then Dunninger's prediction was opened. Success!* The information was the same!

CHAPTER 26

Notes

1. The silliest of these books is *You DO Take it with You,* by R. DeWitt Miller, 1956. The "it" in the title has reference mainly to sex.
2. See Herbert Brean's picture article, "Bridey Murphy Puts Nation in a Hypnotizzy," *Life,* March 19, 1956.
3. From his delightfully written article, "Bridey Murphy: An Irishman's View," *Fantasy and Science Fiction,* Aug., 1956. The article criticizes the book, on internal evidence alone, for picturing "an Ireland that never was, save in the minds of the uninformed and the vulgar." See also editor Anthony Boucher's shrewd comments in the earlier, May, 1956 issue.
4. From Campbell's unfavorable review of *A Scientific Report on "The Search for Bridey Murphy,"* edited by Dr. Milton V. Kline, Julian Press, 1956.
5. The *Chicago American* articles were syndicated by the Hearst papers. In New York City they ran daily in the *Journal American* from June 10 through June 18, 1956. The finding of Bridey was reported in *Time,* June 18, 1956 ("Yes, Virginia, There is a Bridey") and *Life,* June 25, 1956.
6. For an amusing example of wild psychoanalytical-anagrammatical numerological speculation, see the discussion of these and other names in George Devereux's pompous article, "Bridey Murphy: a Psychoanalytic View," *Tomorrow,* Summer, 1956 (an issue devoted primarily to pieces on Bridey). Devereux finds "crushing evidence" that Bridey's imaginary husband, Sean Joseph Brian MacCarthy, is a symbol of Bernstein. The initials of Brian MacCarthy, reversed, are the initials of Morey Bernstein. Two names have a *"y"* ending, and the other two end in "ein" and "ien" (provided we spell Brian "Brien"). If we cross out the letters both names have in common we take care of all except A,C,C,A,H in the husband's name, and O,E,E,N,S in Bernstein's name. Devereux conveniently ignores the excess letters in the husband's name, then goes on to point out that three of the excess letters in Bernstein's name are to be found in "Sean." The remaining two letters, O,E, appear in "Joseph." "Interestingly," Devereux adds, "they are *set apart* from the other letters since they are the *only two vowels* in that name . . ." (italics his).

Shades of Jung and the ancient Cabbalists! My own opinion is that Virginia identified her imaginary husband with George Devereux. If we take the husband's name to be Joseph McCarthy (spelling it with an

"Mc") then we find that "Joseph" and "George" both have six letters and "Devereux" and "McCarthy" both have eight, the total being fourteen which is also the total for "Morey Bernstein." Surely one cannot expect chance to account for this astonishing similarity. Moreover, we find that "George Devereux" and "Joseph McCarthy" have only three letters in common, E,R,O; three of the four letters of "Eros," Greek god of love. Even the missing S is accounted for if we use the French spelling, "Georges."

Devereux also holds the opinion that "Bridie" is the diminutive of "bride," hence Virginia thought of herself unconsciously as Bernstein's "little bride." Apparently it no more occurred to Devereux than to Bernstein that Virginia might have *known* a Bridie Murphy. Nor did he consider the startling possibility that Virginia's memory of scratching paint off her bed could have been a memory (firmly implanted by a strapping) of scratching paint off her bed. ". . . we are dealing here," he writes delicately, "with a so-called screen-memory . . . a fictitious 'memory' made up to blanket a real and less presentable memory of what had, in fact, taken place." I cite only the more plausible parts of Devereux's involved symbolic analysis. It is surprising that the initials of Bridey Murphy did not suggest an anal complex to the learned doctor.

INDEX OF NAMES

Abrams, Albert, 179f. 297
Adamski, George, 288
Adler, Mortimer, 44, 117f.
Albright, William, 286
Alexander, Dan, 298
Alexander, Frederick, 173
Alexandra, Queen of Yugoslavia, 198
Anderson, Margaret, 187
Aquinas, St. Thomas, 243
Aristotle, 44, 129, 213, 247f., 251, 256, 268, 300, 305
Arnold, Kenneth, 46f., 289
Arrhenius, Svante, 104
Atwater, Gordon, 26
Augustine, St., 15, 119–20

Babson, Roger, 78 f.
Bachman, Louis, 104
Bailey, Alfred, 43
Bailey, Herbert, 296
Barker, William, 275, 277
Barrett, Sir William, 87
Barton, Bruce, 169
Bastian, Henry, 100
Bates, William, 201 f., 299

Battell, Joseph, 70f.
Beausoleil, Baron de, 87
Bechamp, Antoine, 166, 171
Bellamy, Hans, 31f., 148–49f.
Belloc, Hilaire, 44, 115, 118
Bender, Peter, 22
Benedict, Ruth, 108
Bennett, Arnold, 188
Bergson, Henri, 123, 269
Beringer, Johann, 106
Berkeley, Bishop, 205
Bernstein, Morey, 275f., 298, 311–12
Besant, Annie, 69, 147
Besterman, Theodore, 87
Bianca, Sonya, 236
Bierce, Ambrose, 273
Birdseye, Clarence, 80, 291
Blau, L. W., 291
Blavatsky, Helena, 101, 146, 147, 160
Block, Edgar, 272
Blondlot, Prosper, 302–303
Boring, Edwin G., 284
Bothezat, Georges de, 71–72
Boucher, Anthony, 311

Bowers, Edwin, 169f., 186
Brady, Mildred, 226
Brean, Herbert, 311
Bridgman, P. W., 307
Brinkley, John, 214f.
Broca, Paul, 137
Brown, Harrison, 282
Brumberg, Abraham, 293
Bryan, William J., 110
Bryant, William C., 165
Bueren, Godfried, 286
Bulwer-Lytton, Edward, 183
Bump, George, 36
Burbank, Luther, 272
Burke, John B., 100
Burton, Sir Richard, 215
Busch, Noel, 299
Butler, Samuel, 123

Calhoun, John, 136
Callahan Rev. Jeremiah, 75f.
Campbell, Jr., John, 231, 239,
    245, 276, 289, 303–304f.,
    311
Canon, Harry, 310
Carmer, Carl, 21
Carrington, Hereward, 192–93,
    208
Carroll, Charles, 136f.
Cayce, Edgar, 188–89f., 276
Cerf, Bennett, 210
Cerminara, Gina, 298
Chamberlain, Houston, 133, 138
Chan, Charlie, 268
Chesterton, Gilbert, 85, 115–16f.,
    161
Chidley, W. J., 300–301
Christopher, Milbourne, 309
Churchward, Col. James, 148f.
Cleaves, Margaret, 297
Cohen, Bernard, 44, 284
Cole, David, 307
Colson, Thomas, 297
Conklin, Groff, 288
Connell, Father Francis, 53, 289

Coover, John, 269
Copen, Bruce, 186
Corbett, Margaret, 210, 300
Corkell, Mrs. Anthony, 277
Cowper, William, 106
Cox, Arthur, 238
Cox, Earnest, 140
Crehore, Albert, 68f.
Cripps, Sir Stafford, 173
Crookshank, Francis, 138f.
Crosse, Andrew, 99f.
Crosse, Mrs. C. H. A., 100
Crowley, Aleister, 188
Curle, James, 140
Cuvier, Baron Georges, 107

Darwin, Charles, 107, 122, 280
Davenport, Walter, 2
Davidson, David, 159f.
Davis, Andrew J., 188
Davis, John, 306
Debs, Eugene, 171–72
De Camp, Sprague, 284, 293
De Casseres, Benjamin, 287
De Ford, Miriam, 287
De Morgan, Augustus, 291
Deri, Susan, 257
Deuel, Wallace, 134
Devereux, George, 310, 311
Dewey, John, 44, 172–73, 248,
    295
Dietrich, Marlene, 166
Donnelly, Ignatius, 29f., 143–44f.,
    269
Douglas, Norman, 301
Doyle, Sir Arthur C., 150, 262,
    269, 271, 291
Drayson, Alfred, 43f.
Dreiser, Theodore, 35, 36, 43
Driesch, Hans, 269
Drown, Ruth, 182f.
Duncan, Charles, 166
Dunninger, Joseph, 272, 310
Durham, Carl, 108
Durovic, Stevan, 296

Eastman, Max, 306
Eddy, Mary B., 171
Eddy, Sherwood, 298
Edgar, John, 158f.
Edgar, Morton, 158f.
Edison, Thomas, 78, 81, 271–72f.
Ehrenwald, Jan, 310
Einstein, Albert, 6, 7, 10, 53, 70–71f., 141, 270, 353
Eisenbud, Jule, 310
Ellis, Havelock, 301
Ellsworth, Oliver, 178
Emerson, Ralph W., 165
Esson, William, 79, 82, 83, 84
Estep, William, 184, 297

Fadiman, Clifton, 26, 286
Faulkner, Edward, 293
Faunce, Cy, 58, 61, 290
Fichte, Johann, 133
Fink, David, 173
Fishbein, Morris, 171–72, 175
Fitzgerald, William, 169f.
Fletcher, Horace, 193
Fliess, Wilhelm, 300
Fodor, Nandor, 269–70, 273
Fonda, Claudia, 308
Ford, Charles de, 9
Fort, Charles, 35f., 52, 56–57, 77, 84, 90, 163, 272–73f., 287
Franklin, T. Bedford, 87
Freud, Sigmund, 269, 300

Galileo, 5, 7, 9, 44, 73, 130
Gall, Francis, 255–56
Garbo, Greta, 198
Gardner, Marshall, 17–18f.
Gardner, Thomas, 289
Garfield, Pres. James, 157
Garner, Richard, 118
Garnier, Col. J., 157
Gauch, Herman, 134
GeBauer, Leo, 289
Ghadiali, Dinshah, 184f., 297

Gillette, George, 73–74f.
Gladstone, William, 107, 145
Gobineau, Comte Joseph de, 133, 138
Goddard, Paulette, 198
Goethe, 69, 129, 290
Gosse, Edmund, 108–109
Gosse, Philip, 107f.
Goudsmit, Samuel, 293
Gould, Charles, 140
Gould, Rupert, 100
Graham, John, 214
Graham, Sylvester, 298–99
Grant, Madison, 138f.
Graydon, Thomas, 73f.
Grimme, Hubert, 151
Gromyko, Andrei, 47
Gross, Henry, 90–91f., 265
Gross, Walter, 142
Günther, Hans, 133f., 138, 140
Gurdjieff, George, 187–88
Gutman, William, 164, 300
Gutowski, Ace, 98

Hackett, Clara, 210, 300
Hahn, Otto, 104–105
Hahnemann, Samuel, 163f., 294
Haldeman-Julius E., 214
Hall, Alexander, 69f.
Halley, Edmund, 17
Haluska, John, 296
Hargrove, Marion, 294
Harland, S. C., 126
Harris, Frank, 215
Hart, Fred, 297
Hauser, Gayelord, 2, 197f., 209f., 299
Hay, William H., 193–94
Hayakawa, S. I., 251
Heard, Gerald, 54–55f., 269, 298
Heaviside, Oliver, 68
Hecht, Ben, 36, 40
Hegel, 40, 161
Heinlein, Robert, 276
Helmholtz, H. von, 290

Hendricks, Wendell, 186
Hering, Daniel, ix
Herodotus, 156
Herschel, John, 56
Herz, Valentin, 286
Hieronymous, Thomas, 304
Hitler, 113, 133–34, 135, 140
Holmes, Oliver W., 179
Holmes, Sherlock, 41, 256
Hooton, Earnest, 257
Hörbiger, Hans, 31f., 149f.
Horsey, Admiral Sir Algernon de, 43
Hoxsey, Harry, 295
Hrdlicka, Ales, 138
Hubbard, L. Ron, 230f., 270, 303f.
Hubbard, Sara, 244
Hutchins, Robert, 44
Huxley, Aldous, 173, 205, 210, 269, 310
Huxley, Julian, 126, 128, 129, 130
Huxley, Thomas, 116

Ivy, Andrew, 282, 296
Izzard, Ralph, 289

Jacobson, Edmund, 173
James, William, 193, 269
Jastrow, Joseph, ix
Jesus, 142, 148, 155, 158, 159, 160
Johnson, Robert, 287
Johnson, Samuel, 7
Joly, Maurice, 141
Jordan, David S., ix
Jung, Carl, 269
Just, Adolph, 166

Kallen, Horace, 26, 286
Kammerer, Paul, 123–24f.
Kaufman, Richard, 267
Kay, Abbott, 183
Kecskemeti, Paul, 306

Keely, John, 83–84
Kellog, John, 166, 294
Kelly, Mayor Edward, 183–84
Kelvin, Lord, 104
Kenealy, Arabella, 212, 213
Keyhoe, Donald, 52f., 288
Kinget, Marian, 260
Kingland, William, 160
Kinnaman, John, 151–52
Kinsey, Alfred, 217
Kleeman, Rita, 297
Kline, Milton, 311
Klyce, Scudder, 295
Kneipp, Sebastian, 166
Koch, William, 186f.
Koestler, Arthur, 219, 269
Kolisko, Lili, 196
Kordel, Lelord, 197
"Koresh" (See Teed, Cyrus)
Korzybski, Count Alfred, 240, 246f., 306
Kritzer, J. Haskell, 168
Kuhne, Louis, 166

Lafleur, Laurence, 8
Lahmann, Heinrich, 166
Lahn, Henry, 168
Lamarck, Jean, 121f.
Langer, Senator William, 187
Larrabee, Eric, 282
Lawrence, David, 47
Lawson, Alfred, 58f., 272, 281, 290
Leadbeater, Charles, 69
Le Blanc, Senator Dudley, 200
Lehrs, Ernst, 290
Leland, Charles, 215
Leslie, Joseph, 145
Levine, J. D., 295
Ley, Willy, 31, 34, 286, 304
Lichtenstein, Rabbi Morris, 171
Lierman, Emily, 207
Lindlahr, Henry, 167, 168
Lipman, Charles, 104
Littlefield, Charles, 101–102f., 272

Locke, John, 29
Locke, Richard, 56
Lombroso, Cesare, 257
Long, Max, 272
Lover, Myron, 80
Lowell, Percival, 55
Lundmark, Knut, 104
Lunn, Arnold, 114–15
Lust, Benedict, 167, 169, 198
Luther, Martin, 15, 107, 132
Lynch, Arthur, 71
Lysenko, Trofim, 121f., 132, 292–93

Maby, J. Cecil, 87
McCabe, Joseph, 114
McCann, Alfred, 115
McDougall, William, 123, 262, 269, 271
Macfadden, Bernarr, 167, 172, 186, 202, 208f., 213, 280, 294
Macfadden, Mary, 295
MacFayden, Ralph, 202
Machover, Karen, 259
McLaughlin, Cmdr. Robert, 52
McNair, Harold, 79
Maeterlinck, Maurice, 101, 269
Mager, Henri, 87
Maltsev, Terenty, 292
Mansfield, Katherine, 187
Mantell, Thomas, 48, 49, 53
Maritain, Jacques, 117
Marriott, Maj. R. A., 43
Marston, Sir Charles, 151
Martin, Morley, 100f.
Martineau, Harriet, 100
Marx, Karl, 161
Masefield, John, 150
Mather, Cotton, 17
Maugham, Somerset, 188
Mayer, Robert, 7
Mencken, H. L., 4, 264, 297
Mendel, Gregor, 124, 128–29
Mendl, Lady Elsie, 198

Mensendieck, Bess, 173
Menzel, Donald, 50
Menzies, Robert, 154
Mesmer, Franz, 6
Messenger, Father Ernest, 119
Michel, Aimé, 289
Michelangelo, 109
Michon, Abbé Jean-Hippolyte, 258
Michurin, I. V., 124f., 293
Miller, Dayton, 72f., 290
Miller, R. DeWitt, 311
Miller, William, 285
Milne, Edward, 71
Mivart, St. George, 116, 292
Moigno, Abbé F., 157
Monboddo, Lord James, 113
Montagu, Ashley, 213, 300
Moody, Senator Blair, 65
Moore, Mary, 82
More, Louis T., 109
More, Paul Elmer, 109
Moreno, Jacob, 252f.
Morse, A. D., 299
Muller, H. J., 124–25f.
Munro, Daniel, 195
Murphy, Bridey, 275f., 311f.
Murphy, Gardner, 261
Murray, Sir Gilbert, 269

Nagel, Ernest, 306
Navarro, Duran, 285
Nelson, Robert, 183
Nelson, Jr., Robert, 183–84
Neupert, Karl, 22
Newton, Silas, 289
Newton, Sir Isaac, 10, 29, 69f., 80, 85, 290
Noyes, Alfred, 7
Noyes, John H., 215–16f.

Ogden, C. K., 310
Ollendorff, Ilse, 301
O'Neill, John J., 26, 68
Orwell, George, 161

Osborn, Henry, 138
Ostoja, Roman, 270
O'Toole, George, 114, 115
Oursler, Fulton, 4, 26
Ouspensky, Peter, 188

Page, Melvin, 194
Palmer, B. J., 176–77f.
Palmer, Daniel D., 175–76
Palmer, Ray, 51f., 296, 298
Pancoast, Seth, 185
Pasteur, Louis, 6, 9
Payne-Gaposchkin, Cecilia, 287
Payne, Rev. Buckner, 136
Peacock, Thomas L., 113
Peakes, Bayard, 284
Peck, John, 285
Peczely, Ignatz, 168
Peppard, Harold, 202, 300
Perkins, Benjamin, 178
Perkins, Elisha, 178f., 182
Petrie, Sir Flinders, 161
Pfeiffer, Ehrenfried, 196–97f.
Plato, 144, 216
Pleasanton, Augustus, 185
Poe, Edgar A., 17, 144
Polanyi, Michael, 290
Poling, Rev. Daniel, 202
Pollack, Philip, 299
Poor, Charles, 71
Poppelbaum, Herman, 196–97
Power, Tyrone, 182
Powys, John C., 36
Pratt, E. H., 173
Price, Cecil, 202
Price, George M., 9, 110f., 292
Price, George R., 307
Priesnitz, Vincenz, 166
Prowitt, Alfred, 285

Ranald, Josef, 258
Rascoe, Burton, 36
Rauch, Oliver, 66
Rawson, Clayton, 268
Read, Allen W., 246f.

Ready, W. B., 276
Rees, Raymond, 302
Reese, Bert, 272
Regardie, Francis, 254
Reich, Wilhelm, 99, 217, 220f.,
	239, 301f., 302
Reichenbach, Karl von, 302
Reynolds, Osborne, 284–85
Rhine, Joseph, 94, 191, 261f.,
	279, 307f.
Riddick, Thomas, 96, 291
Rideout, George, 83, 85
Riesman, David, 301
Riffert, George, 158–60f.
Roberts, Kenneth, 90–92f., 291
Robinson, W. D., 298
Rockefeller, John D., 193
Rodale, Jerome, 195–96f.
Rogers, J. A., 136
Rogers, Loyal D., 166
Romains, Jules, 269, 310
Roman, Klara, 258
Rosenberg, Alfred, 135, 141
Ross, Irwin, 301
Roth, Samuel, 141
Rothschild, Baron Philippe de,
	198
Russell, Bertrand, 157, 250, 285
Russell, Charles T., 158f.
Russell Eric F., 57
Rutherford, Judge J. F., 159f.

Sanderson, Ivan, 113
Sayers, James, 140
Scarne, John, 268
Schell, William, 137
Schemann, Ludwig, 133
Scholfield, Socrates, 304
Schott, Gaspard, 87
Schuchardt, Charlotte, 250
Schuman, Frederick, 232, 305
Scott-Elliott, W., 147f.
Scully, Frank, 54f., 289
Seaborn, Captain, 16–17
Seabrook, William, 303

Seiss, Joseph, 156, 157
Selden, Mrs. Robert, 210
Shakespeare, 29, 256, 269
Shaver, Richard, 51, 150, 288–89
Shaw, George B., 10, 123, 135, 171f., 216, 295
Sheldon, William, 258
Shorey, Paul, 110
Sidgwick, Henry, 269
Silvert, Michael, 302
Sinclair, Upton, 172, 180–81f., 193, 194, 269f., 297, 310
Sinclair, Mrs. Upton, 270f.
Skinner, J. Ralston, 160
Smith, Aaron, 152
Smith, H. Allen, 287
Smith, James, 291
Smith, Kendon, 310
Smyth, Piazzi, 153–54f.
Smyth, William H., 212
Soal, S. G., 307
Spence, Lewis, 145–46, 293
Spengler, Oswald, 161
Spillane, Mickey, 294
Stalin, 125, 126, 128, 292
Standen, Anthony, 44f.
Steig, William, 226, 302
Steiner, Rudolf, 147, 196f., 290
Stekel, W., 269
Stewart, John, 27
Still, Andrew, 173f.
Stockham, Alice B., 217
Stoddard, George, 296
Stoddard, Lothrop, 138f.
Stokes, W. E. D., 140
Stopes, Marie, 217, 300
Stramke, Emil, 295
Streicher, Julius, 134–35, 142
Sugrue, Thomas, 188f., 298
Sweet, H., 305
Swift, Jonathan, 7, 25
Sykes, Egerton, 152
Symmes, John, 16f.
Symonds, John, 188
Székely, Edmund, 298

Szondi, Leopold, 257

Tarkington, Booth, 36
Tastmona, Thothnu, 294
Taylor, John, 152f.
Teed, Cyrus, 19f., 285
Tesla, Nikola, 68
Thackrey, Ted, 26
Thayer, Tiffany, 36, 40, 43, 57, 287, 289
Tighe, Virginia, 275f., 312
Todd, Mabel, 173
Toynbee, Arnold, 161
Trent, Paul, 55
Trillat, Raymond, 306
Tromp, Solcol, 88f.
Tyndall, John, 290

Urbuteit, Fred, 184

Vail, Isaac N., 111
Vavilov, N. I., 125, 128, 293
Velde, T. H. Van de, 216
Velikovsky, Immanuel, 2, 3–4, 8, 23f., 42, 149, 281, 286
Verne, Jules, 17, 150
Vogt, A. E. van, 210, 251–52f., 303
Voliva, Wilbur G., 13f., 285

Walker, Kenneth, 187
Wallace, Alfred R., 256
Warren, William, 143
Wartegg, Ehrig, 260
Washington, George, 178
Wassersug, Joseph, 172
Weaver, Purcell, 298
Webb, Wells Alan, 289
Welles, Orson, 56
Wells, H. G., 43, 56, 78, 115, 118, 149, 199, 249, 269
Weltfish, Gene, 108
Whiston, William, 28f.
White, Ellen G., 112
White, George Starr, 185

White, Rev. Wally, 277
Whitman, Walt, 256
Winchell, Walter, 47
Windsor, Duchess of, 198
Winrod, Gerald, 187
Winter, Joseph, 239–40f.
Wittry, David, 80
Wolfe, Theodore, 223, 226
Wolff, Charlotte, 258
Wolff, Werner, 259, 306
Woltmann, Ludwig, 133

Wood, Robert, 170f., 172
Wood, Robert W., 303
Wollcott, Alexander, 36
Wright, Sr., Cobina, 198
Wynn, Walter, 157
Wyvell, Lois, 302

Yahuda, Joseph, 217
Yogananda, Paramhansa, 272
Young, J. W., 98

Zirkle, Conway, 285